The Ribonucleic Acids

Edited by

P. R. Stewart and D. S. Letham

With Contributions by

G. D. Clark-Walker · L. Dalgarno · A. J. Gibbs
A. J. Howells · D. S. Letham · H. Naora
G. M. Polya · R. Poulson · J. Shine · J. J. Skehel
P. R. Stewart · P. R. Whitfeld

With 56 Figures

Springer-Verlag
New York · Heidelberg · Berlin 1973

ISBN 0-387-06190-8 Springer-Verlag New York Heidelberg Berlin
ISBN 3-540-06190-8 Springer-Verlag Berlin Heidelberg New York

© by Springer-Verlag Berlin · Heidelberg 1973. Library of Congress Catalog Card Number
73-76335. Printed in Germany. Typesetting, printing and bookbinding: Universitätsdruckerei
H. Stürtz AG, Würzburg

Foreword

The central role of the ribonucleic acids (RNA) in mediating the expression of information encoded in DNA in living cells is now well established. Research in this area of biology continues at a remarkable rate, and new and significant information appears almost daily in a wide range of journals, published symposia and specialist reviews.

The scattered nature of this information makes it difficult for the newcomer to the field of ribonucleic acid biochemistry to obtain a general oversight of current activity and new advances. Moreover, the reviews available for the most part are concerned with rather insular aspects of these ubiquitous molecules, or in the case of text-books, the subject is treated as part of a general outline of properties of nucleic acids and thus often tends to be superficial.

With these considerations in mind, a postgraduate course was instituted in the university in Canberra to attempt to provide a comprehensive, though not excessively detailed, outline of the biological roles of RNA. The course was designed for students with a sound undergraduate training in biochemistry, but otherwise with a wide variety of biological interests — plant physiology, virology, organelle biochemistry, genetics.

The chapters in this book represent the matter of that course. The collection is not meant to be a completely comprehensive account of the latest findings in the field. The book is aimed at those already initiated in the basic biology and chemistry of RNA, and who are probably pursuing some aspect of cellular development and differentiation. Hopefully, it provides a grounding in established facts and concepts in most aspects of the function of RNA species. Newer developments that seem to be important and are likely to result in a greater understanding of the role of RNA in living cells are also dealt with.

Canberra, January 1973
P. R. Stewart
D. S. Letham

Contents

CHAPTER   9

Contributors

CLARK-WALKER, G. D., Department of Developmental Biology, Research School of Biological Sciences, Institute of Advanced Studies, Australian National University, Canberra/Australia

DALGARNO, L., Department of Biochemistry, School of General Studies, Australian National University, Canberra/Australia

GIBBS, A. J., Department of Developmental Biology, Research School of Biological Sciences, Institute of Advanced Studies, Australian National University, Canberra/Australia

HOWELLS, A. J., Department of Biochemistry, School of General Studies, Australian National University, Canberra/Australia

LETHAM, D. S., Department of Development Biology, Research School of Biological Sciences, Institute of Advanced Studies, Australian National University, Canberra/Australia

NAORA, H., Molecular Biology Unit, Research School of Biological Sciences, Institute of Advanced Studies, Australian National University, Canberra/Australia

POLYA, G. M., Department of Environmental Biology, Research School of Biological Sciences, Institute of Advanced Studies, Australian National University, Canberra/Australia (present address: Department of Biochemistry, La Trobe University, Bundoora, Victoria/Australia)

POULSON, R., Department of Developmental Biology, Research School of Biological Sciences, Institute of Advanced Studies, Australian National University, Canberra/Australia (present address: Department of Biochemistry, University of British Columbia, Vancouver/Canada)

SHINE, J., Department of Biochemistry, School of General Studies, Australian National University, Canberra/Australia

SKEHEL, J. J., Division of Virology, National Institute for Medical Research, Mill Hill, London, NW7 1AA/Great Britain

STEWART, P. R., Department of Developmental Biology, Research School of Biological Sciences, Institute of Advanced Studies, Australian National University, Canberra/Australia

WHITFELD, P. R., CSIRO Division of Plant Industry, Canberra/Australia

Principal Abbreviations

rDNA	DNA coding for ribosomal RNA
ChRNA	chromosomal RNA
HnRNA	heterogeneous nuclear RNA
LnRNA	low molecular weight nuclear RNA
lRNA	low molecular weight RNA associated with the large rRNA species of eukaryotes
mRNA	messenger RNA
rRNA	ribosomal RNA
tRNA	transfer RNA
A*	adenine or adenosine
C	cytosine or cytidine
G	guanine or guanosine
T	thymine or thymine riboside
U	uracil or uridine
AMP	
CMP	
GMP	} 5'-phosphates of corresponding nucleosides
TMP	
UMP	
polyA	polyadenylic acid
polyU	polyuridylic acid
polyC	polycytidylic acid
Py	pyrimidine
Pu	purine

---

* In current accepted international nomenclature these symbols designate nucleosides. In this book they have been applied to both the base and nucleoside but their meaning is clear from the context in which they are used.

CHAPTER 1

RNA in Retrospect

D.S. LETHAM, P.R. STEWART, and G.D. CLARK-WALKER

The molecular basis of gene expression whereby organisms develop inherited characteristics is of great importance in the broad and expanding domain of modern biological science. Today it is generally accepted that gene expression can be equated with protein formation and function. Growth, differentiation and reproduction all depend on chemical reactions mediated by enzymes, which are a special class of proteins. Although the polypeptide chains of proteins are polymers of 20 different amino acids, all proteins including enzymes have specific sequences and the living cell can reproduce a particular sequence precisely. In protein synthesis, three types of ribonucleic acid (RNA) are intimately involved; these are messenger, ribosomal and transfer. Our present knowledge of the structure and function of RNA is derived from studies involving many disciplines ranging from genetics, through biochemistry and organic chemistry to physics. In this introductory chapter, an attempt is made to record very briefly the development of our knowledge of *basic* structural features common to all RNA species. The key experimental observations which contributed to our understanding of the role played by RNA in protein synthesis are also presented.

Yeast served as the principal source of RNA for the initial structural studies. Hydrolysis of RNA (termed pentose nucleic acid by early workers) with alkali yielded an approximately equimolecular mixture of four nucleotides termed adenylic, guanylic, cytidylic and uridylic acids. In each nucleotide, a heterocyclic base was joined to a pentose sugar monophosphate. Principally as a result of the endeavour of the organic chemist, P.A. LEVENE, and his co-workers in New York over the years 1908–1936, the following conclusions were reached regarding ribonucleotide structure: a) the heterocyclic bases are adenine, guanine, cytosine and uracil; b) the pentose sugar in all nucleotides is D-ribose; c) the sugar moiety has a furanose ring structure and is attached to position 9 of purine bases and to position 3 (position 1 in modern nomenclature) of pyrimidines, but the stereochemical configuration of the linkage remained unknown; d) the phosphate group is attached to either position 2′ or 3′ of the sugar moiety. For a lucid, detailed account of these early achievements the reader is referred to the review by TIPSON (1945). By ion-exchange chromatography, COHN (1950) showed that each nucleotide was a mixture of two isomers which were later identified as the 2′- and 3′-phosphates (BROWN et al., 1954a). These arose by cleavage of a 2′:3′-cyclic phosphate intermediate formed during alkaline hydrolysis. The configuration of the base-sugar linkage was not defined until DAVOLL et al. (1946) demonstrated that the dialdehyde produced by periodate oxidation of adenosine is identical to that obtained by similar treatment of 9-β-D-glucopyranosyladenine, the configuration of which was established by unambiguous synthesis. By similar methods the β configuration of other nucleosides was established.

In the union of ribonucleotides to form polyribonucleotides, several types of linkage are theoretically possible. However, electrometric titration studies by Levene in 1926, and also by later workers, were consistent with the internucleotide bond being a phosphodiester linkage between ribose moieties. These studies also eliminated other likely possibilities, namely, pyrophosphate, phosphoamide and ether linkages. In 1935 Levene and Tipson proposed a 2' to 3' phosphodiester linkage, but the actual location of the bond was not established unequivocally until 1954. Degradation of RNA with snake venom phosphodiesterase yielded the 5'-phosphates of all four nucleosides (Cohn and Volkin, 1953). A spleen nuclease, shown to hydrolyze specifically phosphodiester groups located at the 3' position of ribonucleosides, was found to degrade RNA to the nucleoside 3'-phosphates without formation of a cyclic phosphate intermediate (Brown et al., 1954b; Heppel et al., 1953). Hence enzymic degradation established that the internucleotide bond in RNA is a phosphodiester linkage joining the 3' position of one nucleotide with the 5' position of the adjacent nucleotide. No unequivocal evidence for branching in RNA has been presented; RNA molecules appear to be entirely linear polymers of mononucleotides.

RNA molecules can possess secondary and tertiary structures. The macromolecular properties of polynucleotides are greatly influenced by the negatively charged electrostatic field arising from the phosphodiester groups and by the ability of bases to interact to form helical structures. These two forces tend to oppose each other and consequently conformation depends on ionic strength, temperature and pH (Cox, 1968). Three conformations of single-stranded polyribonucleotides have been characterized and are as follows (Cox, 1968):

a) An amorphous form in which the bases have no preferred orientation with respect to one another.

b) A 'stacked' conformation in which the flat purine and pyrimidine rings tend to pile one upon another to form a single helix.

c) A 'hairpin-loop' conformation stabilized by hydrogen bonding between complementary bases to give a double-helical structure below the loop. In such a conformation, sequences which cannot find appropriate bases for pairing tend to be 'looped out'.

Double helices also form between RNA strands. The resulting structure resembles that of DNA in that it consists of two antiparallel polynucleotide chains stabilized by Watson-Crick type base pairing. Natural double-stranded RNA occurs in certain viruses. Without doubt, conformation greatly affects the functional properties of RNA.

About 1950, as the elucidation of the basic structure of RNA neared completion, biochemists began to study actively the mechanism of protein synthesis. These studies eventually established the role played by RNA in *de novo* synthesis of proteins, the principal function of RNA. RNA may also be involved in memory and learning phenomena (Glassman, 1969) but this topic is beyond the scope of the present volume. Prior to 1950 by use of histochemical techniques, Brachet and Caspersson independently established that cells active in protein synthesis possessed a high RNA content. The suggestion was made that there was an intimate connection between protein synthesis and RNA. It was not until cell-free protein-synthesizing systems were developed, however, that real progress was made in establishing the role of RNA in the synthesis of polypeptide chains. In 1954, ZAMECNIK and associates in Boston had developed such a system from rat liver. Essential components of this cell-free system were amino acids, ATP, GTP, the ribonucleoprotein fraction of the microsomes (i.e. the ribosomes), and factors including enzymes in the 105,000-g supernatant. By use of this system it was soon established that the ribosome was the probable site of synthesis of peptide chains (KELLER et al., 1954; LITTLEFIELD et al., 1955) and that formation of 'activated' amino acids, aminoacyl adenylates, was the first step in the synthesis of protein from amino acids (HOAGLAND et al., 1956). In 1957 and 1958, again by use of the *in vitro* protein-synthesizing system, the ZAMECNIK group made a dramatic advance which established the second step in the biosynthetic pathway. The activated amino acids were shown to become covalently bound to a type of RNA, termed soluble RNA (sRNA), in the 105,000-g supernatant. Next it was demonstrated that aminoacyl-sRNA substituted for free amino acid in the cell-free system and that the transfer of amino acid from sRNA (later termed transfer RNA, tRNA) to the peptide chain on the ribosome was dependent on GTP (HOAGLAND et al., 1958). Evidence soon followed that tRNA was a complex mixture of polynucleotides and that each tRNA species was specific for a particular amino acid.

Although elucidation of the function of sRNA was a great achievement, a considerable gap still existed in our knowledge of protein synthesis in that the link between the genetic information in DNA and the final amino acid sequence in a protein was not established. Prior to the elucidation of the function of tRNA, it was thought by some that DNA might act directly as a template for assembly of amino acids into protein, a concept introduced by GAMOW (1954) after consideration of the Watson-Crick double-helical structure for DNA. Each amino acid was considered to make a direct steric fit with a sequence of bases. Later it was proposed that amino acids were held directly on RNA templates and then linked enzymically. The significance of these early concepts is elegantly discussed by WOESE (1967). Unlike many of his contemporaries, CRICK rejected these concepts and proposed the following (CRICK, 1955):

"Each amino acid would combine chemically, at a special enzyme, with a small molecule which, having a specific hydrogen-bonding surface, would combine specifically with the nucleic acid template".

This became known as the "adaptor hypothesis". The molecules, or adaptors, to which the amino acids became attached were considered to be mediators between the encoded information of the nucleic acid and the extremely variable chemical structures of amino acids. The discovery of tRNA and its specificity completely vindicated the adaptor hypothesis; tRNA possessed all the properties of Crick's adaptor. Complementarity between polynucleotides and amino acids then became a concept of purely historical significance.

By 1958 the ribosome was established as the site of cytoplasmic protein synthesis, and many assumed that ribosomal RNA (rRNA) was a carrier of genetic information transcribed from DNA. It was also considered that rRNA was the template which combined, not with amino acids, but, by hydrogen bonding, with specific groups of bases on the tRNA moiety of aminoacyl-tRNA molecules. By 1961, however, this hypothesis was generally discarded. It was replaced by a new concept supported by the elegant experiments of Brenner et al. (1961). These workers found that after infection of the bacterium *Escherichia coli* with T2-bacteriophage no new ribosomes were formed for synthesis of viral protein, but in fact a new RNA species with a rapid turnover and a base composition complementary to that of phage DNA was produced and attached to pre-existing *E. coli* ribosomes. Such RNA with template function, termed messenger RNA (mRNA), was postulated by Jacob and Monod (1961) in their theory of protein synthesis. RNA species which labelled very rapidly, attached to ribosomes and possessed a DNA-like base composition were soon detected in normal uninfected bacterial cells (Gros et al., 1961).

Further support for the mRNA concept was provided by the classic *in vitro* experiments of Nirenberg and Matthaei (1961) who, using a cell-free protein-synthesizing system from *E. coli*, showed that addition of synthetic polyuridylic acid resulted in the formation of polyphenylalanine. This observation together with similar experiments using other synthetic polymers as templates convincingly established the mRNA concept and also provided important information regarding the now familiar genetic code. In recent years the existence of mRNA has been demonstrated unequivocally and in this connection the work of Lockard and Lingrel (1969) merits special mention. Treatment of mouse reticulocyte polysomes with detergent yielded an RNA species (9S) which sedimented between tRNA and rRNA in sucrose gradients. Although several properties of this RNA were consistent with it being the mRNA for globin, demonstration of ability to direct the synthesis of the globin chains was required to establish this unequivocally. Lockard and Lingrel showed that the 9S RNA from mouse polysomes did direct the formation of mouse globin β-chains in a rabbit reticulocyte cell-free protein-synthesizing system. It has also been demonstrated that an RNA fraction from rabbit reticulocytes can direct synthesis of globin in a cell-free system from *E. coli* (Laycock and Hunt, 1969).

The concept that mRNA carries information in the form of nucleotide sequences from the gene to the protein-synthesizing mechanism is now clearly established. On the ribosome the codons (sequences of three adjacent nucleotides that code for an amino acid) of mRNA pair sequentially with the anticodons of aminoacyl-tRNA molecules, and, since each tRNA is specific for a particular amino acid, amino acid sequence is dictated. Transfer of information from DNA to RNA to protein is an established basic concept, a keystone in the structure of molecular biology. However, the important work of Temin and Mizutani (1970) has established that flow of information from RNA to DNA also occurs. An RNA-dependent DNA polymerase is present in the virions of certain RNA tumour viruses. This polymerase, termed reverse transcriptase, has been used to synthesize *in vitro* DNA copies of mammalian globin mRNA (Verma et al., 1972; Kacian et al., 1972). Our present knowledge of transfer of information between DNA, RNA and protein can be represented diagrammatically (Fig. 1) as suggested by Crick (1970). General transfers which probably occur in all cells are: DNA → DNA, DNA → RNA, and RNA → protein. Mammalian reticulocytes, because they are enucleate, probably lack the first two transfers and may be an exception.

Special transfers which are not of general occurrence are: RNA → RNA, RNA → DNA, and DNA → protein. The first two transfers are known to occur in certain virus-infected cells; the third occurs in a special cell-free system containing single-stranded DNA and a streptomycinoid antibiotic (McCarthy and Holland, 1965).

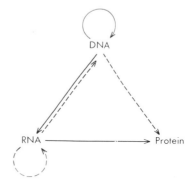

Fig. 1. Flow of information between macromolecules. Solid arrows indicate general transfers; dotted arrows indicate special transfers. (Arrows do not represent flow of matter, but the direction of flow of encoded information)

In eukaryotic cells, the mRNA synthesized on the DNA template passes from the nucleus to the cytoplasm for translation into protein by cytoplasmic ribosomes and transcription and translation are essentially independent processes. In contrast, in bacteria, mRNA appears to be translated by a cluster of ribosomes which follow close behind the RNA polymerase involved in transcription. Thus in bacteria transcription and translation of an mRNA chain may occur concurrently and the two processes may be coupled (Morse et al., 1969; Imamoto and Kano, 1971).

At the beginning of the 20th century, Morgan and others established the chromosomal basis of heredity. Elucidation of the molecular basis of heredity and gene expression then became a challenging concept in the mind of man. Even in 1940, knowledge of this field was meagre indeed. It should be remembered that it was not until 1944 that Avery and co-workers unequivocally established the genetic role of DNA; in 1956 and 1957 experiments with tobacco-mosaic-virus RNA proved the same role for viral RNA (Gierer and Schramm, 1956; Fraenkel-Conrat et al., 1957). Today as evidenced by the contributions to this book concerning RNA, our understanding of the molecular basis of gene expression is very extensive and undoubtedly represents one of the most significant achievements of modern scientific endeavour.

References

Reviews

Tipson, R.S.: in Advan. carbohydrate chemistry (Pigman, W.W., Wolfrom, M.L., eds.), vol. 1, p.193. New York: Academic Press 1945.
Watson, J.D.: Science **140**, 17 (1963).
Zamecnik, P.C.: Cold Spring Harbor Symp. Quant. Biol. **34**, 1 (1969).

Other References

BRENNER, S., JACOB, F., MESELSON, M.: Nature **190**, 576 (1961).

BROWN, D.M., FASMAN, G.D., MAGRATH, D.I., TODD, A.R.: J. Chem. Soc. p.1448 (1954a).

BROWN, D.M., HEPPEL, L.A., HILMOE, R.J.: J. Chem. Soc. p.40 (1954b).

COHN, W.E.: J. Amer. Chem. Soc. **72**, 1471, 2811 (1950).

COHN, W.E., VOLKIN, E.: J. Biol. Chem. **203**, 319 (1953).

COX, R.A.: Quart. Rev. Chem. Soc. p.499 (1968).

CRICK, F.H.C.: A Note for the RNA Tie Club (1955). Quoted by Hoagland, M.B. In: CHARGAFF, E., and DAVIDSON, J.N. (eds.), The nucleic acids, vol.3, p.349. New York and London: Academic Press 1960.

CRICK, F.H.C.: Nature **227**, 561 (1970).

DAVOLL, J., LYTHGOE, B., TODD, A.R.: J. Chem. Soc. p.833 (1946).

FRAENKEL-CONRAT, H., SINGER, B., WILLIAMS, R.C.: Biochim. Biophys. Acta **25**, 87 (1957).

GAMOW, G.: Nature **173**, 318 (1954).

GIERER, A., SCHRAMM, G.: Nature **177**, 702 (1956).

GLASSMAN, E.: Ann. Rev. Biochem. **38**, 605 (1969).

GROS, F., HIATT, H., GILBERT, W., KURLAND, C.G., RISEBROUGH, R.W., WATSON, J.D.: Nature **190**, 581 (1961).

HEPPEL, L.A., MARKHAM, R., HILMOE, R.J.: Nature **171**, 1152 (1953).

HOAGLAND, M.B., KELLER, E.B., ZAMECNIK, P.C.: J. Biol. Chem. **218**, 345 (1956).

HOAGLAND, M.B., STEPHENSON, M.L., SCOTT, J.F., HECHT, L.I., ZAMECNIK, P.C.: J. Biol. Chem. **231**, 241 (1958).

IMAMOTO, F., KANO, Y.: Nature **232**, 169 (1971).

JACOB, F., MONOD, J.: J. Mol. Biol. **3**, 318 (1961).

KACIAN, D.L., SPIEGELMAN, S., BANK, A., TERADA, M., METAFORA, S., DOW, L., MARKS, P.A.: Nature New Biol. **235**, 167 (1972).

KELLER, E.B., ZAMECNIK, P.C., LOFTFIELD, R.B.: J. Histochem. Cytochem. **2**, 378 (1954).

LAYCOCK, D.G., HUNT, J.A.: Nature **221**, 1118 (1969).

LITTLEFIELD, J.W., KELLER, E.B., GROSS, J., ZAMECNIK, P.C.: J. Biol. Chem. **217**, 111 (1955).

LOCKARD, R.E., LINGREL, J.B.: Biochem. Biophys. Res. Commun. **37**, 204 (1969).

MCCARTHY, B.J., HOLLAND, J.J.: Proc. Nat. Acad. Sci. U.S. **54**, 880 (1965).

MORSE, D.E., MOSTELLER, R.D., YANOFSKY, C.: Cold Spring Harbor Symp. Quant. Biol. **34**, 725 (1969).

NIRENBERG, M.W., MATTHAEI, J.H.: Proc. Nat. Acad. Sci. U.S. **47**, 1588 (1961).

TEMIN, H.M., MIZUTANI, S.: Nature **226**, 1211 (1970).

VERMA, I.M., TEMPLE, G.F., FAN, H., BALTIMORE, D.: Nature New Biol. **235**, 163 (1972).

WOESE, C.R.: The genetic code, p.17–31. New York: Harper & Row 1967.

G.M. Polya

Introduction

Transcription is DNA-directed RNA synthesis which yields RNA with a base sequence complementary to that of the DNA template. In this synthesis, RNA polymerase catalyzes the formation of internucleotide 3′ to 5′ phosphodiester bonds and thus plays a vital role in the transfer of information from DNA to RNA.

Major advances have been made in recent years with respect to the molecular basis of the mechanism and control of transcription in bacterial systems. With recent descriptions of the purification of RNA polymerases and transcription-modifying components from eukaryotes, the outlines of transcriptional control mechanisms in higher organisms are beginning to appear.

This chapter deals with the physical and chemical properties of purified RNA polymerases from a variety of sources but discussion of the mechanism of transcription will be limited to the bacterial system. The section on RNA polymerase inhibitors, however, has general relevance to transcription in all cells. Finally, a broad review of mechanisms of control of transcription in a variety of organisms is given. Many excellent reviews and symposia on transcription are available and the reader is referred to these (listed at the end of the chapter) for more detailed discussion of transcription processes.

It should be noted that certain RNA tumour viruses contain an enzyme, termed reverse transcriptase, which synthesizes DNA on RNA templates and thus enables information to flow from RNA to DNA. This DNA polymerase is discussed in Chapter 10.

A. Purification and Physical Properties of RNA Polymerases

1. General Procedural Considerations

a) Assays for RNA Polymerase. RNA polymerases catalyze the DNA-dependent reaction that can be represented as follows:

$$N'ppp + (Np)_n Nppp \rightleftharpoons N'p(Np)_n Nppp + PP_i$$

where Nppp and N′ppp represent ATP, GTP, CTP or UTP, while PP_i denotes inorganic pyrophosphate. The reaction is represented by use of structural formulae in Fig. 1. The nucleoside triphosphate at the 5′ end of the polymeric product is usually a purine nucleotide. RNA polymerase is commonly assayed *in vitro* by

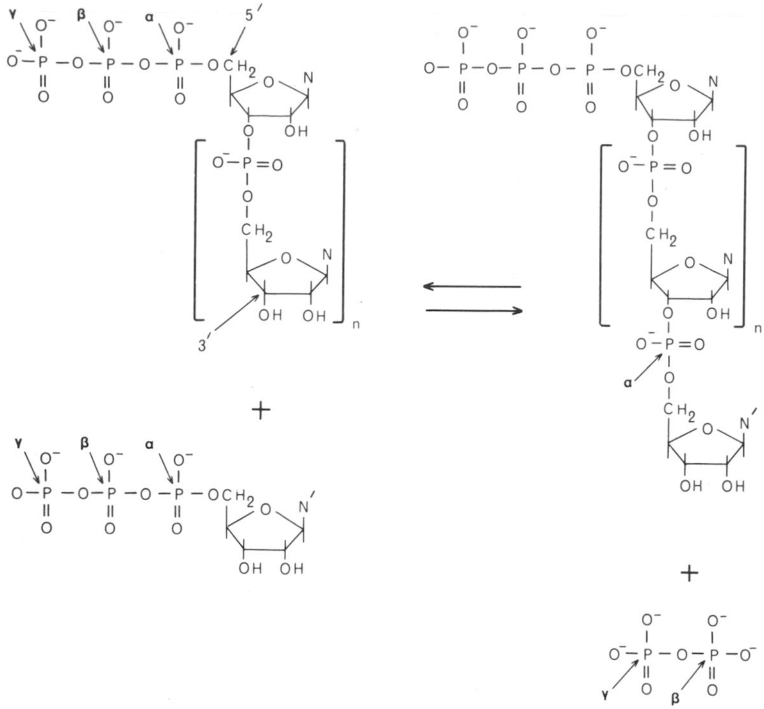

Fig. 1. Structural formulae representing the growth of a polyribonucleotide chain

measuring the amount of labelled nucleoside monophosphate (AMP, GMP, UMP or CMP) incorporated into an acid-insoluble polymer that is collected for counting by precipitation onto Millipore filters or onto filter paper disks. The rate of initiation, as opposed to the rate of elongation, of RNA chains can be determined by measuring the rate of incorporation of γ-^{32}P-labelled purine nucleotides into RNA, since each new chain has a purine nucleoside triphosphate at the 5' end. When an RNA polymerase is of sufficient purity and activity, the activity of the enzyme can be measured by release of inorganic pyrophosphate. RNA polymerase catalyzes a nucleoside triphosphate-^{32}PP$_i$ exchange reaction and this has been used to detect reversibly inactivated RNA polymerase activities. Hybridization techniques have been employed to quantitate the synthesis of specific RNAs *in vitro* and *in vivo*.

   b) *Purity Criteria for RNA Polymerase Preparations.* The specific activity of an RNA polymerase preparation may not be a reliable criterion of purity since many variables (template type and concentration, nucleoside triphosphate concentration, divalent cation type and concentration, ionic strength, temperature and the presence or absence of activating proteins) can affect the rate of RNA synthesis *in vitro* as will become apparent in later sections. Small amounts of certain enzymes or combinations of enzyme impurities (e.g. RNAase, DNAase, nucleoside diphosphokinase, inorganic pyrophosphatase, nucleoside triphosphatase or adenylate kinase) may interfere with correct determination of RNA polymerase activity *in vitro* by utilizing the template, substrates or products of the RNA polymerase reaction. Other enzymes (e.g. polynucleotide phosphorylase, polyphosphate kinase or polyriboadenylate polymerase) may catalyze the formation of acid-insoluble polymers that could make spurious contributions in the appropriate radiochemical assays. In some cases such contaminant activities can be inhibited in the assay by inclusion of appropriate inhibitors, e.g. phosphate inhibits polynucleotide phosphorylase and low levels of ADP inhibit polyphosphate kinase (MAITRA and HURWITZ, 1967). The best estimates of RNA polymerase purity derive from polyacrylamide gel electrophoresis in dissociating conditions (e.g. using gels containing 0.1 % sodium dodecyl sulphate or 8 M urea) (BURGESS, 1969 b).

2. Purification of RNA Polymerases

Purification of RNA polymerases to homogeneity and resolution of component subunits has been achieved in recent years. While only one type of RNA polymerase has been demonstrated in bacterial cells, there are a multiplicity of RNA polymerases in eukaryote cells. These RNA polymerases from fungi, higher plant and mammalian cells have been distinguished in terms of intracellular localization, chromatographic behaviour, divalent cation activation, sensitivity to the inhibitor α-amanitin and, more recently, with respect to subunit compositions. The major types of RNA polymerase from mammalian sources are summarized in Table 1. Isolation and purification of RNA polymerases from bacteria, nuclei and organelles involve some problems peculiar to the enzyme sources. Procedures developed to isolate DNA-dependent RNA polymerases from these different sources are briefly outlined below.

Table 1. Types of mammalian RNA polymerase

Localization	Nomenclatures	Sensitive to α-amanitin	Sensitive to rifampicin
Nucleolar	I a (A I)	−	−
Nucleolar	I b (A II)	−	−
Nucleoplasmic	III (A III)	−	−
Nucleoplasmic	II a (B I)	+	−
Nucleoplasmic	II b (B II)	+	−
Mitochondrial		−	+

RNA polymerases I and II from eukaryote sources have been distinguished in terms of Mg^{2+} or Mn^{2+} concentrations for optimal activity. While both enzymes can function with either cation present, RNA polymerases I and II are preferentially activated by Mg^{2+} and Mn^{2+} respectively (ROEDER and RUTTER, 1970).

a) Purification of Bacterial RNA Polymerases. The first bacterial RNA polymerase purification was achieved by CHAMBERLIN and BERG (1962). This involved precipitation of DNA with streptomycin sulphate, precipitation of RNA polymerase from the supernatant with protamine sulphate and then selective solubilization of the enzyme with 0.1 M ammonium sulphate. The solubilized enzyme was then purified by ammonium sulphate fractionation and DEAE-cellulose chromatography. Further purification procedures were developed subsequently but the best procedure available for purification of *E. coli* RNA polymerase to homogeneity is that of BURGESS (1969 a). To avoid losses of the enzyme as a rapidly sedimenting DNA-RNA polymerase complex in an initial high-speed centrifugation step, the extract was first treated with DNAase, yielding a DNA-free enzyme that was then amenable to classical enzyme purification. Differential streptomycin sulphate or protamine sulphate precipitation of DNA was thus avoided. Subsequent purification by ammonium sulphate fractionation, DEAE-cellulose and phosphocellulose chromatography, and agarose gel filtration or glycerol density gradient centrifugation, yielded a homogeneous RNA polymerase in high yield. Glycerol density gradient centrifugation was performed at low salt concentration and then at high salt concentration taking advantage of the reversible, ionic strength-dependent aggregation of the enzyme to achieve greater purity. As will be discussed below, purification on phosphocellulose results in the separation of a template-specifying factor, σ (sigma), from the remainder of the enzyme. Enzyme lacking σ is termed core enzyme.

b) Purification of Nuclear RNA Polymerases. The first partial purification of an RNA polymerase was the preparation from rat liver of a so-called "aggregate enzyme" in which the RNA polymerase was firmly attached to DNA (WEISS, 1960). The problem of freeing the RNA polymerase from DNA is more severe for eukaryote preparations than in the case of preparations from prokaryotic sources. In addition, the yields of homogeneous RNA polymerase protein from eukaryote sources, plant and animal, have been very much lower than those from prokaryotic cells. A variety of procedures (some derived from bacterial work) have been applied to obtain from eukaryote sources DNA-free RNA polymerase preparations that are then amenable to conventional enzyme purification procedures. These procedures are briefly outlined below:

1) extraction of nuclei with hypotonic buffer,

2) sonication of nuclear extracts at high ionic strength followed by precipitation of DNA at low ionic strength,

3) streptomycin precipitation of DNA,

4) DNAase treatment of extracts followed by selective precipitation of residual DNA or chromatin at low ionic strength,

5) differential phase separation of DNA and RNA polymerase into dextran and polyethyleneglycol layers.

In some cases (e. g. wheat and maize leaves) the above methods are not necessary because soluble, DNA-dependent RNA polymerase activities are present in the supernatants after high-speed centrifugation of the initial tissue homogenates (POLYA and JAGENDORF, 1971; STOUT and MANS, 1967).

c) Purification of Organelle RNA Polymerases. A number of factors interfere with attempts to purify RNA polymerase from mitochondria and chloroplasts. These include: inadequate purification of the organelles from bacteria or other cellular components; contamination by other RNA polymerases released during cell fractionation; presence of organelle DNA to which the RNA polymerase may be attached; small initial amounts of enzyme present in the organelles; and instability of the enzymes after purification. For details the reader is referred to accounts of the purification of chloroplast RNA polymerase (POLYA and JAGENDORF, 1971; BOTTOMLEY et al., 1971) and mitochondrial RNA polymerase (KUNTZEL and SCHAFER, 1971; REID and PARSONS, 1971).

3. Physical Properties of RNA Polymerases

a) Molecular Weights and Subunit Composition. RNA polymerases have now been purified to homogeneity from a variety of eukaryotes and prokaryotes: viruses, bacteriophages, bacteria, mammals, fungi, algae and higher plants. The first RNA polymerase to be purified to homogeneity and characterized with respect to subunit composition was that from *E. coli* (BURGESS, 1969a, b). While much of the discussion in this chapter will refer to the structure and function of this enzyme, the subunit composition and molecular weights of RNA polymerases from other sources are summarized in Table 2 for comparative purposes.

Table 2. Subunit compositions and molecular weights of purified RNA polymerases

Source	Molecular weight	Subunit structure	Subunit MW	Reference
Escherichia coli	400,000	$(\alpha_2\,\beta\,\beta')$	α 39,000	BURGESS (1969b)
			β 155,000	
	490,000	$(\alpha_2\,\beta\,\beta'\,\sigma)$	β' 165,000	
			σ 90,000	
Pseudomonas putida	400,000	$(\alpha_2\,\beta\,\beta')$	α 44,000	JOHNSON et al. (1971)
			β 155,000	
	500,000	$\alpha_2\,\beta\,\beta'\,\sigma$	β' 165,000	
			σ 98,000	
Anacystis nidulans	436,000	$\alpha_2\,\beta\,\beta'\,\sigma$	α 39,000	HERZFELD and
			β 147,000	ZILLIG (1971)
			β' 125,000	
			σ 86,000	
T3	110,000	1 poly-peptide	110,000	DUNN et al. (1971)
T7	107,000	1 poly-peptide	107,000	DUNN et al. (1971);
	110,000		110,000	CHAMBERLIN et al. (1970)
Neurospora crassa mitochondria	variable high MW	1 poly-peptide	64,000	KÜNTZEL and SCHÄFER (1971)
Rat liver mitochondria		1 poly-peptide	64,000–68,000	REID and PARSONS (1971)
Triticum vulgare (wheat leaf, soluble)	$>12\times10^6$ (10 mM Tris) 3×10^6 (200 mM Tris)	1 poly-peptide for minimal enzyme	65,000	POLYA (1973)
Calf thymus and rat liver nucleoplasm	IIa (BI) 400,000	1:1:1:1	190,000 150,000 35,000 25,000	WEAVER et al. (1971) (cf. KEDINGER et al., 1971; CHESTERTON and BUTTERWORTH, 1971; MANDEL and CHAMBON, 1971)
	IIb (BII) 380,000	1:1:1:1	170,000 150,000 35,000 25,000	

It is apparent from Table 2 that RNA polymerases may have quite different degrees of complexity. It should be noted that the mitochondrial and bacterio-phage RNA polymerases are much smaller and simpler proteins (in terms of subunit composition) than the bacterial RNA polymerases or the mammalian nucleoplasmic RNA polymerase II.

The bacterial RNA polymerases (i.e. the holoenzymes) can be reversibly separated into a core enzyme ($\alpha_2\beta\beta'$) and a polypeptide termed sigma (σ) which is involved operationally in initiation-site recognition (BURGESS et al., 1969). Other template-specific activation factors have been found in bacterial systems and in eukaryote systems and are discussed later. Nevertheless when considering the subunit structures given in Table 2 for non-bacterial RNA polymerases, the possibility must be borne in mind that the enzymes may have lost modifying polypeptides not essential for *in vitro* activity. In addition a small protein component, ω, of molecular weight 9,000 and with a variable degree of association with the core enzyme, is found in homogeneous *E. coli* RNA polymerase preparations and may represent a fragment of subunit β', i.e. $\beta' \rightarrow \beta + \omega$ (BURGESS, 1969b). Such specific hydrolysis of a major subunit of mammalian RNA polymerase II(B) has been suggested to account for two types of RNA polymerase II from calf thymus and rat liver in which only one subunit has been modified (WEAVER et al., 1971).

b) Aggregation-Disaggregation Properties. Bacterial RNA polymerases undergo reversible aggregation at low ionic strength. For example, the *E. coli* holoenzyme exists as a monomeric form at high ionic strength (>0.1) but as a dimer at low ionic strength. The core enzyme on the other hand behaves somewhat differently; polymers composed of 6 or more monomeric units may form at low ionic strength [for discussions of these changes see RICHARDSON (1969) and BURGESS (1971)].

The physical state of purified soluble RNA polymerase from wheat seedlings shows a similar dependence on ionic strength. It has an apparent molecular weight of 3×10^6 daltons in 0.2 M Tris, but reversibly aggregates to a $>12 \times 10^6$ molecular weight form in 10 mM Tris (POLYA, 1973). The mitochondrial RNA polymerase from *Neurospora crassa* also shows similar aggregation behaviour (KUNTZEL and SCHAFER, 1971).

B. Initiation of Transcription

As mentioned earlier in this chapter, discussion of the mechanism of transcription, i.e. of initiation, elongation, termination and release of RNA chains, will be almost exclusively concerned with the bacterial system, and more specifically with the RNA polymerase from *E. coli*. The initiation of transcription may be resolved into certain sequential steps. A possible sequence (after BURGESS, 1971) is briefly outlined below:

1) RNA polymerase binds to double-stranded DNA non-specifically and reversibly.

2) In appropriate conditions, RNA polymerase locates and binds to specific initiation sites, probably pyrimidine-rich sites. The physical binding may not

require the σ factor, but this factor appears to be involved in recognition of initiation sites.

3) Local denaturation and strand separation of the DNA occurs at the binding site to form an RNA polymerase-DNA complex, which is highly stable at low ionic strength. This complex can form in the presence of the inhibitor of initiation, rifampicin, and requires the σ factor.

4) A conformational change in the enzyme probably occurs next and yields a rifampicin-resistant preinitiation complex, stabilized by σ.

5) The 5'-terminal nucleoside triphosphate (almost always a purine nucleotide) binds to the enzyme and forms a highly stable initiation complex.

6) A second nucleoside triphosphate binds and the first 3',5'-phosphodiester bond is formed with the elimination of pyrophosphate. Thus the initial condensation yields a 3',5'-nucleotidyl nucleoside-5'-triphosphate (N'pNppp) and polymerization proceeds with this 5' → 3' polarity.

These individual stages are considered in greater detail below.

1. Non-Specific Binding of RNA Polymerase to DNA

The degree of binding of RNA polymerase to DNA can be readily determined experimentally since complexes of RNA polymerase and DNA are retained by membrane filters (JONES and BERG, 1966). The major features of RNA polymerase binding to DNA are as follows:

1) Binding of RNA polymerase to DNA is normally reversible as shown by exchange of labelled enzyme bound to DNA by addition of unlabelled enzyme (STERNBERGER and STEVENS, 1966). However, there may be some sites at which binding is irreversible especially at low ionic strength (BREMER et al., 1966).

2) The binding reaction does not require a divalent cation (RICHARDSON, 1966).

3) The binding reaction may be fast in appropriate conditions; thus a complex formation between RNA polymerase and T4 DNA is detectable within 15 seconds of mixing the components (RICHARDSON, 1969).

4) High ionic strength (e.g. 0.4 M KCl) abolishes complex formation (for a review of ionic strength effects see RICHARDSON, 1969).

5) RNA polymerase binds to both double-stranded and single-stranded DNA and there appear to be more binding sites per unit weight on single-stranded than on double-stranded DNA (RICHARDSON, 1966).

6) The enzyme has some affinity for all regions on DNA and at low ionic strength the enzyme may be tightly bound to DNA in a densely-packed and non-specific manner (PETTIJOHN and KAMIYA, 1967).

2. Specific RNA Polymerase Binding to Initiation Sites

For reasons mentioned above, the number of specific initiation sites on DNA is less than the number of polymerase binding sites. The specific transcriptionally-functional polymerase binding sites are widely separated on DNA with an average of one for every 600 base pairs (VINOGRAD et al., 1965).

In nearly all cases, *in vitro* RNA synthesis is initiated with purine nucleotides (ATP or GTP) (MAITRA and HURWITZ, 1965; BREMER et al., 1965) and there is now considerable evidence suggesting that the initiation sites on DNA may be pyrimidine clusters. RNA homopolymers are able to form stable complexes with some single-stranded DNAs and it is thought that such stable complexes derive from the existence of regions in the DNA containing ten or more adjacent and identical nucleotides. A striking observation is that when transcription is confined to one DNA strand, as in bacteriophages T7 and α, pyrimidine clusters are confined to the same strand (SZYBALSKI et al., 1966). Both strands of λ DNA form complexes with poly G, but the strand that is complementary to the greater number of λ-specific RNA species is also the strand that binds more poly G (TAYLOR et al., 1967). Further evidence for the involvement of pyrimidine clusters in initiation derives from the preferential binding of RNA polymerase to AT-rich sections of double-stranded DNA (SHISHIDO and IKEDA, 1971). RNA polymerase also binds preferentially to T-rich fragments of DNA (SHISHIDO and IKEDA, 1970). Nevertheless nearest neighbour analysis of the 5′-ends of RNA shows that the 5′-purine termini have pyrimidine nearest neighbours indicating that if pyrimidine runs are involved in initiation they are not transcribed (YARUS, 1969).

The σ factor is involved in correct initiation of RNA chains. Both core and holoenzyme initiate nearly exclusively with purine nucleotides. However when transcription is catalyzed by holoenzyme (σ present) the next nucleotide is a pyrimidine, but when core enzyme (σ absent) transcribes, it is a purine which may then be followed by a purine sequence (SUGIURA et al., 1970; OKAMOTO et al., 1970). Thus the σ factor is probably involved in correct initiation but may not be involved in binding to pyrimidine-rich regions. Nevertheless other protein factors to be discussed later in this chapter in the context of control mechanisms can also activate transcription in template-specific ways.

The RNA polymerase dissociates into its monomeric form on binding to single-stranded nucleic acids and can act on double-stranded DNA in conditions in which it exists as a monomer (GEIDUSCHEK and HASELKORN, 1969).

3. DNA Strand Separation as a Consequence of RNA Polymerase Binding

There is considerable evidence that an initial consequence of the binding of RNA polymerase to double-stranded DNA is localized strand-separation. This evidence is briefly summarized below (for further amplification see YARUS, 1969).

1) RNA polymerase can employ single-stranded DNA as a template.

2) RNAase-resistant RNA at the growing (3′) end of RNA has been detected (HAYASHI and HAYASHI, 1968). If this RNA is RNAase-resistant because it exists as a DNA-RNA hybrid, then one DNA strand must have been displaced.

3) Addition of RNA polymerase to DNA results in the enzyme-DNA complex behaving like denatured DNA in its retention on nitrocellulose filters (JONES and BERG, 1966) and in its partition coefficient in an aqueous two-phase polymer system (PETTIJOHN, 1967).

4) A lag in RNA synthesis *in vitro* which is independent of the binding of the polymerase to DNA, is reduced in conditions (higher temperature and lowered

ionic strength) that would favour DNA denaturation. No such lag occurs when single stranded DNA is used as a template (WALTER et al., 1967).

5) RNA polymerase binds more tightly to single-stranded or denaturated DNA than to double-stranded DNA suggesting that RNA polymerase should destabilize native double-stranded DNA (YARUS, 1969).

Transcription of native double-stranded DNA by RNA polymerase is asymmetric, i.e. within a given gene only one of the two complementary strands is transcribed. In some instances, one strand of a double-stranded DNA can serve as template for certain genes, while the remaining genes are transcribed from the other strand. This situation applies to *E. coli* and bacteriophages T4 and λ. However with bacteriophages T7 and α, all transcription is confined to one particular DNA strand.

4. Formation of a Rifampicin-Resistant Preinitiation Complex

The antibiotic rifampicin inhibits bacterial RNA polymerase only if no RNA synthesis has occurred prior to addition of this inhibitor (SIPPEL and HARTMANN, 1968). Rifampicin does not affect elongation, acts before the formation of the first phosphodiester bond and does not affect DNA binding of the enzyme. In appropriate conditions a rifampicin-resistant complex can be formed by the RNA polymerase and DNA in the absence of nucleoside triphosphates (SIPPEL and HARTMANN, 1970; BAUTZ and BAUTZ, 1970). After binding of the enzyme to DNA and local denaturation of DNA, the antibiotic probably prevents a conformational change in the enzyme that is requisite for subsequent initiation of RNA synthesis (see BURGESS, 1971).

5. Binding of the First Nucleotide

In vitro RNA synthesis is initiated almost entirely with purine nucleotides and the relative proportions of chains initiated with ATP or GTP are characteristic of the DNA template used. Since the 5'-terminal triphosphate is initially conserved in the product RNA, initiation can be quantitated radiochemically using γ-^{32}P-labelled nucleotides.

6. Formation of the First Phosphodiester Bond

The nucleotide adjacent to the 5'-terminal nucleotide of a newly formed chain usually contains a pyrimidine, i.e. RNA molecules usually begin with the sequence pppPupPy, although the absence of σ factor can result in a polypurine run. The first phosphodiester bond is formed with the elimination of pyrophosphate. Enzyme-DNA complex stability is probably enhanced by the formation of the first phosphodiester bond although there is some doubt as to the minimum chain length required to stabilize the complex at high ionic strength (see GEIDUSCHEK and HASELKORN, 1969).

C. Elongation and Termination of RNA Chains

1. Elongation of RNA Chains

Chain elongation results in enhanced stability of the RNA polymerase-DNA complex and does not require the σ factor which is released during chain elongation. Several important aspects of RNA chain elongation are amplified below.

a) Direction of Chain Growth. Chain growth of RNA molecules occurs in the 5' to 3' direction. Alkaline hydrolysis of RNA synthesized enzymically *in vitro* yields a nucleoside tetraphosphate (pppNp) from the 5'-end, a nucleoside (N) from the 3'-end and the remaining nucleotides as 2'- and 3'-nucleoside monophosphates. In conditions of elongation and little initiation, added label is incorporated preferentially into the growing 3'-end of the chain as shown by alkaline hydrolysis and separation of product nucleosides and nucleotides (BREMER et al., 1965). Further evidence for $5' \rightarrow 3'$ growth of RNA chains derives from the demonstration that ^{32}P incorporated from γ-^{32}P nucleoside triphosphates cannot be removed by subsequent chain elongation. Accordingly chain growth must proceed from the 3'-end (MAITRA and HURWITZ, 1965). The incorporation of 3'-deoxyadenosine 5'-triphosphate into RNA chains and the resulting termination of chain growth provides further evidence for addition of nucleotides to the 3'-end of growing RNA chains (SHIGEURA and BOXER, 1964).

As a consequence of the 5' to 3' direction of chain growth, growing chains are probably protected from 3'-terminus-specific exonucleases. Since mRNA is translated in the same direction, translation and transcription can occur concurrently.

b) Rates of Chain Elongation. Growth rates of RNA molecules *in vitro* can be quite comparable with growth rates *in vivo*, although there are many variables, notably ionic strength, that markedly affect rates *in vitro*. Table 3 summarizes various estimates of rate of RNA chain growth *in vivo* and *in vitro*.

Since mRNA is synthesized and translated in the same $5' \rightarrow 3'$ direction, and furthermore since some types of RNA are inhibitors of RNA polymerase, it has been suggested that concomitant translation might speed up transcription. This possibility is discussed later in this chapter.

Table 3. RNA chain growth rates

RNA synthesized	Chain growth rate (nucleotides/sec)	Reference
in vivo		
T4 early mRNA	28	BREMER and YUAN (1968)
Trp mRNA	20	BAKER and YANOFSKY (1968)
E. coli rRNA	55	BREMER and BERRY (1971)
E. coli rRNA	38	MANGIAROTTI et al. (1968)
E. coli RNA	15–26	MANOR et al. (1969)
in vitro		
T4 RNA	16	RICHARDSON (1969)
T7 RNA	100	BREMER (1967)
Calf thymus RNA	100	BREMER (1967)
E. coli RNA	15	BREMER and KONRAD (1964)

2. Termination and Reinitiation of RNA Chains

The existence of discrete RNA species such as tRNAs and rRNAs and the distinct sizes of multicistronic mRNAs that have been identified in bacterial systems imply the existence of termination regions on DNA. The nature of such regions remains to be elucidated. Several types of termination mechanism have been established for *in vitro* RNA synthesis and these are outlined below.

   a) Ionic Strength Effects on Termination and Reinitiation. At low ionic strength (e.g. 0.05 M KCl) nonspecific inhibition by product RNA of further elongation of RNA chains occurs; most of the RNA made in these conditions remains attached to the enzyme-DNA complex. At higher ionic strength (e.g. 0.2 M KCl), the release of product RNA and reinitiation of RNA synthesis by the polymerase occurs (RICHARDSON, 1969). The product RNA molecules made in such conditions can be of discrete sizes suggesting specificity of the chain termination (MAITRA and BARASH, 1969).

   b) The Role of ρ Factor in Termination. ROBERTS (1969) has isolated and purified a termination factor termed rho (ρ). This protein is a tetramer, has a molecular weight of 200,000 and at low ionic strength releases RNA chains of discrete sizes that are smaller than RNA molecules made in the absence of ρ. This protein factor binds reversibly to DNA and is displaced by high ionic strength and by denatured DNAs (BECKMAN et al., 1971). In the presence of ρ, transcription of T4 DNA by *E. coli* RNA polymerase yields immediate early RNA* molecules that are about 2,000 nucleotides in length; in the absence of ρ, the product RNA molecules are largely delayed early molecules, 4,000–7,000 nucleotides in length (WITMER, 1971). However PETTIJOHN et al. (1970) have shown that precise *in vitro* transcription can be obtained using a highly-purified *E. coli* RNA polymerase-DNA complex, and that ρ factor may not be involved in this faithful *in vitro* transcription.

D. Inhibitors of RNA Polymerase

A wide variety of compounds are available which inhibit RNA polymerase by protein-specific or DNA-specific mechanisms. An excellent recent review of such compounds is that of GOLDBERG and FRIEDMAN (1971).

1. DNA-Specific Inhibitors of Transcription

A large number of compounds have been characterized that inhibit transcription by binding to DNA and altering its template activity. The different classes of compounds in this category are briefly described below. Apart from phleomycin, for which the nature of the binding is uncertain, these inhibitors form noncovalent complexes with DNA. All act essentially by impairing template function. Two

---

* The "early" RNA transcribed *in vivo* from T4 DNA in *E. coli*, i.e. before T4 DNA replication, can be subdivided into "immediate-early" and subsequently synthesized or "delayed-early" RNA molecules.

compounds, luteoskyrin and kanchanomycin, which may impair RNA polymerase function as well as complexing with the DNA template, are also discussed in this section.

a) *Actinomycins*. The actinomycins are the most intensively studied group of compounds that impair template function by binding to DNA (REICH and GOLDBERG, 1964; GOLDBERG and FRIEDMAN, 1971). Actinomycin D (Fig. 2, I) is the best known of these compounds. A variety of models have been proposed

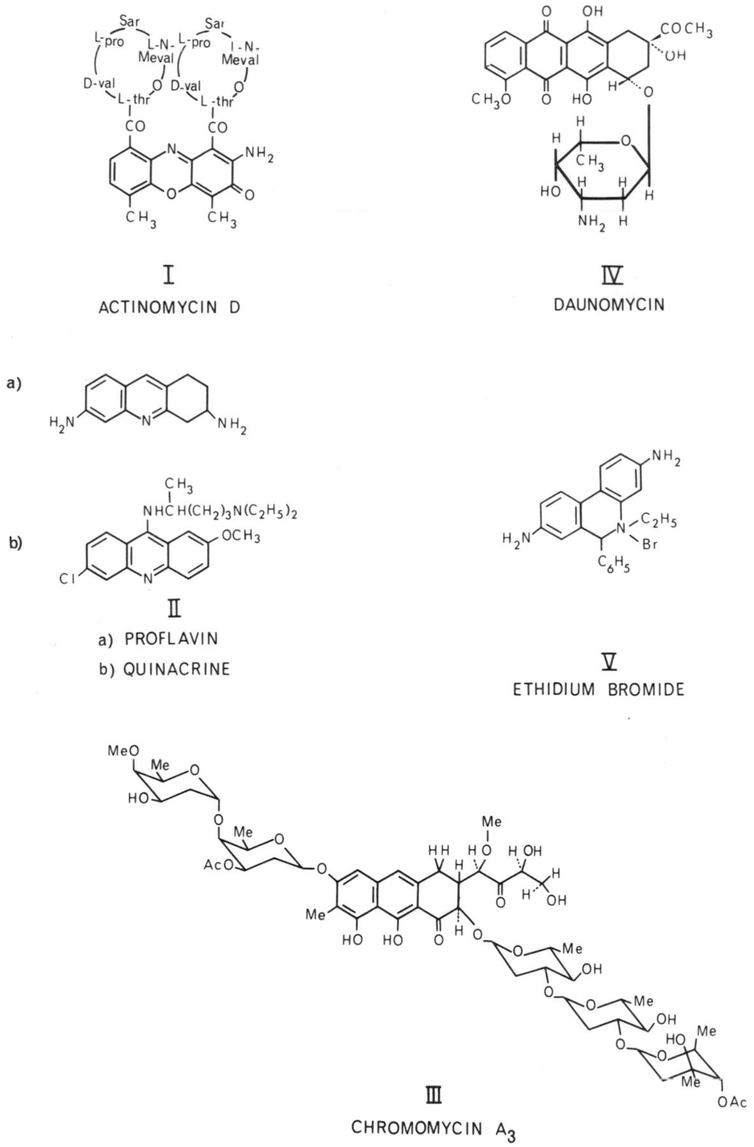

I
ACTINOMYCIN D

IV
DAUNOMYCIN

a)

b)

II

a) PROFLAVIN
b) QUINACRINE

V
ETHIDIUM BROMIDE

III
CHROMOMYCIN A₃

Fig. 2. Structural formulae for inhibitors of transcription which bind to DNA. The following abbreviations were used in the formula for actinomycin D: L-thr = l-threonine; D-val = d-valine; L-pro = l-proline; Sar = sarcosine; L-N-Meval = N-methyl-l-valine

for the binding of actinomycin D. One of the more probable is that of MULLER and CROTHERS (1968) in which the actinomycin D phenoxazone ring intercalates between base pairs adjacent to any G—C base pair while the peptide rings project into the minor groove of the DNA. SOBELL et al. (1971) have established that the phenoxazone ring intercalates into the DNA helix and that both cyclic peptides can hydrogen bond to dG residues. It is clear that the presence of guanine and helicity is mandatory for stable binding of actinomycin D to DNA and that binding occurs in the minor groove of DNA (GOLDBERG and FRIEDMAN, 1971). Actinomycin D at very low concentrations (1 μM) has little effect on the binding of the RNA polymerase to DNA (RICHARDSON, 1966) or on initiation (MAITRA et al., 1967) but preferentially blocks elongation; at higher concentrations (10 μM) actinomycin D also inhibits initiation. The inhibition of elongation by actinomycin D can be overcome by adding more DNA but not by adding more RNA polymerase. Differential effects of actinomycin D on RNA synthesis in different organisms and organelles *in vivo*, and on the synthesis of different types of RNA *in vivo*, may be related largely to membrane permeability, and to the operon size of the relevant DNA regions respectively (GOLDBERG and FRIEDMAN, 1971). RNA synthesis is considerably more sensitive to actinomycin D than is DNA synthesis.

   b) Proflavine. Proflavine (Fig. 2, IIa) intercalates with DNA and thereby alters the secondary structure of DNA (LERMAN, 1961). Proflavine inhibits the binding of RNA polymerase to DNA and inhibits initiation but has little effect on elongation of RNA chains (MAITRA et al., 1967; RICHARDSON, 1966). Other acridines such as quinacrine (Fig. 2, IIb) and acriflavine also intercalate with DNA (LERMAN, 1961, 1963), thereby inhibiting RNA synthesis.

   c) Chromomycin A_3, Mithramycin, Echinomycin and Olivomycin. These closely related antibiotics (chromomycin A_3: Fig. 2, III) are potent inhibitors of RNA synthesis (GOLDBERG, 1965). They require the presence of a divalent cation to bind to DNA and it is likely that a Mg^{2+}-antibiotic complex is the actual inhibitor (WARD et al., 1965). As is the case with actinomycin D, binding of these antibiotics to DNA requires the presence of guanine and helicity in the DNA (BEHR et al., 1969). These compounds probably do not intercalate with DNA (WARING, 1970). With some exceptions, RNA synthesis in mammalian and bacterial cells is more sensitive to these compounds than is DNA synthesis.

   d) Anthracyclines. The anthracycline antibiotics nogalomycin and daunomycin (Fig. 2, IV) bind noncovalently to DNA and are thought to intercalate between adjacent base pairs of DNA (KERSTEN et al., 1966; WARD et al., 1965).

   While the binding of anthracycline antibiotics to DNA increases with increasing G+C content of the DNA, G+C content is not mandatory for binding (KERSTEN et al., 1966). RNA and DNA polymerase activities are similarly sensitive to daunomycin, nogalomycin and other anthracycline antibiotics. Like actinomycin D, nogalomycin acts as an inhibitor of chain elongation rather than initiation (RICHARDSON, 1966).

   e) Rubiflavin, Hedamycin and Pluramycin. These antibiotics, as yet not fully characterized structurally, are inhibitors of both RNA and DNA synthesis in microorganisms and mammalian cells (NAGAI et al., 1967). *In vivo*, DNA synthesis is more sensitive to these compounds than RNA synthesis but *in vitro* RNA synthesis can be as sensitive as DNA synthesis (TANAKA et al., 1965; JOEL and GOLDBERG, 1970). Hedamycin acts as an inhibitor of RNA chain elongation (JOEL and GOLDBERG, 1970).

f) Distamycin A. Distamycin A (Fig. 3, I) inhibits DNA-dependent DNA and RNA synthesis through binding to the DNA (BUSCHENDORF et al., 1971). This antibiotic binds preferentially to AT rich DNA and inhibits the binding of *E. coli* RNA polymerase to DNA. In this context it is worth noting that there is evidence for preferential binding of the *E. coli* RNA polymerase to AT rich sequences of DNA (SHISHIDO and IKEDA, 1971).

g) Ethidium Bromide. Ethidium bromide (Fig. 2, V) intercalates with DNA thereby impairing template function. Of particular interest is the differential

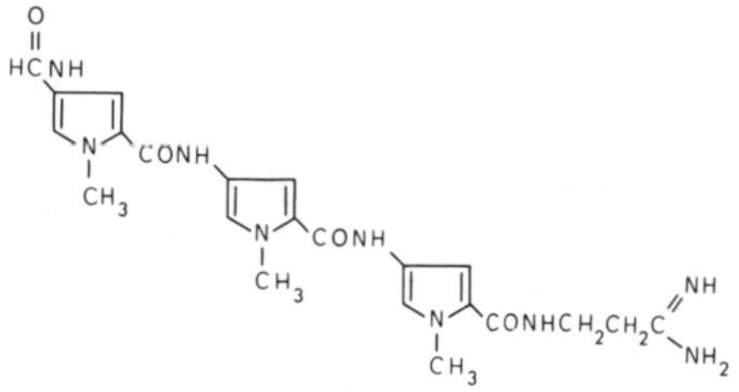

I

DISTAMYCIN A

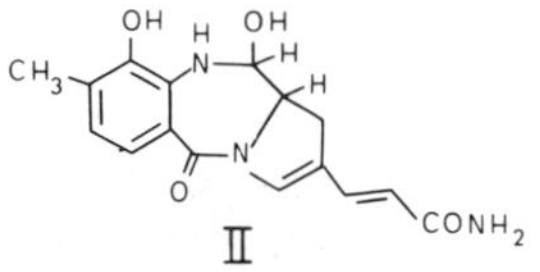

II

ANTHRAMYCIN

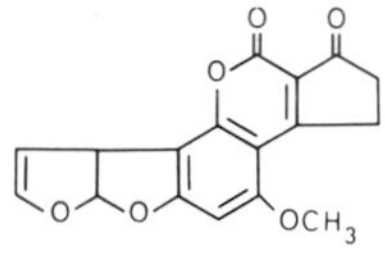

V

AFLATOXIN B1

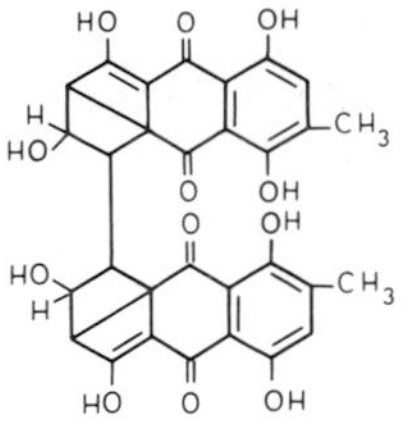

III

LUTEOSKYRIN

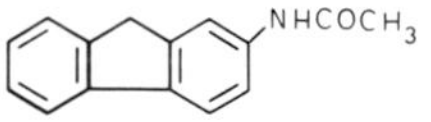

VI

2-ACETYLAMINOFLUORENE

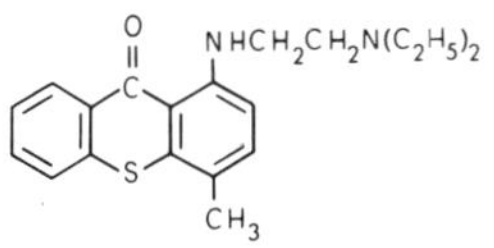

IV

MIRACIL D

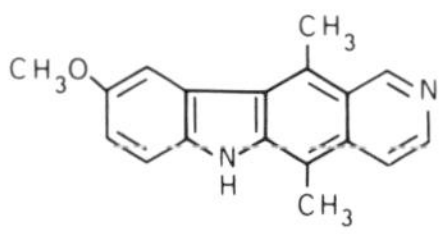

VII

METHOXY-9-ELLIPTICINE

Fig. 3. Structural formulae for inhibitors of transcription which bind to DNA binding of ethidium bromide to circular supercoiled as compared to linear DNA which permits use of ethidium bromide as a selective inhibitor of mitochondrial DNA-dependent processes (ZYLBER et al., 1969; KNIGHT, 1969; WARING, 1966; RADLOFF et al., 1967).

h) Phleomycin. Phleomycin is a mixture of copper-chelating, peptide-containing antibiotics as yet not completely defined structurally (TAKITA et al., 1969). Phleomycin inhibits DNA-dependent RNA and DNA synthesis as well as $Q\beta$-RNA polymerase (WATANABE and AUGUST, 1968). Phleomycin binds well to DNA with high $A+T$ content (FALASCHI and KORNBERG, 1964), but the type of binding to DNA has not been established.

i) Anthramycin. Anthramycin (Fig. 3, II) inhibits RNA and DNA synthesis in bacterial and mammalian cells (ADAMSON et al., 1968; KOHN et al., 1968; HORWITZ and GROLLMAN, 1969). This compound binds preferentially to native helical DNA, more weakly to denatured DNA and not at all to RNA. Anthramycin binds far more firmly to DNA than the antibiotics considered above but the exact nature of its binding is uncertain (GOLDBERG and FRIEDMAN, 1971).

j) Luteoskyrin. Luteoskyrin (Fig. 3, III), from *Penicillium islandicum*, is like chromomycin, mithramycin and olivomycin, in forming an antibiotic-Mg^{2+} complex that subsequently binds to DNA (OHBA and FROMAGEOT, 1967, 1968). Luteoskyrin forms two types of reversible complexes with DNA and RNA. Complex I is formed rapidly with purine-containing single-stranded DNA or RNA and contains antibiotic, magnesium ion and purine base in equimolar amounts. Complex II is formed slowly with double-stranded DNA or RNA or apurinic single-stranded nucleic acids, and possibly represents less specific aggregations of luteoskyrin-Mg^{2+} with the polynucleotide. Luteoskyrin inhibits the *E. coli* RNA polymerase reaction more effectively when denatured rather than native DNA is used as the template. Increasing DNA concentration does not overcome the inhibition, suggesting that luteoskyrin may also impair RNA polymerase function (SENTENAC et al., 1967).

k) Kanchanomycin. This antibiotic of unknown structure resembles luteoskyrin in forming two types of complexes with polynucleotides in the presence of stoichiometric amounts of Mg^{2+} (FRIEDMAN et al., 1969 a, b). Kanchanomycin inhibits both DNA-dependent RNA and DNA synthesis. Inhibition of RNA synthesis by kanchanomycin can be overcome by addition of more RNA polymerase or by addition of DNA that has not been treated with the antibiotic (JOEL et al., 1970). Kanchanomycin may thus have a dual effect of impairing both template and RNA polymerase functions.

l) Miracil D. Miracil D (Fig. 3, IV) binds to DNA, inhibits total cellular RNA synthesis in *E. coli* and also inhibits RNA polymerase *in vitro* (WEINSTEIN et al., 1965).

m) Aflatoxin. This toxic compound and powerful carcinogen from *Aspergillus fucus* (Fig. 3, V) is a potent inhibitor of DNA and RNA synthesis *in vivo* (DE RECONDO et al., 1966) and probably acts through impairing template activity by binding to DNA (for review see SCHOENTAL, 1967).

n) 2-Acetylaminofluorene. This hepatocarcinogen (Fig. 3, VI) binds to DNA and can abolish its template activity for RNA polymerase as well as decreasing its primer activity for DNA polymerase (TROLL et al., 1968).

o) Polyamines. Various polyamines bind to DNA and inhibit RNA polymerase activity. These inhibitors are reviewed by TABOR (1965).

p) Other Intercalation Compounds. A variety of other compounds inhibit RNA synthesis by intercalation with DNA. These include methoxy-9-ellipticine (FESTY et al., 1971) (Fig. 3, VII) and various substituted p-nitroanilines (GABBAY et al., 1972).

2. Protein-Specific Inhibitors of Transcription

Relatively few enzyme-specific inhibitors of transcription are known but the antibiotics of this kind that have been found are of considerable importance as selective inhibitors of prokaryotic, eukaryotic organelle or eukaryotic nucleoplasmic RNA synthesis. These inhibitors have been very useful in distinguishing between different modes of transcription in eukaryotic organisms. Rifampicin has been of considerable use in the definition of transcription processes.

a) Rifamycin and Its Derivatives. The rifamycins are a group of compounds that are potent inhibitors of bacterial RNA polymerase. Rifampicin (Fig. 4, Ib), a synthetic derivative of rifamycin B (Fig. 4, Ia; from *Streptomyces mediterranei*),

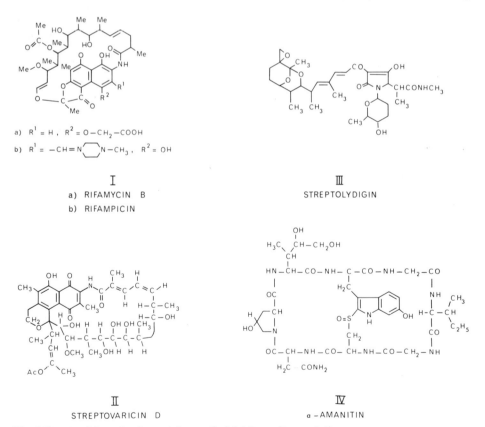

Fig. 4. Structural formulae for protein-specific inhibitors of transcription

is mostly widely used. Rifampicin binds noncovalently to the bacterial RNA polymerase (WEHRLI et al., 1968). One molecule of rifampicin is bound per molecule of the core enzyme ($\alpha_2 \beta \beta'$), and since a rifampicin-resistant mutant has been shown to contain a modified β chain (HEIL and ZILLIG, 1970), this subunit is probably involved in binding the inhibitor. Rifampicin is a highly specific inhibitor of bacterial RNA polymerase (50% inhibition at 2×10^{-8} M) and does not inhibit DNA polymerase. The initiation reaction only is inhibited by rifampicin, and not chain elongation (SIPPEL and HARTMANN, 1968). Preincubation of enzyme, DNA and nucleoside triphosphates provide a complex relatively insensitive to rifampicin. The primary step at which rifampicin acts is prior to the binding of the first nucleotide since rifampicin-resistant complexes of enzyme plus DNA have been demonstrated (SIPPELL and HARTMANN, 1970; BAUTZ and BAUTZ, 1970). Rifampicin does not inhibit nuclear RNA polymerases from mammalian, fungal or higher plant sources. There have been conflicting reports on the rifampicin-sensitivity of mitochondrial and chloroplast RNA polymerases, possibly due to membrane permeability effects and contamination of the preparations (see KUNTZEL and SCHÄFER, 1971). However solubilized mitochondrial RNA polymerases from animals and fungi are rifampicin-sensitive (KUNTZEL and SCHÄFER, 1971; REID and PARSONS, 1971). While RNA polymerases from *Chlamydomonas reinhardii* plastids (SURZYCKI, 1969) and from the blue-green alga *Anacystis nidulans* (HERZFELD and ZILLIG, 1971) are inhibited by rifampicin, partially-purified RNA polymerases from wheat and maize chloroplasts are insensitive to the antibiotic (BOTTOMLEY et al., 1971; POLYA and JAGENDORF, 1971). RNA polymerase III from the aquatic fungus *Blastocladiella emersonii* is sensitive to rifampicin (HORGEN and GRIFFIN, 1971). The T3 and T7 phage RNA polymerases are insensitive to rifampicin and streptolydigin (DUNN et al., 1971).

   b) Streptovaricin. Streptovaricin is chemically related to the rifamycins and like the rifamycins is an enzyme-specific inhibitor of initiation of bacterial RNA synthesis (MIZUNO et al., 1968). Again, like rifamycins, streptovaricin does not inhibit eukaryotic nuclear RNA polymerases. Streptovaricin is a mixture of related compounds; the structure of streptovaricin D is shown in Fig. 4 (formula II).

   c) Streptolydigin. Streptolydigin (Fig. 4, III) inhibits RNA chain elongation by bacterial RNA polymerase (SCHLEIF, 1969). Streptolydigin, like rifampicin, inhibits the core enzyme rather than interacting with the σ subunit. The concentration for 50% inhibition of *E. coli* RNA polymerase is 7×10^{-6} M (CASSANI et al., 1971).

   d) α-Amanitin. This polypeptide toxin (Fig. 4, IV), isolated from the mushroom *Amanita phalloides* (WIELAND, 1968), selectively inhibits the nucleoplasmic Mn^{2+}- and ammonium sulphate-activated RNA polymerase (RNA polymerase II or RNA polymerase B) of mammalian cells (JACOB et al., 1970a, b). α-Amanitin does not inhibit bacterial, eukaryote nucleolar (RNA polymerase I), chloroplast, or mitochondrial RNA polymerases, but does inhibit fungal nuclear RNA polymerases at relatively high concentrations (DEZELEE et al., 1970; HORGEN and GRIFFIN, 1971). One of the soluble RNA polymerases of maize is sensitive to α-amanitin (STRAIN et al., 1971). A variety of active analogues of α-amanitin exist (BUKU et al., 1971; FIUME and WIELAND, 1970).

3. Polyanions as Inhibitors of RNA Polymerase

Many polyanions such as polynucleotides, heparin and polyethylenesulphonate inhibit bacterial RNA polymerases by binding to the enzyme and thereby prevent binding of the enzyme to DNA. tRNA inhibits RNA polymerase in this fashion but the inhibition is negligible in reaction mixtures containing 0.2 M KCl. Re-initiation by bacterial RNA polymerase is inhibited by RNA product at low ionic strength (0.05 M KCl), but at higher ionic strength (e.g. 0.2 M KCl) the product RNA is released from an enzyme-DNA-RNA complex, allowing re-initiation to take place (for a review of these effects see RICHARDSON, 1969).

4. Nucleotide Analogues as Inhibitors of RNA Polymerase

A variety of nucleotide analogues act as inhibitors of RNA polymerase *in vivo* or *in vitro* by altering template characteristics through incorporation into DNA or by acting as analogues of RNA polymerase substrates (SUHADOLNIK, 1970). Studies of effects of analogues of the ribonucleoside triphosphate substrates have yielded information on structural requirements for hydrogen-bonding between complementary bases (KAHAN and HURWITZ, 1962) and for nearest neighbour incorporation (for brief reviews of this work see CRAMER et al., 1971; GERARD et al., 1971). Several such analogues act as RNA chain growth terminators. 3'-Deoxyadenosine 5'-triphosphate is incorporated into RNA chains in a reaction catalyzed by RNA polymerase but inhibits further elongation because of the absence of a 3'-hydroxyl (SHIGEURA and BOXER, 1964). 3'-Amino-3'-deoxyadenosine 5'-triphosphate acts in a similar fashion (SHIGEURA et al., 1966). 2'-O-Methyl-adenosine 5'-triphosphate also acts as a substrate for RNA polymerase but despite the presence of a 3'-hydroxyl, incorporation of the next nucleotide is considerably inhibited (GERARD et al., 1971).

5. Physiological Inhibitors of RNA Polymerase

Pyrophosphate, the product of phosphodiester bond formation in RNA synthesis, inhibits RNA polymerase. In T3-infected *E. coli* cells, an inhibitory protein is found which antagonizes the σ component of the host RNA polymerase. While this protein strongly inhibits the host holoenzyme, the core enzyme and the T3 RNA polymerase are not significantly inhibited (MAHADIK et al., 1971). A lipid inhibitor of mammalian and bacterial RNA synthesis is introduced into HeLa cells by poliovirus infection (HO and WASHINGTON, 1971).

E. Control of Prokaryote Transcription

The last decade has seen the isolation of many protein components that can specifically control transcription in bacterial and bacteriophage systems. This section briefly examines the major mechanisms of control that have been elucidated.

There are some excellent recent reviews dealing with control of prokaryote transcription (BURGESS, 1971; TRAVERS, 1971; PASTAN and PERLMAN, 1971; GEIDUSCHEK and HASELKORN, 1969; YARUS, 1969).

1. Repressors

An important prediction of the genetic regulation model of JACOB and MONOD (1961) requires that repressor proteins in the cell interact with and regulate the transcription of DNA. Certain repressor proteins have been isolated (GILBERT and MULLER-HILL, 1966; PTASHNE, 1967), and recent work on their characterization is reviewed by EPSTEIN and BECKWITH (1968) and YARUS (1969). A brief summary of the properties of repressor proteins and their interaction with RNA polymerase is given below.

   a) The Lac Repressor. The *lac* repressor, isolated by GILBERT and MULLER-HILL (1966), is coded for by the lactose (*lac*) operon regulatory gene and binds to operator DNA. The *lac* repressor, which is a tetrameric protein with a molecular weight of 150,000 (RIGGS and BOURGEOIS, 1968), binds a gratuitous inducer isopropylthiogalactoside (dissociation constant 1.3×10^{-6} M) and binds tightly to operator DNA (dissociation constant 3×10^{-10} M in 0.15 M KCl and 0.01 M $MgCl_2$; 3×10^{-12} M at lower ionic strength). The inducer greatly reduces the affinity of the *lac* repressor for DNA (GILBERT and MULLER-HILL, 1966). As will be discussed later, the *lac* repressor blocks transcription of the *lac* operon by *E. coli* RNA polymerase *in vitro*.

   b) The λ Bacteriophage Repressor. The λ repressor, isolated by PTASHNE (1967), prevents the production of λ-specific RNA by *E. coli* RNA polymerase using λ DNA as template (ECHOLS et al., 1968). The λ repressor is an acidic protein of molecular weight 30,000 which binds very tightly to λ DNA (dissociation constant 10^{-10} M). As with the *lac* repressor, affinity for the appropriate DNA is decreased by increasing ionic strength and binding is abolished by denaturation of the DNA (PTASHNE, 1967).

   c) Other Repressor Proteins. The galactose (*gal*) operon repressor protein has been partially purified using affinity chromatography. The *gal* repressor binds to *gal* operon DNA (dissociation constant 10^{-12} M). Fucose and galactose (inducers of the galactose operon *in vivo*) reduce binding of the *gal* repressor by one half at 5×10^{-5} M (PARKS et al., 1971). GREENBLATT and SCHLEIF (1971) have demonstrated regulation of transcription of the arabinose operon *in vitro* by the positive regulator *ara C* protein.

2. Specific Transcription Activation Factors

A variety of transcription activation factors have been found for the bacterial system in recent years including the σ, C, M and ψ factors and the catabolite gene-activating protein. The properties of these factors are outlined below.

   a) σ-Factors. The bacterial RNA polymerase holoenzyme can be separated into core enzyme and σ-factor by phosphocellulose chromatography (BURGESS et al., 1969). The material already presented in this chapter indicates that σ plays a vital rôle in the initiation process. Further information regarding the regulatory function of σ is now presented. The core enzyme retains catalytic activity but its ability to transcribe T4 DNA is much less than that of holoenzyme. σ stimulates initiation at specific sites, functions catalytically and is released from an enzyme-DNA-nascent RNA complex for re-use in initiation. σ, which does not itself bind to DNA, is also dissociated from the holoenzyme by single-stranded poly-ribonucleotides suggesting that the elongation of RNA chains leads to the release of σ soon after initiation (KRAKOV and FRONK, 1969). As discussed earlier, the presence of σ restricts *E. coli* RNA polymerase to transcription of early cistrons on T4 and T7 DNA while in the absence of σ transcription is much less specific. SUGIURA et al. (1970) have shown that while the holoenzyme transcribes 3 discrete RNA chains from one strand of the fd phage replicative form, the core enzyme transcribes both strands and produces heterogeneous RNAs with a variety of initial sequences. Thus the stimulation of transcription by σ is due to stimulation of specific initiation.

b) M Factor and H Factor. M factor, isolated from *E. coli* ribosomes by DAVISON et al. (1969), is a protein of about 5S that greatly stimulates transcription of T4 and λ DNA by *E. coli* core and holoenzyme RNA polymerase. The stimulation of transcription by M factor is remarkably template-specific. Although M factor is distinct from σ, it stimulates initiation at the same sites.

H factor, a thermostable substance from *E. coli* lysate supernatant, stimulates *in vitro* transcription in the presence of excess DNA, and may inhibit when enzyme is present in excess. In this respect H factor resembles M factor since M factor stimulation is decreased when the enzyme to DNA ratio is increased. H factor, which has been purified to the point of electrophoretic homogeneity, has a molecular weight of 10,000 and probably acts at the stage of formation of the pre-initiation complex (JUEQUET et al., 1971).

c) Factor C. Like M factor, this factor is released from ribosomes by high salt extraction. Factor C stimulates transcription as well as promoting formyl-methionyl-tRNA binding to 30S subunits (REVEL et al., 1968; REVEL and GROS, 1967).

d) ψ-Factor and the Stringent Response. While synthesis of rRNA can account for up to 40% of RNA synthesized in rapidly growing *E. coli* (although the rRNA cistrons represent only 0.2–0.4% of the genome), this proportion was not found by TRAVERS et al. (1970) when *E. coli* DNA was transcribed by the RNA polymerase *in vitro*; less than 0.2% of the RNA thus made was rRNA. These workers have isolated a factor called ψ ($ψ_r$) which apparently stimulates rRNA synthesis several hundred-fold *in vitro* so that rRNA accounts for a substantial proportion of RNA synthesized *in vitro* with *E. coli* holoenzyme and *E. coli* DNA. $ψ_r$ activity is also exhibited by Qβ replicase subunits coded for by *E. coli* DNA. The nucleoside tetraphosphate ppGpp, which increases when stringent strains of *E. coli* are starved of amino acids, inhibits *in vitro* ψ-stimulated rRNA synthesis rather specifically (TRAVERS et al., 1970). This provides an explanation for the fact that preferential inhibition of rRNA synthesis results from amino acid starvation of stringent *E. coli* (STAMATO and PETTIJOHN, 1971; EDLIN and BRODA, 1968). Nevertheless a more critical examination of this system by HASELTINE

(1972) shows that rRNA synthesis *in vitro* may represent 7–14% of RNA synthesis; that ψ_r, while stimulating transcription of *E. coli* DNA by the RNA polymerase holoenzyme, does not stimulate rRNA synthesis; and that ppGpp does not inhibit rRNA synthesis in the absence or presence of ψ_r. Thus the regulatory rôle of ψ_r is at present unclear.

Ribosomes stimulate *in vitro* transcription (MORRIS and GOULD, 1971; JONES et al., 1968), and a further ribosome-derived factor, in addition to factors C, M and ψ, has been demonstrated by MAHADIK and SRINIVASAN (1971). This factor, derived from RC^{rel} strains of *E. coli* that continue to make RNA in conditions of amino acid starvation, stimulates core RNA polymerase 20-fold; the holoenzyme activity is increased only twofold in the same conditions.

e) Catabolite Gene-activating Protein and Catabolite Repression. Catabolite repression of genes involved in sugar metabolism, such as those of the *gal*, *lac* and arabinose operons, is associated with a decrease in the 3′,5′-cyclic AMP concentration in bacterial cells and repression of these operons is relieved by cyclic AMP (for a review see PASTAN and PERLMAN, 1971). A protein called catabolite gene-activating protein (CAP) or cyclic AMP receptor protein (CR protein) has been isolated (ZUBAY et al., 1970; EMMER et al., 1970). This protein binds cyclic AMP and is required for cyclic AMP-dependent β-galactosidase synthesis *in vitro* (ZUBAY et al., 1970; EMMER et al., 1970). CAP with bound cyclic AMP attaches to *lac* DNA, stimulates β-galactosidase synthesis in a DNA-dependent protein-synthesizing system (ZUBAY et al., 1970) and is required for *in vitro* synthesis of *lac*-specific mRNA (see below). VARMUS et al. (1971) have established that the cyclic AMP regulation of the expression of the *lac* operon *in vivo* results from direct control of transcription, not translation. CAP has a molecular weight of 45,000 and is composed of 2 identical polypeptides of molecular weight 22,500 (ANDERSON et al., 1971).

f) Specific in vitro Transcription. Highly resolved *in vitro* transcription systems have been shown to require many of the factors discussed above. *In vitro* transcription of the *gal* operon (attached to λ DNA) by *E. coli* RNA polymerase requires CAP and cyclic AMP for maximal specific synthesis of *gal* mRNA (NISSLEY et al., 1971). DE CROMBRUGGHE et al. (1971a–c) have developed an *in vitro* system for transcription of the *lac* operon (attached to λ DNA). Transcription requires the RNA polymerase holoenzyme, cyclic AMP, and CAP and is inhibited by the *lac* repressor protein. Isopropylthiogalactoside and methylthiogalactoside both reverse this inhibition, caused by inability to form a rifampicin-resistant preinitiation complex. Cyclic AMP and CAP are required for appropriate binding of the RNA polymerase holoenzyme to the promoter region to form a preinitiation complex. The *lac* repressor protein and the RNA polymerase bind independently to *lac* DNA *in vitro* (CHEN et al., 1971).

g) Possible Role of Cyclic AMP-dependent Protein Kinases. MARTELO et al. (1970) have shown that cyclic AMP-dependent protein kinases from rabbit muscle or rabbit erythrocytes stimulate *E. coli* RNA polymerase activity on T4 DNA through phosphorylation of the σ-factor. This raises the possibility of a more general control of transcription in bacterial cells by cyclic AMP.

h) Possible Control of Transcription by Translation. Since RNA molecules are synthesized in the same (5′→3′) direction as they are translated, and since product RNA can inhibit transcription, there is a possibility that translation of mRNA might control the rate of concomitant transcription. This was proposed by STENT (1964, 1966) who suggested that the movement of ribosomes along the messenger may be required to release the RNA from its DNA template. Lengyel has proposed that initiation of protein synthesis controls the rate of mRNA synthesis (SHIH et al., 1966).

Some of the major lines of evidence supporting these hypotheses can be summarized as follows:

1) There is a close association of nascent RNA and ribosomes (MANGIAROTTI and SCHLESSINGER, 1967) and mRNA can be used translationally before its synthesis is complete (ALPERS and TOMKINS, 1966; LEIVE and KOLLIN, 1967).

2) The mRNA from operons containing a gene with a strong polar amber mutation (which terminates translation permaturely) can be shorter than mRNA from the normal operon (IMAMOTO and YANOFSKY, 1967a, b).

3) The stringent response of RC^{str} strains of *E. coli* can be mimicked by a temperature shift from 30° to 44°, resulting in a transient decrease in RNA synthesis as well as a transient inactivation of initiation of protein synthesis. This phenomenon, which is independent of amino-acid presence or absence, is not observed with RC^{rel} strains (PATTERSON and GILLESPIE, 1971). A temperature-sensitive mutant defective for protein chain elongation at 44° shows no such "stringent response" after a shift to 44°.

4) When translation in a temperature-sensitive *E. coli* mutant is inhibited by elevation of temperature, the rate of synthesis of tryptophan mRNA decreases proportionally. However synthesis of rRNA and tRNA is not impaired (IMAMOTO and KANO, 1971).

However there is evidence that the hypotheses of Stent and Lengyel are not generally applicable. Rates of RNA synthesis *in vitro* in the absence of translation can be quite comparable to rates *in vivo* (see Table 3; for a detailed discussion see GEIDUSCHEK and HASELKORN, 1969). *Lac* mRNA synthesized during recovery from potassium deficiency or from chloramphenicol inhibition is not translated into functional protein and there is no obligatory coupling of transcription of the *lac* operon to translation (ARTMAN and ENNIS, 1972). Furthermore, amino-acid starvation of stringent strains of bacteria preferentially inhibits rRNA synthesis (STAMATO and PETTIJOHN, 1971), and while net RNA synthesis is completely inhibited in these conditions, synthesis of mRNA continues at its former rate (MORRIS and KJELDGAARD, 1968).

3. Control of Bacteriophage Transcription

The ordered transcription of bacteriophage DNA after infection of *E. coli* is now partly explicable in terms of bacterial RNA polymerase specificities, changes to the bacterial RNA polymerases, bacterial and phage transcription-specifying factors and synthesis of phage RNA polymerases. Some of the major observations are summarized below (for a review see BURGESS, 1971).

a) Role of Bacterial Transcription Factors. In vitro the *E. coli* σ factor restricts the RNA polymerase to transcription on those regions of T4 and T7 DNA transcribed *in vivo* immediately after bacteriophage infection. No such specificity is shown by the core enzyme (for references see BURGESS, 1971). As mentioned previously, the presence or absence of σ has a considerable effect on the nature and length of T4 DNA transcripts *in vitro* (WITMER, 1971).

   b) Phage Transcription Factors. After early and late T4 phage infection σ-like factors have been detected. These specify transcription of certain T4 DNA regions not normally transcribed by the *E. coli* holoenzyme (TRAVERS, 1969, 1970). Similar factors appear to be coded for by λ DNA.

   c) Phage RNA Polymerases. Both T3 and T7 bacteriophages code for RNA polymerases that are highly specific for T3- and T7-DNAs respectively (CHAMBERLIN et al., 1970; DUNN et al., 1971; MAITRA, 1971). Both these RNA polymerases are small proteins (110,000 molecular weight) composed of one polypeptide.

   d) Modification of Host RNA Polymerase. After T4 phage infection modification of all subunits of the *E. coli* RNA polymerase has been shown to occur. Soon after T4 infection label can be shown to be incorporated into an ω-like subunit of the RNA polymerase but not into subunits, α, β or β' (STEVENS, 1970). Nevertheless subsequent modification of subunits α, β and β' of the RNA polymerase has been demonstrated (SCHACHNER and ZILLIG, 1971).

F. Control of Eukaryote Transcription

A variety of transcription-modifying factors have been isolated from eukaryote systems and the outlines of eukaryote transcriptional control are now appearing. In many cases factors analogous in function to some bacterial transcription-modifying factors have been isolated.

1. Specific Inhibitor of rRNA Synthesis

CRIPPA (1970) has isolated a protein which appears to function as a specific repressor of rRNA synthesis in *Xenopus* oocytes. While at stage 1 of *Xenopus* oocyte development more than 97% of RNA synthesized is rRNA, by stage 6 rRNA synthesis has decreased greatly. This inhibition of rRNA synthesis is attributable to the appearance of a protein which has been shown to bind to rRNA cistrons and to inhibit rRNA synthesis *in vitro*.

2. Transcription Activation Factors

Protein factors which stimulate *in vitro* RNA synthesis by eukaryote RNA polymerase have been isolated from several sources. A stimulation factor (termed S) from calf thymus stimulates calf thymus RNA polymerase II activity 10-fold only when native DNA is used as template. Factor S does not enhance the activity of the *E. coli* RNA polymerase and is thus a protein- and template-specific activation factor (STEIN and HAUSEN, 1970). A factor (factor C) requisite for activity of RNA polymerases A and B from coconut milk nuclei has been demonstrated (Mondal et al., 1970).

3. The Role of Chromosomal RNA and Histones in Gene Repression

Eukaryote chromosomal DNA is largely complexed with basic proteins called histones which prevent transcription of regions of the DNA. When histone-free regions of chromatin are removed, the template activity of the chromatin is decreased greatly (Bonner and Huang, 1963) and removal of histones from chromatin greatly enhances its template activity. Thus it appears that only histone-free regions of chromatin are available for transcription (Marushige and Bonner, 1966). Competition-hybridization experiments have shown that RNA synthesized *in vitro* by bacterial RNA polymerase using eukaryote chromatin as template corresponds to the RNA synthesized *in vivo* (Paul and Gilmore, 1966; Bekhor et al., 1969) although only a small proportion of the total chromatin DNA is complementary to this RNA (Bekhor et al., 1969).

According to Bonner and co-workers, chromosomal RNA confers specificity on DNA-histone interactions. Concepts of repression involving chromosomal RNA are not generally accepted at present and the existence of this RNA has been questioned (see Chapter 3 for a critical assessment of this subject).

Histones interact with transcription components in a variety of ways. Histone III inhibits RNA synthesis by inactivating the RNA polymerase, whereas histone I inhibits RNA synthesis by binding to DNA (Spelsberg et al., 1969). Histones may be involved in crosslinking DNA strands in chromatin (Littau et al., 1965) and certain histones are required for supercoiling of chromatin DNA (Richards and Pardon, 1970). Histones are methylated, phosphorylated and acetylated *in vivo* and in some cases such modifications can be induced by hormones and correlated with increased RNA synthesis. Nevertheless the precise mechanisms of specific histone control of transcription remain obscure. For an excellent recent review of histone structure and function see De Lange and Smith (1971).

4. Acidic Nuclear Proteins and Cyclic AMP Control of Transcription

Acidic nuclear proteins (ANPs), derived from chromatin, enhance rates of transcription *in vitro* in template-specific fashion. ANPs from rat kidney increase the rate of transcription of rat liver and rat kidney DNA by *E. coli* RNA polymerase by nearly 100% but have no effect when calf thymus DNA is used as template (Teng et al., 1970). ANPs bind to DNA of the same species and in some cases to the DNA of closely related species, and binding to DNA correlates with enhancement of RNA synthesis (Teng et al., 1971). ANPs are present in the chromatin of metabolically active tissues, have high rates of turnover, and are located in the regions of chromatin most active in RNA synthesis. Many of these proteins are phosphoproteins and phosphorylation increases at times of gene activation by drugs or hormones (Ruddon and Rainey, 1971). It has been suggested that ANPs may be eukaryote analogues of the bacterial σ factors (TENG et al., 1971). RUDDON and RAINEY (1971) have proposed a model for gene activation as a result of hormone or carcinogen treatment of mammalian cells. It is suggested such "inducers" bind to ANPs thereby making them substrates for cyclic AMP-activated kinases. The phosphorylated ANPs bind to DNA, promoting RNA polymerase binding and transcription of naked regions of DNA. This model is consistent with stimulation of the synthesis and phosphorylation of ANPs by a variety of hormones and drugs, e.g. nicotine, phenobarbital, oestradiol, 3-methylcholanthrene and hydrocortisol (RUDDON and RAINEY, 1971).

5. Effects of Hormones on Higher Plant Transcription

Application of some plant hormones to plant tissue increases rates of RNA synthesis (KEY, 1969); this response has also been demonstrated with isolated plant nuclei (JOHRI and VARNER, 1968; CHERRY, 1967; MATTHYSSE and PHILLIPS, 1969). MATTHYSSE and ABRAMS (1970) have shown that a species-specific cyto-kinin-reactive protein is required for kinetin stimulation of in vitro transcription of pea bud DNA by E. coli RNA polymerase. VENIS (1971) has isolated a protein which appears to bind to auxin and stimulates transcription of the pea and corn chromatin by E. coli RNA polymerase. Indol-3-ylacetate has been shown to promote incorporation of labelled adenine into a nucleotide which appears to be 3',5'-cyclic AMP (AZHAR and MURTI, 1971). Hence cyclic AMP may be involved in activation of RNA synthesis in plants as well as in bacterial and animal cells. As with animal systems, application of certain plant hormones can also increase the amount of chromatin-associated RNA polymerase activity though it is not clear whether these effects derive from template or enzyme modification (for references and amplification see LEFFLER et al., 1971).

6. Effects of Hormones on Transcription in Animal Systems

Animal hormones may activate transcription in a variety of ways but many hormone effects may be attributed primarily to activation of adenyl cyclase (JOST and RICKENBERG, 1971) and in turn to activation of 3',5'-cyclic AMP-dependent protein kinases. Some of the other major hormonal responses are discussed below.

   a) Changes in RNA Polymerases. The relative activities of the multiple RNA polymerases of animal cells change during embryo development (ROEDER and RUTTER, 1970) and as a result of hormone treatment. Hypophysectomy (removal of pituitary gland) of rats increases the level of RNA polymerase C in rat liver nuclei; subsequent treatment with growth hormone increases enzyme A and decreases the level of C (SCHMUCKLER and TATA, 1971). Phytohaemagglutinin increases human lymphocyte RNA synthesis some 20-fold and increases RNA polymerase A 4–5 fold (COOKE et al., 1971). On the other hand both Mn^{2+}- and Mg^{2+}-dependent RNA polymerases from mammalian thymus decrease in activity after treatment of the animals with the thymolytic hormone cortisol

(Kehoe et al., 1969; Nakagawa and White, 1971). Oestradiol administration to calf uteri results in activation of adenyl cyclase and phosphorylation of a 5S receptor protein which in turn activates transcription. The hormone also induces modification of nucleolar RNA polymerase so that the polymerase becomes a better substrate for a cyclic AMP-dependent kinase (Arnaud et al., 1971).

b) Specificity of Transcriptional Activation by Hormones. In view of the observations presented above, it is not surprising to find that certain hormones stimulate synthesis of specific RNA molecules. For example, oestrogen administration to ovariectomized rats increases the synthesis of 28S and 18S RNA as well as precursor RNA for these species (Luck and Hamilton, 1972). Phytohaemagglutinin increases synthesis of 45S RNA in lymphocytes (Cooper, 1970).

c) Modification of Histones and Templates. Chromatin may be modified by polyadenosine diphosphoribosylation during the human cell cycle with consequent inhibition of its template activity (Smulson et al., 1971). Histones, which may modify DNA conformation, can in turn be modified by acetylation and methylation and there is a considerable literature on phosphorylation of histones (as well as of ANPs) concomitant with hormonal activation of RNA synthesis (for an extensive review see de Lange and Smith, 1971).

References

Bautz, E. K. F.: In Molecular genetics (Taylor, J. H., ed.), part II, p. 213. New York: Academic Press 1967.
Burgess, R. R.: Ann. Rev. Biochem. **40**, 711 (1971).
De Lange, R. J., Smith, E. L.: Ann. Rev. Biochem. **40**, 279 (1971).
Epstein, W., Beckwith, J. R.: Ann. Rev. Biochem. **37**, 411 (1968).
Geiduschek, E. P., Haselkorn, R.: Ann. Rev. Biochem. **38**, 647 (1969).
Goldberg, I. H.: Am. J. Med. **39**, 722 (1965).
Goldberg, I. H., Friedman, P. A.: Ann. Rev. Biochem. **40**, 775 (1971).
Gross, P. R.: Ann. Rev. Biochem. **37**, 631 (1968).
Jost, J. P., Rickenberg, H. V.: Ann. Rev. Biochem. **40**, 741 (1971).
Martin, R. G.: Ann. Rev. Genet. **3**, 181 (1969).
Pastan, I., Perlman, R. L.: Nature New Biol. **229**, 5 (1971).
Reich, E., Goldberg, I. H.: Progr. Nucl. Acid Res. Mol. Biol. **3**, 183 (1964).
Richardson, J. P.: Progr. Nucl. Acid Res. Mol. Biol. **9**, 75 (1969).
Silvestri, L.: First Lepetit Colloqium on RNA Polymerase. Amsterdam: North-Holland 1970.
Suhadolnik, R. J.: Nucleoside Antibiotics. New York: Wiley-Interscience 1970.
Travers, A.: Nature New Biol. **229**, 69 (1971).
Yarus, M.: Ann. Rev. Biochem. **38**, 841 (1969).

Other References
Adamson, R. H., Hart, L. G., De Vita, N. T., Oliverio, V. T.: Cancer Res. **28**, 343 (1968).
Alpers, D. H., Tompkins, G. M.: J. Biol. Chem. **241**, 4434 (1966).
Anderson, W. B., Schneider, A. B., Emmer, M., Perlman, R. L., Pastan, I.: J. Biol. Chem. **246**, 5929 (1971).
Arnaud, M., Beziat, Y., Borgna, J. L., Guilleux, J. C., Mousseron-Caret, M.: Biochim. Biophys. Acta **254**, 241 (1971).
Artman, M., Ennis, H. L.: J. Bacteriol. **110**, 652 (1972).
Azhar, S., Krishna Murti, C. R.: Biochem. Biophys. Res. Commun. **43**, 58 (1971).

BAKER, R.F., YANOFSKY, C.: Proc. Nat. Acad. Sci. U.S. **60**, 313 (1968).

BAUTZ, E.K.F., BAUTZ, F.A.: Nature **226**, 1219 (1970).

BECKMAN, J.S., DANIEL, V., TICHAUER, Y., LITTAUER, U.Z.: Biochem. Biophys. Res. Commun. **45**, 806 (1971).

BEHR, W., HONIKEL, K., HARTMANN, G.: Eur. J. Biochem. **9**, 82 (1969).

BEKHOR, I., KUNG, G.M., BONNER, J.: J. Mol. Biol. **39**, 35 (1969).

BONNER, J., HUANG, R.C.: J. Mol. Biol. **6**, 169 (1963).

BOTTOMLEY, W., SMITH, H.J., BOGORAD, L.: Proc. Nat. Acad. Sci. U.S. **68**, 2412 (1971).

BREMER, H.: Mol. Gen. Genetics **99**, 362 (1967).

BREMER, H., BERRY, L.: Nature New Biol. **234**, 81 (1971).

BREMER, H., KONRAD, M.W.: Proc. Nat. Acad. Sci. U.S. **51**, 807 (1964).

BREMER, H., KONRAD, M.W., BRUNER, R.: J. Mol. Biol. **16**, 104 (1966).

BREMER, H., KONRAD, M.W., GAINES, K., STENT, G.S.: J. Mol. Biol. **13**, 540 (1965).

BREMER, H., YUAN, D.: J. Mol. Biol. **34**, 527 (1968).

BUKU, A., CAMPADETTI-FIUME, G., FIUME, L., WIELAND, T.: FEBS Lett. **14**, 42 (1971).

BURGESS, R.R.: J. Biol. Chem. **244**, 6160 (1969a).

BURGESS, R.R.: J. Biol. Chem. **244**, 6168 (1969b).

BURGESS, R.R., TRAVERS, A.A., DUNN, J.J., BAUTZ, E.K.F.: Nature **221**, 43 (1969).

BUSCHENDORF, B., PETERSEN, E., WOLF, H., WERDRAU, H.: Biochem. Biophys. Res. Commun. **43**, 617 (1971).

CASSANI, G., BURGESS, R.R., GOODMAN, H.M., GOLD, L.: Nature New Biol. **230**, 197 (1971).

CHAMBERLIN, M., BERG, P.: Proc. Nat. Acad. Sci. U.S. **48**, 81 (1962).

CHAMBERLIN, M., MCGRATH, J., WASKELL, L.: Nature **228**, 227 (1970).

CHEN, B., DE CROMBRUGGHE, B., ANDERSON, W.B., GOTTESMAN, M.E., PASTAN, I., PERLMAN, R.L.: Nature New Biol. **233**, 67 (1971).

CHERRY, J.H.: Ann. N.Y. Acad. Sci. **144**, 154 (1967).

CHESTERTON, C.J., BUTTERWORTH, P.H.W.: FEBS Lett. **15**, 181 (1971).

COOKE, A., KAY, J.I., COOPER, J.L.: Biochem. J. **125**, 74 P (1971).

COOPER, H.L.: Nature **227**, 1105 (1970).

CRAMER, F., GOTTSCHALK, E.M., MATZURA, H., SCHEIT, K.-H., STERNBACH, H.: Eur. J. Biochem. **19**, 379 (1971).

CRIPPA, M.: Nature **227**, 1138 (1970).

DAVISON, J., PILANSKI, L.M., ECHOLS, H.: Proc. Nat. Acad. Sci. U.S. **63**, 168 (1969).

DE CROMBRUGGHE, B., CHEN, B., ANDERSON, W., NISSLEY, P., GOTTESMAN, M., PASTAN, I., PERLMAN, R.: Nature New Biol. **231**, 139 (1971a).

DE CROMBRUGGHE, B., CHEN, B., ANDERSON, W.B., GOTTESMAN, M.E., PERLMAN, R.L., PASTAN, I.: J. Biol. Chem. **246**, 7343 (1971b).

DE CROMBRUGGHE, B., CHEN, B., GOTTESMAN, M., PASTAN, I., VARMUS, H.E., EMMER, M., PERLMAN, R.L.: Nature New Biol. **230**, 37 (1971c).

DE RECONDO, A.M., FRAYSSINET, C., LAFARGE, C., LE BRETON, E.: Biochim. Biophys. Acta **119**, 322 (1966).

DEZELEE, S., SENTENAC, A., FROMAGEOT, P.: FEBS Lett. **7**, 220 (1970).

DUNN, J.J., BAUTZ, F.A., BAUTZ, E.K.: Nature New Biol. **230**, 95 (1971).

ECHOLS, H., PILARSKI, L., CHENG, P.Y.: Proc. Nat. Acad. Sci. U.S. **59**, 1016 (1968).

EDLIN, G., BRODA, P.: Bacteriol. Rev. **32**, 206 (1968).

EMMER, M., DE CROMBRUGGHE, B., PASTAN, I., PERLMAN, R.L.: Proc. Nat. Acad. Sci. U.S. **66**, 480 (1970).

FALASCHI, A., KORNBERG, A.: Fed. Proc. **23**, 940 (1964).

FESTY, B., POISSON, J., PAOLETTI, C.: FEBS Lett. **17**, 321 (1971).

FIUME, L., WIELAND, T.: FEBS Lett. **8**, 1 (1970).

FRIEDMAN, P.A., JOEL, P.B., GOLDBERG, I.H.: Biochemistry **8**, 1535 (1969a).

FRIEDMAN, P.A., LI, T.-K., GOLDBERG, I.H.: Biochemistry **8**, 1545 (1969b).

GABBAY, E.J., DE STEFANO, R., SANFORD, K.: Biochem. Biophys. Res. Commun. **46**, 155 (1972).

GERARD, G.F., ROTTMAN, F., BOEZI, J.A.: Biochemistry **10**, 1974 (1971).

GILBERT, W., MULLER-HILL, B.: Proc. Nat. Acad. Sci. U.S. **56**, 1891 (1966).

GREENBLATT, J., SCHLEIF, R.: Nature New Biol. **233**, 166 (1971).

HASELTINE, W.A.: Nature **235**, 329 (1972).

HAYASHI, M. N., HAYASHI, M.: Proc. Nat. Acad. Sci. U. S. **61**, 1107 (1968).
HEIL, A., ZILLIG, W.: FEBS Lett. **11**, 165 (1970).
HERZFELD, F., ZILLIG, W.: Eur. J. Biochem. **24**, 242 (1971).
HO, P. P. K., WASHINGTON, A. L.: Biochemistry **10**, 3646 (1971).
HORGEN, P. A., GRIFFIN, D. H.: Proc. Nat. Acad. Sci. U. S. **68**, 338 (1971).
HORWITZ, S. B., GROLLMAN, A. P.: In Antimicrobial agents and chemotherapy—1968 (HOBBY, G. L., ed.), p. 21. Bethesda: Amer. Soc. Microbiol. 1969.
IMAMOTO, F., KANO, Y.: Nature New Biol. **232**, 169 (1971).
IMAMOTO, F., YANOFSKY, C.: J. Mol. Biol. **28**, 1 (1967a).
IMAMOTO, F., YANOFSKY, C.: J. Mol. Biol. **28**, 25 (1967b).
JACOB, F., MONOD, J.: J. Mol. Biol. **3**, 318 (1961).
JACOB, S. T., SAJDEL, E. M., MUNRO, H. N.: Biochem. Biophys. Res. Commun. **38**, 765 (1970a).
JACOB, S. T., SAJDEL, E. M., MUNRO, H. N.: Nature **225**, 60 (1970b).
JACQUET, M., CUBIER-KAHN, R., PLA, J., GROS, F.: Biochem. Biophys. Res. Commun. **45**, 1597 (1971).
JOEL, P. B., FRIEDMAN, P. A., GOLDBERG, I. H.: Biochemistry **9**, 4421 (1970).
JOEL, P. B., GOLDBERG, I. H.: Biochim. Biophys. Acta **224**, 361 (1970).
JOHNSON, C. J., DE BACKER, M., BOEZI, J. A.: J. Biol. Chem. **246**, 1222 (1971).
JOHRI, M. M., VARNER, J. E.: Proc. Nat. Acad. Sci. U. S. **59**, 269 (1968).
JONES, O. W., BERG, P.: J. Mol. Biol. **22**, 199 (1966).
JONES, O. W., DIECKMAN, M., BERG, P.: J. Mol. Biol. **31**, 177 (1968).
KAHAN, F. M., HURWITZ, J.: J. Biol. Chem. **237**, 3778 (1962).
KEDINGER, C. P., NURET, P., CHAMBON, P.: FEBS Lett. **15**, 169 (1971).
KEHOE, J. M., LUST, G., BEISEL, W. R.: Biochim. Biophys. Acta **174**, 761 (1969).
KERSTEN, W., KERSTEN, H., SZYBALSKI, W.: Biochemistry **5**, 236 (1966).
KEY, J. L.: Ann. Rev. Plant Physiol. **20**, 449 (1969).
KNIGHT, E.: Biochemistry **8**, 5089 (1969).
KOHN, K. W., BONO, V. H., KANVI, H. E.: Biochim. Biophys. Acta **155**, 121 (1968).
KRAKOV, J. S., FRONK, E.: J. Biol. Chem. **244**, 5988 (1969).
KÜNTZEL, H., SCHÄFER, K. P.: Nature New Biol. **231**, 265 (1971).
LEFFLER, H. R., O'BRIEN, T. J., GLOVER, D. V., CHERRY, J. H.: Plant Physiol. **48**, 43 (1971).
LEIVE, L., KOLLIN, V.: J. Mol. Biol. **24**, 247 (1967).
LERMAN, L. S.: J. Mol. Biol. **3**, 18 (1961).
LERMAN, L. S.: Proc. Nat. Acad. Sci. U. S. **49**, 94 (1963).
LITTAU, V. C., BURDICK, C. J., ALLFREY, V. G., MIRSKY, A. E.: Proc. Nat. Acad. Sci. U. S. **54**, 1204 (1965).
LUCK, D. N., HAMILTON, T. H.: Proc. Nat. Acad. Sci. U. S. **69**, 157 (1972).
MAHADIK, S. P., DHARMGRONGARTAMA, B., SRINIVASAN, P. R.: Proc. Nat. Acad. Sci. U. S. **69**, 162 (1971).
MAHADIK, S. P., SRINIVASAN, P. R.: Proc. Nat. Acad. Sci. U. S. **68**, 1898 (1971).
MAITRA, U.: Biochem. Biophys. Res. Commun. **43**, 443 (1971).
MAITRA, U., BARASH, F.: Proc. Nat. Acad. Sci. U. S. **64**, 779 (1969).
MAITRA, U., HURWITZ, J.: Proc. Nat. Acad. Sci. U. S. **54**, 815 (1965).
MAITRA, U., HURWITZ, J.: J. Biol. Chem. **242**, 4897 (1967).
MAITRA, U., NAKATA, Y., HURWITZ, J.: J. Biol. Chem. **242**, 4908 (1967).
MANDEL, J. L., CHAMBON, P.: FEBS Lett. **15**, 175 (1971).
MANGIAROTTI, G., APIRION, D., SCHLESSINGER, D., SILENGO, L.: Biochemistry **7**, 456 (1968).
MANGIAROTTI, G., SCHLESSINGER, D.: J. Mol. Biol. **29**, 395 (1967).
MANOR, H., GOODMAN, D., STENT, G. S.: J. Mol. Biol. **39**, 1 (1969).
MARTELO, O. J., WOO, S. C. L., REIMANN, E. M., DAVIE, E. W.: Biochemistry **9**, 4807 (1970).
MARUSHIGE, K., BONNER, J.: J. Mol. Biol. **15**, 160 (1966).
MATTHYSSE, A. G., ABRAMS, M.: Biochim. Biophys. Acta **119**, 511 (1970).
MATTHYSSE, A. G., PHILLIPS, C.: Proc. Nat. Acad. Sci. U. S. **63**, 897 (1969).
MIZUNO, S., YAMAZAKI, H., NITTA, K., UMEZAWA, H.: Biochim. Biophys. Acta **157**, 322 (1968).
MONDAL, H., MANDAL, R. K., BISWAS, B. B.: Biochem. Biophys. Res. Commun. **40**, 1194 (1970)
MORRIS, D. W., KJELDGAARD, N. O.: J. Mol. Biol. **31**, 145 (1968).
MORRIS, M. E., GOULD, H.: Proc. Nat. Acad. Sci. U. S. **68**, 481 (1971).
MÜLLER, W., CROTHERS, D. M.: J. Mol. Biol. **35**, 251 (1968).
NAGAI, K., YAMAKI, H., TANAKA, N., UMEZAWA, H.: J. Biochem. (Tokyo) **62**, 321 (1967).

NAKAGAWA, S., WHITE, A.: Biochem. Biophys. Res. Commun. **43**, 239 (1971).

NISSLEY, S. P., ANDERSON, W. B., GOTTESMAN, M. E., PERLMAN, R. L., PASTAN, I.: J. Biol. Chem. **246**, 4671 (1971).

OHBA, Y., FROMAGEOT, P.: Eur. J. Biochem. **1**, 147 (1967).

OHBA, Y., FROMAGEOT, P.: Eur. J. Biochem. **6**, 98 (1968).

OKAMOTO, T., SUGIURA, M., TAKANAMI, M.: Biochemistry **9**, 3533 (1970).

PARKS, J. S., GOTTESMAN, M., SHIMADA, K., WEISBERG, R., PERLMAN, R. L., PASTAN, I.: Proc. Nat. Acad. Sci. U. S. **68**, 1891 (1971).

PATTERSON, D., GILLESPIE, D.: Biochem. Biophys. Res. Commun. **45**, 476 (1971).

PAUL, J., GILMOUR, R.: J. Mol. Biol. **16**, 242 (1966).

PETTIJOHN, D.: Eur. J. Biochem. **3**, 25 (1967).

PETTIJOHN, D., KAMIYA, T.: J. Mol. Biol. **29**, 275 (1967).

PETTIJOHN, D. E., STONINGTON, O. G., KOSSMAN, C. R.: Nature **228**, 235 (1970).

POLYA, G. M.: Arch. Biochem. Biophys., in press (1973).

POLYA, G. M., JAGENDORF, A. T.: Arch. Biochem. Biophys. **146**, 635 (1971).

PTASHNE, M.: Proc. Nat. Acad. Sci. U. S. **57**, 306 (1967).

RADLOFF, R., BAUER, W., VINOGRAD, J.: Proc. Nat. Acad. Sci. U. S. **57**, 1514 (1967).

REID, B. D., PARSONS, P.: Proc. Nat. Acad. Sci. U. S. **68**, 2830 (1971).

REVEL, M., GROS, F.: Biochem. Biophys. Res. Commun. **27**, 12 (1967).

REVEL, M., HERZBERG, M., BECAREVIC, A., GROS, F.: J. Mol. Biol. **33**, 231 (1968).

RICHARDS, B. M., PARDON, J. F.: Exp. Cell Res. **62**, 184 (1970).

RICHARDSON, J. P.: J. Mol. Biol. **21**, 83 (1966).

RIGGS, A. D., BOURGEOIS, S.: J. Mol. Biol. **34**, 361 (1968).

ROBERTS, J. W.: Nature **224**, 1168 (1969).

ROEDER, R. G., RUTTER, W. J.: Biochemistry **9**, 2543 (1970).

RUDDON, R. W., RAINEY, C. H.: FEBS Lett. **14**, 170 (1971).

SCHACHNER, M., ZILLIG, W.: Eur. J. Biochem. **22**, 513 (1971).

SCHLEIF, R.: Nature **223**, 1068 (1969).

SCHMUCKLER, E. A., TATA, J. R.: Nature New Biol. **234**, 37 (1971).

SCHOENTAL, R.: Ann. Rev. Pharmacol. **7**, 343 (1967).

SENTENAC, A., RUET, A., FROMAGEOT, P.: Bull. Soc. Chim. Biol. **49**, 247 (1967).

SHIGEURA, H. T., BOXER, G. E., MELONI, M. L., SAMPSON, S. D.: Biochemistry **5**, 994 (1966).

SHIGEURA, J. T., BOXER, G. E.: Biochem. Biophys. Res. Commun. **17**, 758 (1964).

SHIH, A.-Y., EISENSTADT, J., LENGYEL, P.: Proc. Nat. Acad. Sci. U. S. **56**, 1599 (1966).

SHISHIDO, K., IKEDA, Y.: J. Biochem. (Tokyo) **68**, 881 (1970).

SHISHIDO, K., IKEDA, Y.: Biochem. Biophys. Res. Commun. **44**, 1420 (1971).

SIPPEL, A. E., HARTMANN, G.: Biochim. Biophys. Acta **157**, 218 (1968).

SIPPEL, A. E., HARTMANN, G.: Eur. J. Biochem. **16**, 152 (1970).

SMULSON, M., HENRIKSEN, O., RIDEAU, C.: Biochem. Biophys. Res. Commun. **43**, 1266 (1971).

SOBELL, H. M., JAIN, S. C., SAKORE, T. D., NORDMAN, C. E.: Nature New Biol. **321**, 200 (1971).

SPELSBERG, T. C., TANKERSLEY, S.: Proc. Nat. Acad. Sci. U. S. **62**, 1218 (1969).

STAMATO, T. D., PETTIJOHN, D. E.: Nature New Biol. **234**, 99 (1971).

STEIN, H., HAUSEN, P.: Eur. J. Biochem. **14**, 270 (1970).

STENT, G. S.: Science **144**, 816 (1964).

STENT, G. S.: Proc. Roy. Soc. (London) Ser. B **164**, 181 (1966).

STERNBERGER, N., STEVENS, A.: Biochem. Biophys. Res. Commun. **24**, 937 (1966).

STEVENS, A.: Biochim. Biophys. Res. Commun. **41**, 367 (1970).

STOUT, E. R., MANS, R. J.: Biochim. Biophys. Acta **134**, 327 (1967).

STRAIN, G. C., MULLINIX, L., BOGORAD, L.: Proc. Nat. Acad. Sci. U. S. **68**, 2647 (1971).

SUGIURA, M., OKAMOTO, T., TAKANAMI, M.: Nature **225**, 598 (1970).

SURZYCKI, S. J.: Proc. Nat. Acad. Sci. U. S. **63**, 1327 (1969).

SZYBALSKI, W., KUBINSKI, H., SHELDRICK, P.: Cold Spring Harbor Symp. Quant. Biol. **31**, 123 (1966).

TABOR, H., TABOR, C. W.: Pharmacol. Rev. **16**, 245 (1965).

TAKITA, T., MAEDA, K., UMEZAWA, H., OMOTO, S., UMEZAWA, S.: J. Antibiot. (Tokyo) **22**, 237 (1969).

TANAKA, N., NAGAI, K., YAMAGUCHI, H., UMEZAWA, H.: Biochem. Biophys. Res. Commun. **21**, 328 (1965).

TAYLOR, K., HRADECNA, Z., SZYBALSKI, W.: Proc. Nat. Acad. Sci. U. S. **57**, 1618 (1967).

Teng, C. T., Teng, C. S., Allfrey, V. G.: Biochem. Biophys. Res. Commun. **41**, 690 (1970).
Teng, C. S., Teng, C. T., Allfrey, V. G.: J. Biol. Chem. **246**, 3597 (1971).
Travers, A. A.: Nature **223**, 1107 (1969).
Travers, A. A.: Nature **225**, 1009 (1970).
Travers, A. A., Kamen, R. I., Schleif, R. F.: Nature **328**, 748 (1970).
Troll, W., Belman, S., Berkowitz, E., Chmielewicz, Z. F., Ambrus, J. L., Bardos, T. J.: Biochim. Biophys. Acta **157**, 16 (1968).
Varmus, H. E., Perlman, R. L., Pastan, I.: Nature New Biol. **230**, 41 (1971).
Venis, M. A.: Proc. Nat. Acad. Sci. U. S. **68**, 1824 (1971).
Vinograd, J., Lebowitz, J., Radloff, R., Watson, R., Laipis, P.: Proc. Nat. Acad. Sci. U. S. **53**, 1104 (1965).
Walter, G., Billig, W., Palm, P., Fuchs, E.: Eur. J. Biochem. **3**, 194 (1967).
Ward, D. C., Reich, E., Goldberg, I. H.: Science **149**, 1259 (1965).
Waring, M. J.: Biochem. Biophys. Acta **114**, 234 (1966).
Waring, M. J.: J. Mol. Biol. **54**, 247 (1970).
Watanabe, M., August, J. T.: J. Mol. Biol. **33**, 21 (1968).
Weaver, R. F., Blatti, S. P., Rutter, W. J.: Proc. Nat. Acad. Sci. U. S. **68**, 2994 (1971).
Wehrli, W., Knüsel, F., Schmid, K., Staehelin, M.: Proc. Nat. Acad. Sci. U. S. **61**, 667 (1968).
Weinstein, B., Chernoff, R., Finkelstein, I., Hirschberg, E.: Mol. Pharmacol. **1**, 297 (1965).
Weiss, S. B.: Proc. Nat. Acad. Sci. U. S. **46**, 1020 (1960).
Wieland, T.: Science **159**, 946 (1968).
Witmer, H. J.: J. Biol. Chem. **246**, 5220 (1971).
Zubay, G., Schwartz, D., Beckwith, J.: Proc. Nat. Acad. Sci. U. S. **66**, 104 (1970).
Zylber, E., Vesco, C., Penman, S.: J. Mol. Biol. **44**, 195 (1969).

CHAPTER 3

Nuclear RNA

H. Naora

Introduction

The nucleus is the main centre for storage of genetic information and for regulation of gene expression. Virtually all of the cellular DNA is now known to be nuclear. Chloroplasts and mitochondria contain their own DNA, but the informational content and amount of DNA in these organelles is very small relative to the nuclear genome. Control of transcription of nuclear DNA is of fundamental importance in the growth and development of cells and is discussed in Chapter 2. Precursor rRNA, rRNA and tRNA, are major species of nuclear RNA. However these are considered in detail elsewhere in this book and are not discussed in the present chapter which is devoted to other RNA species that occur in the nucleus.

A. Preparation of Nuclear RNA

A better understanding of the experimental results summarized in this chapter is possible by considering briefly certain aspects of the experimental techniques used in the preparation of nuclear RNA. It should be understood from the outset that satisfactory preparations of nuclei and nuclear RNA are to some extent dependent on the source of experimental material and its physiological condition. Each situation presents particular problems, and for comprehensive description, the reader should refer to earlier publications (STEELE and BUSCH, 1967; KRUH, 1967; NOLL and STUTZ, 1968; WANG, 1968; STERN, 1968; KIRBY, 1968).

1. Isolation of Nuclei

In most experiments, nuclear RNA has been prepared from isolated nuclei or nuclear subfractions. To be effective, the isolation procedure ideally ought to meet the following requirements: 1) isolated nuclei should be free of cytoplasmic RNA; 2) nuclear RNA species should be preserved without loss or degradation during isolation; 3) no redistribution of nuclear RNA should take place within the compartments of the nucleus.

A number of significant advances in isolation of nuclei have contributed to a partial satisfaction of these requirements. For instance, clean preparations of nuclei from liver, kidney and tumor cells are obtained using high density sucrose solutions. This method was first employed by CHAUVEAU et al. (1956), and takes advantage of the high density of nuclei (1.29–1.31) compared with whole cells and other cellular structures. The procedure is simple, and provides a clean preparation of nuclei though in low yield. There is some release of material from nuclei in the high osmolarity conditions used in this procedure. Furthermore centrifugation of liver nuclei in high density sucrose solution results in a partial fractionation of the nuclear population (FISHER et al., 1963; NIEHAUS and BARNUM, 1964) since liver cells are heterogeneous in ploidy (NAORA, 1957; BUCHER, 1963). Thus nuclei isolated by this means are not always representative of the whole nuclear population of liver cells. There are other procedures to isolate nuclei, but no single method has been reported which gives satisfactory preparations of nuclei from all kinds of tissues. More detailed procedures for isolation of nuclei are described by SIEBERT (1967), BUSCH (1967), and WANG (1967).

2. Preparation of RNA from Isolated Nuclei

General details of RNA extraction methods are given in Chapter 11. Procedures specifically adopted with nuclei are reviewed by STEELE and BUSCH (1967) and GEORGIEV (1967). The sodium dodecyl sulphate (SDS)-phenol method is probably the most commonly used (SIBATANI, 1966; STEELE and BUSCH, 1967; GEORGIEV, 1967). The concentration of SDS used is critical to achieve effective extraction of RNA without excessive contamination by DNA. High concentrations of SDS may result in an increased yield of RNA and greater inhibition of nuclease activity, but at the same time more DNA is released. The concentration should be empirically determined with each type of nuclear preparation used.

3. Differential Extraction of Cellular RNA

Apart from the type and concentration of detergent used, temperature and ionic strength also determine the molecular species of the RNA that are extracted from cells or nuclei by phenol (SIBATANI, 1966; GEORGIEV, 1967). In the absence of detergent, for example, nuclear material, which is rich in DNA-like RNA, remains at the interface after phenol extraction at low temperature.

By appropriate variation of temperature and detergent concentration, differential extraction of cytoplasmic and high molecular weight nuclear RNA has been demonstrated using whole lymphocyte cells (COOPER and KAY, 1969). In order to extract only cytoplasmic RNA, cells are suspended in acetate buffer containing salt and bentonite, without detergent, and extracted with phenol at 20°. This is an efficient way of extracting virtually all of the cytoplasmic rRNA and tRNA (MACH and VASSALLI, 1965; COOPER and KAY, 1969), though some low molecular weight nuclear RNA is also removed (HELLUNG-LARSEN and FREDE-RIKSEN, 1972). High molecular weight RNA remains in the residue. Since RNAase remains active during treatment with phenol alone, extraction without detergent may result in some degradation of RNA. High molecular weight nuclear and cytoplasmic RNA are extracted with 0.5% SDS-phenol at 40°, but with a reduced amount of nuclear heterogeneous RNA. Treatment with 0.5% SDS-phenol at 62° results in the extraction of an almost complete spectrum of nuclear and cytoplasmic RNA species from cells. It should be mentioned, however, that DNA is always present in this type of preparation, and is the most troublesome contaminant. Small amounts of DNA may be removed by extraction at high salt concentration (STEELE and BUSCH, 1967), by CsCl centrifugation (SIBATANI, 1972) and/or by treatment with DNAase. The former two procedures are very effective at removing DNA, but have limited use because partial fractionation and aggregation of RNA occur.

B. The Concept of Nucleus-Specific RNA and Migrating RNA

A now classical series of autoradiographic and chemical experiments have revealed that when cells are exposed to a radioactive RNA precursor, the RNA in the cell nucleus is labelled initially, and radioactivity appears subsequently in the cytoplasm. Radioactivity continues to increase in cytoplasmic RNA after a plateau of activity has been reached in the nuclear RNA (WATTS and HARRIS, 1959; HARRIS, 1959; GOLDSTEIN and MICOU, 1959; PERRY et al., 1961). From these results together with those obtained from nuclear transplantation (GOLD-STEIN and PLAUT, 1955) and enucleation experiments (NAORA et al., 1960; PRE-SCOTT, 1960), it has been concluded that some species of nuclear RNA are transported to the cytoplasm where they function in a variety of roles associated with protein synthesis. Careful studies have further demonstrated that a regulated but substantial breakdown of RNA occurs in the nucleus before it is transported to the cytoplasm. Part of newly synthesized RNA is found only in the nucleus; this RNA is referred to as "nucleus-specific RNA" and the RNA that moves to the cytoplasm as "migrating RNA". Further evidence for the occurrence of nucleus-specific RNA comes from pulse-chase and hybridization experiments.

The presence of these two classes of RNA, one restricted to the nucleus, the other moving to the cytoplasm, is commonly recognised today, although the functions of some of this RNA remain obscure. Recent experiments with amoebae expand the idea of migrating RNA to a "two-way traffic" (GOLDSTEIN et al., 1969; GOLDSTEIN and TRESCOTT, 1970). Several hours after a nucleus containing ^{3}H-RNA is transplanted to an unlabelled cell, ^{3}H-RNA moves to the host-cell nucleus. An interesting observation is that a certain species of ^{3}H-RNA is accumulated in the host-cell nucleus to a concentration of the order of 12 times the cytoplasmic concentration of ^{3}H-RNA. Further experiments with actino-mycin D have shown that this appearance of radioactivity in the host nucleus does not result from re-use of the radioactive degradation-products of the ^{3}H-RNA grafted, nor is it due to simple diffusion of cytoplasmic ^{3}H-RNA to the cell nucleus.

From the evidence so far accumulated, nuclear RNA may be classified in the following way:
  a) Migrating RNA.
    1) RNA that is transported only from the cell nucleus to the cytoplasm.
    2) RNA that migrates back and forth between the cell nucleus and cytoplasm.
  b) Nucleus-specific (non-migrating) RNA.

The relative proportion of these classes of RNA in a given cell is subject to variation with the physiological and pathological condition of the cell (see for example, CHURCH and McCARTHY, 1970). Some RNA molecules restricted to the nucleus in normal cells are transported to the cytoplasm in regenerating liver (CHURCH and McCARTHY, 1967) and tumour cells (DREWS et al., 1968; CHURCH et al., 1969; MENDECKI et al., 1969; SHEARER and SMUCKLER, 1972). These observations indicate that selective transport of RNA to the cytoplasm may be closely involved in regulatory mechanisms affecting cell function.

rRNA, tRNA and mRNA are migrating RNAs. This chapter is concerned with heterogeneous nuclear RNA (HnRNA), chromosomal RNA (ChRNA) and low molecular weight nuclear RNA (LnRNA). One other species of nuclear RNA merits mention here but will not be discussed at length. It has been reported that double-stranded RNA (dsRNA) occurs in normal (uninfected) animal cells (MONTAGNIER, 1968; COLBY and DUESBERG, 1969; STOLLAR and STOLLAR, 1970; STERN and FRIEDMAN, 1971; HAREL and MONTAGNIER, 1971; MONCKTON and NAORA, 1972). This type of RNA had previously been thought to appear only in DNA virus-infected animal cells, in DNA phage-infected bacterial cells and in RNA virus-infected cells during replication of viral RNA. It now appears that dsRNA may play an important role in fundamental functions in normal cells. dsRNA has been characterized and possible functions have been proposed (DE MAEYER et al., 1971; HUNT and EHRENFELD, 1971; KIMBALL and DUESBERG, 1971). The nuclear origin of this RNA has now been established (JELINEK and DARNELL, 1972; MONCKTON and NAORA, 1972).

C. Heterogeneous Nuclear RNA (HnRNA)

1. Synthesis and Degradation of HnRNA

HnRNA is identified by rapid incorporation of radioactive RNA precursors, and by polydisperse size distribution (10–100 S). HnRNA is DNA-like in base composition (SOEIRO et al., 1968) and is probably confined to the nuclei of eukaryotic cells. It is of interest to note that HnRNA is actively metabolized even in cells which are not multiplying and which are fully differentiated in specialized function, for example avian erythroblasts (SCHERRER et al., 1966; ATTARDI et al., 1966).

In HeLa cells with a doubling time of 24 hours, the half life of mRNA has been estimated to be at least 3–4 hours (PENMAN et al., 1963). In contrast, the average half life of HnRNA in these cells has been calculated to be 3 minutes, based on the assumption that HnRNA is synthesized from a common nucleotide pool (SOEIRO et al., 1968). On the other hand, PENMAN et al. (1968) found that the kinetics of labelling of HnRNA in HeLa cells indicated a mean lifetime of 60 minutes. The lifetime of HnRNA on polytene chromosomes was also calculated to be not more than 45 minutes (DANEHOLT and SVEDHEM, 1971). Another observation, indicating that HnRNA is rapidly labelled, is that after pulse-labelling of tissue-cultured monkey kidney cells for 2 minutes, the incorporated ^{3}H-uridine is already localized in the nucleoplasm (FAKAN and BERNHARD, 1971), where HnRNA is solely located (PENMAN et al., 1968).

The rapid turnover of HnRNA is also indicated by the base composition of rapidly labelled, high molecular weight nuclear RNA. The average guanine plus cytosine (G+C) content of HeLa cell HnRNA is about 44%, which is similar to that of HeLa cell DNA (42–47%) (SOEIRO et al., 1968; DARNELL, 1968). 45 S rRNA precursor, by contrast, possesses a high G+C content (70%). The high molecular weight nuclear RNA labelled during short periods of incubation has a low G+C content; longer periods result in an increase in labelling of RNA with higher G+C content (DARNELL, 1968; SOEIRO et al., 1968).

Kinetic labelling experiments indicate that not more than 10–20% of HnRNA migrates to the cytoplasm after processing (SCHERRER et al., 1966; ATTARDI et al., 1966; SOEIRO et al., 1968; PENMAN et al., 1968). These observations suggest that degradation of HnRNA occurs in the nucleus and the majority of the products return to an acid-soluble pool (HOUSSAIS and ATTARDI, 1966; WARNER et al., 1966). This type of nuclear degradation is also suggested in the experiments with salivary gland cells of *Chironomus tentans* (DANEHOLT and SVEDHEM, 1971). Some HnRNA molecules synthesized on chromosomes are selectively degraded either on the chromosomes or shortly after they have been released from the chromosomes.

However in sea urchin embryos a different situation appears to exist. ARONSON (1972) has reported that almost all of the HnRNA synthesized in the nuclei of sea urchin embryos appears to be transferred to the cytoplasm without any processing in the nucleus. ARONSON also suggests that HnRNA immediately on passing through the nuclear membrane may be loaded onto ribosomes present on the outer nuclear membrane and associated with stable polysome structures. An endonuclease present in the cytoplasm, but not in the nucleus, is found to be involved in the degradation of the RNA transported to the cytoplasm. The RNA is degraded first to "4S" RNA products by the cytoplasmic endonuclease unless it associates with ribosomes, in which case it is protected from degradation. Further degradation of the "4S" RNA products may occur within lysosomes. An interesting observation is that the cytoplasmic endonuclease is not active on poly A or poly C. Since, as will be described later, HnRNA and mRNA contain poly A sequences, it is possible that such a sequence may protect critical regions of the transported HnRNA from the endonuclease. It is unclear, however, whether all of the nucleotide sequences present in HnRNA of sea urchin embryos remain in the RNA molecules which appear in the cytoplasm. It also remains to be determined whether this picture of transport of HnRNA to the cytoplasm is unique to sea urchin embryos.

2. Is HnRNA a Precursor of Cytoplasmic mRNA?

There are observations outlined below which suggest that HnRNA is a precursor of cytoplasmic mRNA. It should be noted here, however, that conclusive experiments for or against this possibility are still lacking.

   a) *DNA-Like Base Composition*. HnRNA from avian erythroblasts has a
DNA-like base composition (Attardi et al., 1966; Scherrer et al., 1966) which
is very different from rRNA. Similar analysis has been carried out with HnRNA
from tissue-cultured cells in which rRNA synthesis is inhibited by low concentrations of actinomycin D. As mentioned above, the base composition of
HeLa cell HnRNA is very similar to that of cellular DNA (42–47% G+C),
whereas rRNA has over 60% G+C. This G+C content of HnRNA is an average
of all molecules of HnRNA synthesized in the nucleus. Examination of fractions
of HnRNA shows it to be heterogeneous in base composition from chromosome
to chromosome, and also from segment to segment within one chromosome
(Danehold and Svedhem, 1971).

   b) *HnRNA and mRNA Are Not Methylated*. Neither mRNA (Srinivasan and
Borek, 1964) nor HnRNA from rat liver are methylated (Muramatsu and
Fujisawa, 1968). rRNA and its precursors, on the other hand, are methylated
(see Chapter 6).

   c) *Hybridization*. If HnRNA is a precursor of cytoplasmic mRNA, HnRNA
should contain nucleotide sequences which are present in cytoplasmic mRNA.
Hybridization of HnRNA with nuclear DNA has been shown in a variety of
types of cells (Hoyer et al., 1963; Whitely et al., 1966; Birnboim et al., 1967;
Greenberg and Perry, 1971). Hybridization-competition experiments clearly
show that all of the nucleotide sequences present in cytoplasmic mRNA are
present in HnRNA, but many sequences present in HnRNA are absent from
cytoplasmic mRNA (Georgiev, 1967; Shearer and McCarthy, 1967; Drews
et al., 1968; Church et al., 1969; Church and McCarthy, 1970; Soeiro and
Darnell, 1970). This observation is consistent with, though not proof of, the idea
that HnRNA is a precursor of cytoplasmic mRNA. More supporting evidence for
this idea has been obtained in experiments on the kinetics of hybridization and
on thermal stability of the hybrids formed. In L cells, HnRNA and mRNA are
transcribed both from highly reiterated DNA sequences, involving as many as
10,000 copies per genome, and from DNA sequences with a low degree of repetition,
one or a few copies per genome. The degree of reiteration of DNA sequences for
HnRNA is similar to that for cytoplasmic mRNA. The proportion of RNA transcribed from highly reiterated DNA sequences is found to be less in mRNA than in
HnRNA. This observation implies that some of the highly reiterated sequences may
be used only for transcribing RNA which is restricted to the nucleus. However, the
RNA transcribed from highly reiterated sequences appears to be only a minor fraction of the non-migrating HnRNA and much of the HnRNA is transcribed from
unique or relatively rare DNA sequences (Greenberg and Perry, 1971). If these
latter RNA species pass to the cytoplasm as mRNA, cytoplasmic mRNA should
be the product transcribed from unique or rare DNA sequences. Indeed, cytoplasmic mRNA, such as haemoglobin mRNA is found to be transcribed from
DNA sequences which have low, if any, reiteration (Bishop et al., 1972).

   Another attempt to examine the possibility of a direct precursor-product
relationship between HnRNA and mRNA has been carried out using RNA-RNA
hybridization. A highly labelled RNA (anti-mRNA) complementary to 10S
haemoglobin mRNA from anaemic duck erythrocytes can be synthesized using
Micrococcus lysodeikticus RNA polymerase. Approximately 10% of heavier

HnRNA and 20% of the lighter fraction of HnRNA from duck erythrocytes hybridize with the labelled anti-mRNA, indicating the presence of sequences complementary to the anti-mRNA in HnRNA (MELLI and PEMBERTON, 1972).

d) Insensitivity to Low Doses of Actinomycin D. The selective inhibition of rRNA synthesis by low doses of actinomycin D without significant effect on the synthesis of other species of RNA including HnRNA is well documented (PERRY, 1963; COZZONE and MARCHIS-MOUREN, 1967; PENMAN et al., 1968; NAORA and KODAIRA, 1969). The molecular mechanism for this specific inhibition is not known. Since the rRNA precursor contains a high $G+C$ content (70%) (WILLEMS et al., 1968; JEANTEUR et al., 1968), it has been thought that actinomycin D selectively interacts with rDNA. The binding of actinomycin D to DNA is discussed in Chapter 2. Other possible explanations for this differential effect may lie in the difference in structure or organization of ribosomal and HnRNA genomes, or in differences in the repetitive sequences in the respective genomes (PERRY and KELLY, 1970).

Incubation of HeLa or L cells with 0.04–0.08 µg/ml of actinomycin D for 30 minutes before labelling with radioactive precursors results in complete inhibition of rRNA synthesis, with virtually no effect on the synthesis of HnRNA or cytoplasmic mRNA (PENMAN et al., 1968; PERRY and KELLY, 1970). However, the synthesis of both HnRNA and cytoplasmic mRNA is completely suppressed by high doses (approximately 10 µg/ml) of actinomycin D, lending further support to the idea of a precursor-product relation between HnRNA and cytoplasmic mRNA. However toxic or non-specific effects of actinomycin D cannot be excluded when actinomycin D is present at high concentration (HONIG and RABINOWITZ, 1965; KORN et al., 1965; LASZLO et al., 1966; SOEIRO and AMOS, 1966).

These differential inhibitory effects of actinomycin D are also observed with plasma cell tumour of mice (KEMPF and MANDEL, 1969) and with liver cells of rats (REVEL and HIATT, 1964; COZZONE and MARCHIS-MOUREN, 1967; NAORA and KODAIRA, 1969).

e) Presence of Poly A Sequences in Both HnRNA and Cytoplasmic mRNA Molecules. Adenylic acid rich sequences have been found in a variety of mRNAs of animal cells (LIM and CANELLAKIS, 1970; LEE et al., 1971; EDMONDS et al., 1971; DARNELL et al., 1971 a, b; BURR and LINGREL, 1971; SHELDON et al., 1972; MENDECKI et al., 1972; PEMBERTON and BALGLIONI, 1972), and of DNA virus-infected cells (KATES, 1970; PHILLIPSON et al., 1971). Rous sarcoma viral RNA (LAI and DUESBERG, 1972), murine sarcoma and avian myeloblastosis viral RNA (GREEN and CARTAS, 1972) and non-oncogenic viral RNA (ARMSTRONG et al., 1972) also contain poly A sequences. Since T4 mRNA has no poly A sequences (SHELDON et al., 1972), poly A in mRNA seems to be unique to eukaryotic cells although more information is required before a final conclusion on this subject can be reached. The length of poly A sequence appears to be quite variable. For instance, the histone messengers do not contain large poly A sequences (see SCHOCHETMAN and PERRY, 1972). An average size of poly A sequences in whole mRNA of HeLa and mouse ascites tumour cells and of duck immature erythrocytes is estimated to be approximately 150–200 residues (see for example MENDECKI et al., 1972). Mammalian haemoglobin mRNA contains stretches of only 7 to 8 adenylic acid residues (BURR and LINGREL, 1971), but these are linked together with

GMP, UMP and CMP residues (Hunt et al., 1972). Poly A sequences are also present in HnRNA, and comprise 1.8%, 1.1% and 0.2% of 10–20S, 20–30S and larger than 35S HnRNA molecules, respectively (Ryskov et al., 1972). Kinetics of the labelling of nuclear and cytoplasmic poly A sequences indicates that these sequences in HnRNA molecules become labelled prior to labelling in cytoplasmic mRNA (Darnell et al., 1971a). A further suggestive observation in this respect is that the length of poly A sequences in HeLa cell HnRNA is similar to that in cytoplasmic mRNA (Edmonds et al., 1971). Poly A sequences are probably located at the 3'-end of mRNA and HnRNA (Kates, 1970; Burr and Lingrel, 1971; Mendecki et al., 1972), though they have also been reported to be present at the 5'-end of HnRNA (Ryskov et al., 1972). Competition-hybridization of DNA and HnRNA labelled in the 3'-end reveals that the 5'-end of HnRNA is preferentially degraded during processing in the nucleus (Coutelle et al., 1970).

The presence of poly A sequences in HnRNA indicates that poly A is synthesized in the nucleus, probably by an enzyme similar to that in calf thymus nuclei (Edmonds and Abrams, 1960), and is linked to HnRNA after transcription in the nucleus (Darnell et al., 1971a; Mendecki et al., 1972). The fact that cellular and adenoviral DNA do not contain a stretch of more than 50 TMP residues supports the suggested post-transcriptional addition of poly A sequences to HnRNA molecules.

A specific function for poly A sequences present in HnRNA and mRNA has not yet been demonstrated. It has been proposed that poly A may be involved in the selective processing of HnRNA in the nucleus, and the subsequent transport of the processed HnRNA to the cytoplasm (Lee et al., 1971; Edmonds et al., 1971; Darnell et al., 1971a). When poly A synthesis in the nucleus is inhibited by cordycepin (3'-deoxyadenosine), the labelling of cytoplasmic mRNA is more significantly inhibited than the synthesis of HnRNA (Darnell et al., 1971a; Mendecki et al., 1972). This could be interpreted as meaning that inhibition of poly A synthesis causes the interruption of transport of newly synthesized HnRNA, without interfering with HnRNA synthesis. A transport role, however, is probably not the sole function of poly A sequences, since vaccinia mRNA, synthesized in the cytoplasm contains poly A sequences (Kates, 1970). Other possible functions are considered elsewhere (see Chapter 4).

f) Virus-specific Sequences in HnRNA. It is now established that virus-specific RNA species are covalently linked to HnRNA of the host. This finding is further evidence in support of the idea of a precursor-product relationship between HnRNA and mRNA.

Cells transformed by SV40 contain viral DNA covalently integrated into cellular chromosomes (Sambrook et al., 1968). Several species of SV40-specific RNA are synthesized and probably code for characteristic virus-specific proteins. Using hybridization techniques, SV40-specific sequences are detected in both HnRNA and polysomal RNA (cytoplasmic mRNA). Although the molecular weight of double-stranded SV40 DNA is 3×10^6 daltons (Green, 1970), HnRNA which contains SV40-specific sequences has a molecular weight of $1.5–4 \times 10^6$ daltons (Lindberg and Darnell, 1970; Jaenisch, 1972). The molecules transcribed may thus be longer than a single strand of SV40 DNA or of SV40-specific viral mRNA found in the cytoplasm [0.57×10^6 daltons (Tonegawa et al., 1970)].

These large molecules are transcribed from covalently linked cellular and SV40 templates, as shown by hybridization experiments with SV40 DNA and DNA from SV40-transformed and untransformed parental cells (TONEGAWA et al., 1970; WALL and DARNELL, 1971; JAENISCH, 1972). This finding implies that virus-specific mRNA in transformed cells is first transcribed as a high molecular weight HnRNA. This contains both viral and cellular sequences and is supplied to the cytoplasm after the selective cleavage of the molecule with concomitant degradation of nucleus-specific sequences.

Similar observations have been made with adenovirus-transformed and herpes virus-infected cells (GREEN et al., 1970; ROIZMAN et al., 1970). Recent experiments have shown further that the sequences of mouse mammary tumour viral RNA are present both in nuclear RNA and in cytoplasmic mRNA from tumour cells (AXEL et al., 1972). Thus the presence of virus-specific sequences in HnRNA covalently linked to normal sequences may be a general phenomenon.

g) Interaction of HnRNA and mRNA with Ribosomes. Selective recognition of RNA by ribosomes associated with binding-factors is observed in experiments with whole nuclear RNA, tissue-specific HnRNA and cytoplasmic mRNA of rat liver, kidney, spleen and thymus cells (NAORA and KODAIRA, 1968, 1969, 1970 and 1972). Protein factors associated with ribosomes or present in the post-ribosomal fraction are apparently responsible for this selective recognition only when these factors are bound to ribosomes (NAORA et al., 1971; NAORA and PRITCHARD, 1971; WHITELAM and NAORA, 1972). Direct involvement of initiation factors in recognizing particular species of RNA is now known in bacterial ribosome systems (see Chapter 7). Similar observations have been reported in the case of animal ribosomes (HEYWOOD, 1970; MATHEWS, 1970; LEADER et al., 1972). The translation of heterologous mRNA (see Chapter 4) *in vitro* may be a special situation since heterologous mRNA is usually added in saturating amounts to the isolated ribosomes with no or insufficient homologous mRNA (NAORA, 1972). The specific recognition of HnRNA and mRNA by ribosomes suggests that HnRNA contains specific ribosome binding site(s) which may be similar to those of mRNA.

3. Some Comments on HnRNA

All of the findings described above are consistent with a direct precursor-product relationship between HnRNA and cytoplasmic mRNA, but they do not definitely establish it. Indeed some findings have raised questions as to this relationship. For example, when cells are exposed to labelled uridine in the presence of low concentrations of actinomycin D, the incorporation of the uridine into HnRNA reaches a plateau after two hours. Cytoplasmic mRNA is almost linearly labelled during the period of incubation (PENMAN et al., 1968). The kinetics of labelling indicates that if the HnRNA is an obligatory precursor to cytoplasmic mRNA, such a relationship would not be simple.

Another question is raised by differential inhibition of synthesis of HnRNA and cytoplasmic mRNA by cordycepin. Addition of this inhibitor to a culture of HeLa cells results in a complete inhibition of synthesis of mitochondrial

RNA (PENMAN et al., 1970), fragmentation of rRNA precursor in the nucleolus, and significant reduction in the appearance of label in cytoplasmic mRNA (PENMAN et al., 1970; DARNELL et al., 1971a). Little if any effect on HnRNA synthesis is seen. Since cordycepin has no effect on transport of mature mRNA molecules to the cytoplasm, PENMAN and co-workers have interpreted this to mean that HnRNA and cytoplasmic mRNA are transcribed by two separate enzymes with different sensitivities to cordycepin. As mentioned earlier, cordycepin inhibits the incorporation of adenosine into poly A sequences of HnRNA. An alternative explanation of this result may thus be that the analogue by preferentially inhibiting the synthesis or addition of poly A sequences to HnRNA causes the production of incomplete HnRNA molecules, products of which cannot pass to the cytoplasm. If this interpretation is the case, the differential effect of cordycepin may not be inconsistent with the idea of the precursor-product relationship.

It should also be pointed out that all HnRNA does not necessarily give rise to cytoplasmic mRNA. As mentioned earlier, a large fraction of HnRNA synthesized is degraded within the cell nucleus and never enters the cytoplasm. Several possibilities can be considered to account for this state of affairs. First of all, many extra copies of HnRNA might be transcribed, but only a limited number enter the cytoplasm after processing; the other copies may then be degraded within the nucleus. This wastage may be a necessary aspect of some regulatory mechanism to maintain active cell functions. Similar wastage has been observed in rRNA synthesis in lymphocytes (COOPER, 1969, 1971; YAMASHITA and NAORA, 1972). The second possibility is that HnRNA may be a mixed population of at least two different species; one of non-translatable molecules which do not migrate to the cytoplasm, the other of RNA molecules containing partially or largely translatable segments which are transferred to the cytoplasm usually after "trimming". If this is the case, HnRNA molecules of the former class are entirely unrelated to cytoplasmic mRNA and may have other, unknown functions. Thirdly, it is possible that HnRNA consists of molecules containing both untranslatable and translatable sequences, and after removal of most, but not all of the untranslatable segments, the molecule moves into the cytoplasm. These untranslatable segments may be transcripts of DNA at regulatory loci, which are recognized by RNA polymerase, repressors or derepressors and other regulator molecules such as chromosomal proteins (GEORGIEV, 1969). Since the RNA chain is formed from the 5'-end, these untranslatable segments, complementary to the sequences of regulatory loci, should be located near the 5'-end of HnRNA. That this is likely is shown by the observation that the translatable segments which migrate to the cytoplasm are present near the 3'-end of HnRNA (COUTELLE et al., 1970).

D. Low Molecular Weight Nuclear RNA (LnRNA)

Recent evidence has indicated the existence of a class of nuclear RNA species with sedimentation coefficients ranging from 4 to 12S and with characteristics entirely different from HnRNA. This RNA is termed low molecular weight nuclear RNA (LnRNA) in this chapter. LnRNA species other than nuclear tRNA comprise approximately 0.4% of total cellular RNA in amoebae (GOLD-STEIN and TRESCOTT, 1970), 0.5% in HeLa cells (WEINBERG and PENMAN, 1969) and 0.2–0.7% in Ehrlich ascites tumour cells (HELLUNG-LARSEN and FREDERIKSON, 1972). The total amounts of LnRNA correspond to approximately 11% of nuclear RNA in Ehrlich ascites tumour cells and a similar proportion in Novikoff hepatoma cell nuclei (MORIYAMA et al., 1970). In addition to the cell types described above, LnRNA has been detected in a wide variety of animal cells. The occurrence of LnRNA species in eukaryotic cells of diverse genera and in both normal and cancerous cells suggests that they are probably involved in fundamental nuclear events. However LnRNA has not been reported to occur in plant cells.

1. Physical and Chemical Properties

Little is known about the physical and chemical nature of this new class of nuclear RNA. Unlike HnRNA which is polydisperse, LnRNA yields at least 11 well-defined bands when subjected to gel electrophoresis (WEINBERG and

Table 1. Molecular species of LnRNA

Species of LnRNA			Esti-mated number of nu-cleotides per molecule [a]	Base composition [b]				Methy-lation
WEINBERG and PENMAN (1968); HELLUNG-LARSEN and FREDERIKSEN (1972)	BUSCH et al. (1971)	ZAPISEK et al. (1969)		A	U	G	C	
L			260					Yes
K			220					Yes
	8S (28S)							
A	U3		180	20.2	24.6	29.6	25.3	Yes
B			170					
C	U2	H, VII	165	22.0	27.1	30.1	20.7	Yes
D	U1b U1c	G, VI	150					Yes
E	U1a	F, V	140					No
F			125					No
G	5S RNA$_I$ 5S RNA$_{II}$ 5S RNA$_{III}$	E	121	19.2 20.7 24.0	25.2 23.4 31.3	30.6 30.0 24.5	25.0 25.9 20.1	Yes Yes No
G'		D, IV	121					Yes
H	4.5S RNA$_I$ 4.5S RNA$_{II}$ 4.5S RNA$_{III}$	C, III	100	24.8 25.4 26.7	24.4 26.0 25.6	26.0 26.0 29.0	24.0 22.7 19.8	No No Yes
I	4S 4S$_{nucleolar}$	B, II	80	19.6 19.5	20.7 25.8	32.6 27.7	27.0 26.8	Yes

[a] These values were obtained with LnRNA from HeLa cells.
[b] These results were obtained with LnRNA from Novikoff hepatoma cells.

Penman, 1968; Busch et al., 1971). LnRNA species have been classified according to their electrophoretic mobilities and are listed in Table 1 where base compositions and estimated chain lengths are also given. Different systems of nomenclature have been used by different groups of workers. In Table 1 an attempt has been made to inter-relate the species of LnRNA reported by three groups. However to avoid any ambiguity in discussion of the work of a particular group, the nomenclature of that group is used.* Two species of 4S RNA are listed in Table 1. One species is confined to the nucleolus and does not accept amino acids, while the other is tRNA probably identical to cytoplasmic tRNA. However its occurrence in nuclear RNA is not simply a consequence of cytoplasmic contamination. Three species of 5S tRNA are recorded in Table 1. Two species (5S RNA_I and 5S RNA_{II}) are ribosomal type; the third (5S RNA_{III}) is nucleus specific. Nuclear tRNA, 5S RNA_I and 5S RNA_{II} are not discussed in this chapter.

LnRNA species in a mobility group can often be resolved by column chromatography. Thus by fractionation on DEAE-Sephadex, 4.5S LnRNA can be separated into three subfractions termed 4.5S RNA_I, 4.5S RNA_{II} and 4.5S RNA_{III} (Busch et al., 1971).

The electrophoretic patterns of LnRNA from mouse L and 3T3 cells are identical, and those for human W138 tissue-cultured cells (non-malignant fibroblasts) are indistinguishable from those for HeLa cells (Rein and Penman, 1969). However, LnRNA from both mouse 3T3 cells and mouse Ehrlich tumour cells differ electrophoretically from LnRNA of HeLa cells. HeLa cells and the Ehrlich tumour cells yield A species of LnRNA which differ in mobility; the K species present in HeLa cells is absent in Ehrlich tumour cells (Weinberg and Penman, 1969; Hellung-Larsen and Frederiksen, 1972). Because of observations of this type, it has been suggested that the nature of LnRNAs is determined by the species of animal from which the cells were derived rather than by cell type or karyotype, though in at least one case LnRNAs from cells of two species are apparently identical (Ro-Choi et al., 1970; Hellung-Larsen and Frederiksen, 1972). No convincing evidence for tumour-specific LnRNA has been presented.

Species of LnRNA differ both in chain length (range 80–260 nucleotides) and also in nucleotide sequence. Pancreatic and T_1 RNAase digests of 4.5S RNA_I,

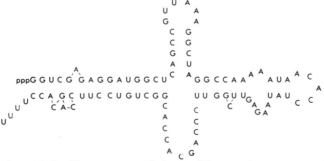

Fig. 1. Nucleotide sequence and the possible secondary structure of 4.5S RNA_I prepared from Novikoff hepatoma cells. The sequence was determined by Ro-Choi et al. (1972)

---

* Expressions such as A, A(U3) and H(4.5S RNA_I) are used in this article. "A(U3)" denotes both the A species of Penman's group and the U3 species of Busch's group, but "A" means only the A species of the former group.

4.5S RNA$_{III}$, U1a RNA and U2 RNA have been characterized by Sanger's methods. The resulting "finger-prints" are all different indicating that these species of LnRNA vary in nucleotide sequence (BUSCH et al., 1971). The complete sequence for 4.5S RNA$_I$ has recently been determined (Fig. 1). Each LnRNA species has a unique base composition and some contain methylated bases (Table 1). Thus A, C(U2), D and H (4.5S RNA$_{III}$) contain 3, 6, 3 and 4 methylated nucleotides respectively per molecule although their position has not been determined (WEINBERG and PENMAN, 1968; ZAPISEK et al., 1969; BUSCH et al., 1971; HELLUNG-LARSEN and FREDERIKSEN, 1972). However as indicated in Table 1, five species lack methylated bases (see ZAPISEK et al., 1969; BUSCH et al., 1971).

2. Intracellular Distribution

Species of LnRNA which are localized in the nucleolus are 8S, A(U3), B, F and a species of 4S RNA which does not accept amino acids. 8S LnRNA is hydrogen bonded by base pairing to 28S nucleolar RNA of Novikoff hepatoma cells. U3 is also hydrogen bonded to 28S nucleolar RNA when this is extracted at 25°, but U3 is readily released when 28S nucleolar RNA is extracted at 65°. No evidence has been presented that U3 is associated with 45S or 35S nucleolar RNA or with 28S rRNA of ribosomes (BUSCH et al., 1971). A(U3) species are possibly very firmly associated with nucleolar ribonucleoprotein particles, but this association may be dependent on nucleolar function. If cells are treated with a low dose (0.04 µg/ml) of actinomycin D to inhibit the synthesis of 45S rRNA precursor in the nucleolus, virtually all of A RNA can be released from the nuclei by treatment with 0.2 M salt, which is normally ineffective for extraction of A species (WEINBERG and PENMAN, 1969).

LnRNA species which do not occur in the nucleolus are C(U2), D(U1), 5S RNA$_{III}$ and 4.5S RNA (three species). It is not known whether these RNAs are soluble in the nucleoplasm or bound to chromatin (WEINBERG and PENMAN, 1969; BUSCH et al., 1971).

Although LnRNAs (4S tRNA, 5S RNA$_I$ and 5S RNA$_{II}$ excepted) are nucleus-specific in the interphase cell, certain species, notably A, C and D, appear in the cytoplasm during mitosis. After completion of mitosis, however, they return to the nucleus and reassociate with chromatin (GOLDSTEIN and TRESCOTT, 1970; REIN, 1971).

3. LnRNA-Nucleoprotein Particles

In interphase nuclei of HeLa cells, LnRNA is not released from chromatin by treatment with dilute salt solutions (0.01 M NaCl). However, with the exception of the A species, LnRNA can be easily extracted from interphase nuclei by salt concentrations of more than 0.2 M. The released LnRNA molecules occur as nucleoprotein particles with sedimentation coefficients less than 15S (WEINBERG and PENMAN, 1969). Although RNA-binding proteins are present in the cells

(BALTIMORE and HUANG, 1970; HEIBERG et al., 1971; STEPANOV et al., 1971; WHITELAM and NAORA, 1972), this association of LnRNA with proteins is probably not entirely an artifact (ENGER and WALTERS, 1970; REIN, 1971). It is known that high concentrations of salt are effective in releasing proteins from ribonucleoprotein complexes (ATSMON et al., 1969; CLEGG and ARNSTEIN, 1970; NAORA and PRITCHARD, 1971). Hence the high salt concentration which releases the LnRNA-protein particles from the nucleus may cause a partial removal of proteins from these particles. If this is the case, the size of the particles present in the nucleus would be larger than those extracted. In fact, 30 to 180S ribonucleoprotein particles, obtained from mitotic HeLa cells, are found to contain A, C and D species of LnRNA. These particles are converted to smaller complexes (less than 30S) by treatment with $0.3 M$ $NaCl$–$0.03 M$ $MgCl_2$ (REIN, 1971). Large LnRNA-containing particles are also observed in growing Chinese hamster cells (ENGER and WALTERS, 1970). It is not known, however, whether these large particles are artifacts produced during extraction by association of proteins with small particles.

4. Synthesis and Metabolic Stability

In contrast to the instability of HnRNAs, LnRNAs are metabolically very stable. B, C and H species are less stable than the A, D and G' species. In HeLa cells whose generation time is 22–24 hours, the half lives of the former 3 species of LnRNA are 20, 25 and 30 hours respectively measured relative to the stability of cytoplasmic rRNA (WEINBERG and PENMAN, 1969). This observation suggests that these species of LnRNA are probably synthesized only once during the cell cycle. A similar situation has been observed in hamster kidney cells (CLASON and BURDON, 1969). A, D(U1) and G' species are extremely stable and remain associated with the nuclei over several cell generations (WEINBERG and PENMAN, 1968, 1969; MORIYAMA et al., 1969). Long-lived RNA (half life at least 30 days) has also been reported in several tissues of mice, in which rRNA has an approximate half life of 5 to 6 days (MENZIES et al., 1969). This long-lived RNA is possibly LnRNA.

LnRNA can be synthesized throughout the cell cycle and its synthesis is quite independent of DNA synthesis (CLASON and BURDON, 1969; REIN, 1971). The LnRNA synthesis takes place on reiterated DNA sequences, since U1a and U2 species can be hybridized with 0.14 and 0.10% of the DNA, respectively. The hybridization plateaus for 4.5S RNA_I and 4.5S RNA_{III} are higher than the percentage of the total DNA that hybridizes with rRNA (BUSCH et al., 1971). The less stable LnRNA may be broken down throughout the cell cycle, uncoordinated with specific cellular events (WEINBERG and PENMAN, 1969).

5. The Relationship between LnRNA and Other RNA Species

Both 45S rRNA precursor and HnRNA are processed in the nucleus and only parts of these molecules migrate to the cytoplasm. However LnRNAs are not cleavage products of precursor rRNA or HnRNA. The labelling kinetics for HnRNA and LnRNA suggest that a simple precursor-product relationship does not exist between these RNA species (WEINBERG and PENMAN, 1969). While LnRNA synthesis is partially inhibited by actidione, cordycepin and low levels of actinomycin D, HnRNA synthesis is almost unaffected (WEINBERG and PENMAN, 1969). Low concentrations of actinomycin D, which completely inhibit 45S rRNA precursor synthesis, only partially inhibit LnRNA synthesis (BUSCH et al., 1971). It has also been established that LnRNA is not formed by degradation of rRNA nor is it derived from ribosomes (HELLUNG-LARSEN and FREDERIKSEN, 1972).

Since the amino acid acceptor activity of nuclear tRNA is essentially the same as that of cytoplasmic tRNA, the majority of I(4S) LnRNA is probably matured tRNA. Evidence has been presented that LnRNA is unlikely to be the precursor of cytoplasmic tRNA. The molecular features of tRNA precursor seem to be different from those of matured tRNA and LnRNA. Labelling of cytoplasmic tRNA is not inhibited by low concentrations of certain inhibitors which affect the synthesis of various species of LnRNAs (WEINBERG and PENMAN, 1969). Kinetic labelling studies also indicate that LnRNAs are not precursors of cytoplasmic tRNA (BERNHARDT and DARNELL, 1969; WEINBERG and PENMAN, 1969). Nucleolar 4S RNA is also unlikely to be a precursor of cytoplasmic tRNA or of non-nucleolar 4S tRNA (NAKAMURA et al., 1968; BUSCH et al., 1971).

6. Possible Functions of LnRNA

There has been much speculation as to the functions of LnRNAs. They may be involved in the basic structure of chromosomes or in the regulation of repression and derepression of gene activity. Some LnRNAs may possibly have a role in DNA synthesis or in protein synthesis. Future attempts to determine the functions of LnRNA may well yield new concepts concerning regulatory mechanisms. However with the limited knowledge of today we can do little more than speculate regarding function of LnRNA. Metabolic stability, monodisperse nature and nuclear localization (including association with chromatin) favour the speculation that LnRNAs are concerned with regulation of expression of the genome. The number of LnRNA molecules per cell is compatible with this idea. Assuming that there is no loss of LnRNA from nuclei during isolation, the estimated amount of LnRNA in rat liver cells is 6×10^{-12} g per nucleus (BUSCH et al., 1971). Since the estimated molecular weights of LnRNAs are 24,000 to 78,000, the rat liver cell nucleus contains approximately 10^6 to 10^7 molecules of LnRNA (PRESTAYKO and BUSCH, 1968; BUSCH et al., 1971). This number of molecules approximates the total number of potentially functional genes in cells and LnRNAs may be associated with genes on a one-to-one basis. It is particularly notable that approximately 10^6 molecules of D(U1b, U1c) species are present in the non-nucleolar fraction, and are probably associated with chromatin. Estimates of numbers of molecules of LnRNA species are listed in Table 2 together with the amounts of other cellular RNA species.

Table 2. Estimated numbers of molecules of various types of RNA of HeLa cells

Species of RNA	Estimated numbers of molecules per cell		
	Cytoplasm	Nucleoplasm	Nucleolus
A (U3)	—	—	2×10^5
B	—	—	0.1×10^5
C (U2)	—	4×10^5	1×10^5
D (U1b, U1c)	—	10×10^5	0.7×10^5
E (U1a)	50×10^5	0.5×10^5	0.1×10^5
F	—	—	0.3×10^5
G (5S) and G'	50×10^5	3×10^5	3×10^5
H (4.5S)	—	2×10^5	1×10^5
I (4S tRNA)	1×10^8	2×10^5	—
18S	50×10^5	0.6×10^5	—
28S	50×10^5	0.7×10^5	—
45S	—	—	0.1×10^5

From WEINBERG and PENMAN, 1968.

The suggestion that LnRNAs act as gene regulators does not exclude the possibility that some species of LnRNA have other nuclear regulatory functions. For instance, the A(U3) species is exclusively located in the nucleolus and is thought to be transiently associated with 28S rRNA in the nucleolus. Since the nucleolus is the principal site for processing rRNA precursor and for maturation of ribonucleoprotein particles in ribosome formation, it has been suggested that the A(U3) RNA may play a particular role in the processing of rRNA and/or in maturation or transport of particles (BUSCH et al., 1971).

E. Chromosomal RNA (ChRNA)

A species of nuclear RNA which is covalently associated with chromosomal proteins has been reported (HUANG and BONNER, 1965). This species of RNA has been termed chromosomal RNA (ChRNA).

The general properties of ChRNA are unique. It has been suggested that ChRNA may have a key role in regulating gene function by conferring specificity on the DNA-chromosomal protein interaction (BONNER and WIDHOLM, 1967; BEKHOR et al., 1969; HUANG and HUANG, 1969). These results have been obtained only in Bonner's and Huang's laboratories and recently doubts have been raised with regard to the occurrence and significance of ChRNA. A number of investigators have attempted to repeat the isolation and characterization of ChRNA from chromatin of various tissues. All have failed to isolate a ChRNA similar to that prepared and characterized by HUANG et al. and BONNER et al. They have concluded that the RNA found in the ChRNA fractions is not covalently bound to chromosomal proteins and is probably an artifact produced by degradation of cellular RNA. Because of the opposing views concerning ChRNA, it will be discussed only briefly in this chapter.

1. The Properties of ChRNA as Described by the Bonner and Huang Groups

In the preparative procedure used by BONNER and coworkers, a solution of chromatin in 4M CsCl was centrifuged yielding a chromosomal protein fraction at the solution surface. ChRNA was prepared from this protein by the following sequential steps: washing with ethanol, digestion with pronase, deproteinization with phenol, DEAE-Sephadex column chromatography, dialysis and precipitation with ethanol.

This procedure, with slight modifications depending upon the source of material, has been used to isolate ChRNA from pea bud (HUANG and BONNER, 1965), pea cotyledon (BONNER and WIDHOLM, 1967), chick embryo (HUANG and HUANG, 1969), rat ascites tumour cells (DAHMUS and McCONNELL, 1969; JACOBSON and BONNER, 1971) and calf thymus (SHIH and BONNER, 1969). More recently, ChRNA was isolated by treatment of chromatin with deoxycholate, but without CsCl centrifugation (MAYFIELD and BONNER, 1971).

According to the BONNER and HUANG groups, ChRNA has properties which readily distinguish it from other RNA species.

About 1–2% of total cellular RNA is ChRNA. Elution patterns from DEAE-Sephadex or DEAE-cellulose exhibit a single, sharp symmetrical peak (BONNER and WIDHOLM, 1967; SHIH and BONNER, 1969; DAHMUS and McCONNELL, 1969; JACOBSON and BONNER, 1971). Sedimentation coefficients range from 3.2–3.8S depending on origin. The chain lengths of ChRNA as determined by the end-group method are 42 and 45 nucleotides for pea (BONNER and WIDHOLM, 1967) and calf thymus (SHIH and BONNER, 1969) respectively. This RNA is essentially a linear unfolded molecule.

ChRNA is characterized by a relatively high content of dihydropyrimidine, 8.1–9.6 mole% depending upon the species of organism. Apart from this similarity, the base composition of ChRNAs from different organisms varies considerably (Table 3). ChRNA is methylated to a certain extent, but much less than tRNA (DAHMUS and McCONNELL, 1969).

It is notable that ChRNA is covalently associated with chromosomal proteins (HUANG and HUANG, 1969; JACOBSON and BONNER, 1971) and in native chromatin the ChRNA is resistant to RNAase action (BONNER et al., 1961; BEKHOR et al.,

Table 3. Base composition (mole %) of ChRNA from various sources

Sources of ChRNA	References	A	U	Dihydro-pyrimidine	G	C
Calf thymus	SHIH and BONNER (1969)	11.1	17.5	8.5	35.0	27.9
Rat ascites	JACOBSON and BONNER (1968)	17.4	19.4	8.1	30.9	24.2
Chick embryo	ref. from DAHMUS and McCONNELL (1969)	27.6	12.8	9.6	24.6	25.6
Pea bud	HUANG and BONNER (1965)	39.8	19.2	8.5	19.3	13.1

1969). The proteins covalently linked to this RNA are acidic in nature (HUANG and HUANG, 1969) and have a sedimentation coefficient of 2.7S (JACOBSON and BONNER, 1971). Since a protein-bound RNA-histone complex is found in pea (HUANG and BONNER, 1965) and in chick embryo (HUANG and HUANG, 1969), histone may be associated with protein-bound RNA through hydrogen bonding (HUANG and HUANG, 1969).

The idea that ChRNA may be involved in gene regulation has been suggested by the finding that the sequence-specific reconstitution of chromatin can be achieved only in the presence of ChRNA. Since the sequence-specific reconstitution is achieved only under conditions favourable for annealing of RNA to DNA, sequence specificity is thought to be achieved by base-pairing (BEKHOR et al., 1969; HUANG and HUANG, 1969). The presence of tissue-specific sequences in ChRNA also favours this idea (MAYFIELD and BONNER, 1971).

ChRNA hybridizes to about 4–5% of total nuclear DNA of pea (BONNER and WIDHOLM, 1967; SIVOLAP and BONNER, 1971), ascites tumour (DAHMUS and McCONNELL, 1969) and chick embryo (HUANG and HUANG, 1969). The repetitive fractions obtained from pea DNA hybridize with ChRNA extensively, more than 13%. The high level of hybridization is in contrast to that of rRNA and tRNA. According to the BONNER group, there is no sequence homology in ChRNA to tRNA, rRNA and mRNA (BONNER and WIDHOLM, 1967; DAHMUS and McCONNELL, 1969).

2. The Critical View of ChRNA

At least five laboratories have endeavoured to confirm the findings of BONNER and HUANG but all have failed (COMMERFORD and DELIHAS, 1966; DE FILIPPES, 1970; HEYDEN and ZACHAU, 1971; ARTMAN and ROTH, 1971; SZESZAK and PIHL, 1972). No evidence was found for the presence of a special class of nuclear RNA which is covalently bound to chromosomal protein.

Perhaps the most obvious conclusion that can be drawn from the experiments of other workers in this field is that ChRNA possibly consists of degraded cellular RNA (SZESZAK and PIHL, 1972; ARTMAN and ROTH, 1971) or degraded tRNA (HEYDEN and ZACHAU, 1971).

Nevertheless, if it is accepted that ChRNA is an artifact arising from the preparative procedures used, certain properties remain to be explained, for example the extensive hybridization to DNA and the high content of dihydro-pyrimidine. These differences cannot be immediately reconciled. Very recently BONNER and coworkers have reaffirmed their belief that ChRNA is a special class of RNA and have presented evidence that it is not a degradation product of rRNA or tRNA (HOLMES et al., 1972). It would be premature at present to draw a final conclusion as to whether ChRNA is a special class of nuclear RNA with important regulatory functions, or is an artifact of preparation. However, the experiments of BONNER, HUANG and co-workers will not really support the former concept until they can be repeated by others. More rigorous experiments are required to understand the significant of ChRNA.

References

ARMSTRONG, J.A., EDMONDS, M., NAKAZATO, H., PHILLIPS, B.A., VAUGHAN, M.H.: Science **176**, 526 (1972).
ARONSON, A.I.: Nature New Biol. **235**, 40 (1972).
ARTMAN, M., ROTH, J.S.: J. Mol. Biol. **60**, 291 (1971).
ATSMON, A., SPITNIK-ELSON, P., ELSON, D.: J. Mol. Biol. **45**, 125 (1969).
ATTARDI, G., PARNAS, H., HWANG, M.I.H., ATTARDI, B.: J. Mol. Biol. **20**, 145 (1966).
AXEL, R., SCHLOM, J., SPIEGELMAN, S.: Proc. Nat. Acad. Sci. U.S. **69**, 535 (1972).
BALTIMORE, D., HUANG, A.S.: J. Mol. Biol. **47**, 263 (1970).
BEKHOR, I., KUNG, G.K., BONNER, J.: J. Mol. Biol. **39**, 351 (1969).
BERNHARDT, D., DARNELL, J.E.: J. Mol. Biol. **42**, 43 (1969).
BIRNBOIM, H.C., PENE, J.J., DARNELL, J.E.: Proc. Nat. Acad. Sci. U.S. **58**, 320 (1967).
BISHOP, J.O., PEMBERTON, R., BAGLIONI: Nature New Biol. **235**, 231 (1972).
BONNER, J., HUANG, R.C.C., MAHESHWARI, N.: Proc. Nat. Acad. Sci. U.S. **47**, 1548 (1961).
BONNER, J., WIDHOLM, J.: Proc. Nat. Acad. Sci. U.S. **57**, 1397 (1967).
BUCHER, N.L.R.: Intern. Rev. Cytol. **15**, 245 (1963).
BURR, H., LINGREL, J.B.: Nature New Biol. **233**, 41 (1971).
BUSCH, H.: In Methods in enzymology (GROSSMAN, L., and MOLDAVE, K., eds.), vol. 12, part A, p. 421. New York: Academic Press 1967.
BUSCH, H., RO-CHOI, T.S., PRESTAYKO, A.W., SHIBATA, H., CROOKE, S.T., EL-KHATIB, S.M., CHOI, Y.C., MAURITZEN, C.M.: Perspectives Biol. Med. **15**, 117 (1971).
CHAUVEAU, J., MOULE, Y., ROUILLER, C.: Exp. Cell Res. **11**, 317 (1956).
CHURCH, R.B., LUTHER, S.W., MCCARTHY, B.J.: Biochim. Biophys. Acta **190**, 30 (1969).
CHURCH, R.B., MCCARTHY, B.J.: Proc. Nat. Acad. Sci. U.S. **58**, 1547 (1967).
CHURCH, R.B., MCCARTHY, B.J.: Biochim. Biophys. Acta **199**, 103 (1970).
CLASON, A.E., BURDON, R.H.: Nature **223**, 1063 (1969).
CLEGG, J.C.S., ARNSTEIN, H.R.V.: Eur. J. Biochem. **13**, 149 (1970).
COLBY, C., DUESBERG, P.H.: Nature **222**, 940 (1969).
COMMERFORD, S.L., DELIHAS, N.: Proc. Nat. Acad. Sci. U.S. **56**, 1759 (1966).
COOPER, H.L.: J. Biol. Chem. **244**, 1946 (1969).
COOPER, H.L.: Nature New Biol. **234**, 272 (1971).
COOPER, H.L., KAY, J.E.: Biochim. Biophys. Acta **174**, 503 (1969).
COUTELLE, C., RYSKOV, A.P., GEORGIEV, G.P.: FEBS Lett. **12**, 21 (1970).
COZZONE, A., MARCHIS-MOUREN, G.: Biochemistry **6**, 3911 (1967).
DAHMUS, M.E., MCCONNELL, D.J.: Biochemistry **8**, 1524 (1969).
DANEHOLT, B., SVEDHEM, L.: Exp. Cell Res. **67**, 263 (1971).
DARNELL, J.E.: Bacteriol. Rev. **32**, 262 (1968).
DARNELL, J.E., PHILIPSON, L., WALL, R., ADESNIK, M.: Science **174**, 507 (1971a).
DARNELL, J.E., WALL, R., TUSHINSKI, R.J.: Proc. Nat. Acad. Sci. U.S. **68**, 1321 (1971b).
DE FILIPPES, F.M.: Biochim. Biophys. Acta **199**, 562 (1970).
DE MAEYER, E., DE MAEYER-GUIGNARD, J., MONTAGNIER, L.: Nature New Biol. **229**, 109 (1971).
DREWS, J., BRAWERMAN, G., MORRIS, H.P.: Eur. J. Biochem. **3**, 284 (1968).
EDMONDS, M., ABRAMS, R.: J. Biol. Chem. **235**, 1142 (1960).
EDMONDS, M., VAUGHAN, M.H., NAKAZATO, H.: Proc. Nat. Acad. Sci. U.S. **68**, 1336 (1971).
ENGER, M.D., WALTERS, R.A.: Biochemistry **9**, 3551 (1970).
FAKAN, S., BERNHARD, W.: Exp. Cell Res. **67**, 129 (1971).
FISHER, R.F., HOLBROOK, D.J., IRVIN, J.L.: J. Cell Biol. **17**, 231 (1963).
GEORGIEV, G.P.: Progr. Nucl. Acid Res. Mol. Biol. **6**, 259 (1967).
GEORGIEV, G.P.: J. Theoret. Biol. **25**, 473 (1969).
GOLDSTEIN, L., MICOU, J.: J. Biophys. Biochem. Cytol. **6**, 1 (1959).
GOLDSTEIN, L., PLAUT, W.: Proc. Nat. Acad. Sci. U.S. **41**, 874 (1955).
GOLDSTEIN, L., RAO, M.V.N., PRESCOTT, D.M.: Ann. Embryo Morph., Suppl. **1**, 189 (1969).
GOLDSTEIN, L., TRESCOTT, O.H.: Proc. Nat. Acad. Sci. U.S. **67**, 1367 (1970).
GREEN, M.: Ann. Rev. Biochem. **39**, 701 (1970).
GREEN, M., CARTAS, M.: Proc. Nat. Acad. Sci. U.S. **69**, 791 (1972).

GREEN, M., PARSONS, J.T., PIÑA, M., FUJINAGA, K., CAFFIER, H., LANDGRAF-LEURS, I.: Cold Spring
    Harbor Symp. Quant. Biol. 35, 803 (1970).
GREENBERG, J.R., PERRY, R.P.: J. Cell Biol. 50, 774 (1971).
HAREL, L., MONTAGNIER, L.: Nature New Biol. 229, 106 (1971).
HARRIS, H.: Biochem. J. 73, 362 (1959).
HEIBERG, R., OLSNES, S., PIHL, A.: FEBS Lett. 18, 169 (1971).
HELLUNG-LARSEN, P., FREDERIKSEN, S.: Biochim. Biophys. Acta 262, 290 (1972).
HEYDEN, H.W., ZACHAU, H.G.: Biochim. Biophys. Acta 232, 651 (1971).
HEYWOOD, S.M.: Proc. Nat. Acad. Sci. U.S. 67, 1782 (1970).
HOLMES, D.S., MAYFIELD, J.E., SANDER, G., BONNER, J.: Science 177, 72 (1972).
HONIG, G., RABINOWITZ, M.: Science 149, 1504 (1965).
HOUSSAIS, J.F., ATTARDI, G.: Proc. Nat. Acad. Sci. U.S. 56, 616 (1966).
HOYER, B.H., McCARTHY, B.J., BOLTON, E.T.: Science 140, 1408 (1963).
HUANG, R.C., BONNER, J.: Proc. Nat. Acad. Sci. U.S. 54, 960 (1965).
HUANG, R.C.C., HUANG, P.C.: J. Mol. Biol. 39, 365 (1969).
HUNT, J.A., PETER, O.F., JOHNSON, B.E.: Proc. Aust. Biochem. Soc. 5, 34 (1972).
HUNT, T., EHRENFELD, E.: Nature 230, 91 (1971).
JACOBSON, R.A., BONNER, J.: Biochem. Biophys. Res. Commun. 33, 716 (1968).
JACOBSON, R.A., BONNER, J.: Arch. Biochem. Biophys. 146, 557 (1971).
JAENISCH, R.: Nature New Biol. 235, 46 (1972).
JEANTEUR, P., AMALDI, F., ATTARDI, G.: J. Mol. Biol. 33, 757 (1968).
JELINEK, W., DARNELL, J.E.: Proc. Nat. Acad. Sci. U.S. 69, 2537 (1972).
KATES, J.: Cold Spring Harbor Symp. Quant. Biol. 35, 743 (1970).
KEMPF, J., MANDEL, P.: Bull. Soc. Chim. Biol. 51, 1121 (1969).
KIMBALL, P.C., DUESBERG, P.H.: J. Virol. 7, 697 (1971).
KIRBY, K.S.: In Methods in enzymology (GROSSMAN, L. and MOLDAVE, K., eds.), vol. 12, part B, p. 87.
    New York and London: Academic Press 1968.
KORN, D., PROTASS, J.J., LEIVE, L.: Biochem. Biophys. Res. Commun. 19, 473 (1965).
KRUH, J.: In Methods in enzymology (GROSSMAN, L., and MOLDAVE, K., eds.), vol. 12, part A, p. 609.
    New York and London: Academic Press 1967.
LAI, M.M.C., DUESBERG, P.H.: Nature 235, 383 (1972).
LASZLO, J., MILLER, D.S., McCARTY, K.S., HOCHSTEIN, P.: Science 151, 1007 (1966).
LEADER, D.P., KLEIN-BREMHAAR, H., WOOL, I.G., FOX, A.: Biochem. Biophys. Res. Commun. 46, 215
    (1972).
LEE, S.Y., MENDECKI, J., BRAWERMAN, G.: Proc. Nat. Acad. Sci. U.S. 68, 1331 (1971).
LIM, L., CANELLAKIS, E.S.: Nature 227, 710 (1970).
LINDBERG, U., DARNELL, J.E.: Proc. Nat. Acad. Sci. U.S. 65, 1089 (1970).
MACH, B., VASSALLI, P.: Science 150, 662 (1965).
MATHEWS, M.B.: Nature 228, 661 (1970).
MAYFIELD, J.E., BONNER, J.: Proc. Nat. Acad. Sci. U.S. 68, 2652 (1971).
MELLI, M., PEMBERTON, R.E.: Nature New Biol. 236, 172 (1972).
MENDECKI, J., LEE, S.Y., BRAWERMAN, G.: Biochemistry 11, 792 (1972).
MENDECKI, J., MINC, B., CHORAZY, M.: Biochem. Biophys. Res. Commun. 36, 494 (1969).
MENZIES, R.A., PRESS, G.D., GOLD, P.H., HENDLEY, D.D., STREHLER, B.L.: Cell and Tissue Kinetics 2,
    133 (1969).
MONCKTON, R.P., NAORA, H.: In preparation (1972).
MONTAGNIER, L.: Compt. Rend. Ser. D. 267, 1417 (1968).
MORIYAMA, Y., HODNETT, J.L., PRESTAYKO, A.W., BUSCH, H.: J. Mol. Biol. 39, 335 (1969).
MORIYAMA, Y., IP, P., BUSCH, H.: Biochim. Biophys. Acta 209, 161 (1970).
MURAMATSU, M., FUJISAWA, T.: Biochim. Biophys. Acta 157, 476 (1968).
NAKAMURA, T., PRESTAYKO, A.W., BUSCH, H.: J. Biol. Chem. 243, 1368 (1968).
NAORA, H.: J. Biophys. Biochem. Cytol. 3, 949 (1957).
NAORA, H.: In preparation (1972).
NAORA, H., KODAIRA, K.: Biochim. Biophys. Acta 161, 276 (1968).
NAORA, H., KODAIRA, K.: Biochim. Biophys. Acta 182, 469 (1969).
NAORA, H., KODAIRA, K.: Biochim. Biophys. Acta 209, 196 (1970).
NAORA, H., KODAIRA, K.: Unpublished observations (1972).

NAORA, H., KODAIRA, K., PRITCHARD, M.J.: Biochim. Biophys. Acta **246**, 280 (1971).
NAORA, H., NAORA, H., BRACHET, J.: J. Gen. Physiol. **43**, 1083 (1960).
NAORA, H., PRITCHARD, M.J.: Biochim. Biophys. Acta **246**, 269 (1971).
NIEHAUS, W.G., BARNUM, C.P.: J. Biol. Chem. **239**, 1198 (1964).
NOLL, H., STUTZ, E.: In Methods in enzymology (GROSSMAN, L., and MOLDAVE, K., eds.), vol. 12, part B, p.129. New York and London: Academic Press 1968.
PEMBERTON, R.F., BAGLIONI, C.: J. Mol. Biol. **65**, 531 (1972).
PENMAN, S., FAN, H., PERLMAN, S., ROSBASH, M., WEINBERG, R., ZYLBER, E.: Cold Spring Harbor Symp. Quant. Biol. **35**, 561 (1970).
PENMAN, S., SCHERRER, K., BECKER, Y., DARNELL, J.E.: Proc. Nat. Acad. Sci. U.S. **49**, 654 (1963).
PENMAN, S., VESCO, C., PENMAN, M.: J. Mol. Biol. **34**, 49 (1968).
PERRY, R.P.: Exp. Cell Res. **29**, 400 (1963).
PERRY, R.P., ERRERA, M., HELL, H., DURWALD, H.: J. Biophys. Biochem. Cytol. **11**, 1 (1961).
PERRY, R.P., KELLEY, D.E.: J. Cell Physiol. **76**, 127 (1970).
PHILLIPSON, L., WALL, R., GLICKMAN, R., DARNELL, J.E.: Proc. Nat. Acad. Sci. U.S. **68**, 2806 (1971).
PRESCOTT, D.M.: Exp. Cell Res. **19**, 29 (1960).
PRESTAYKO, A., BUSCH, H.: Biochim. Biophys. Acta **169**, 327 (1968).
REIN, A.: Biochim. Biophys. Acta **232**, 306 (1971).
REIN, A., PENMAN, S.: Biochim. Biophys. Acta **190**, 1 (1969).
REVEL, M., HIATT, H.H.: Proc. Nat. Acad. Sci. U.S. **51**, 810 (1964).
RO-CHOI, T.S., MORIYAMA, Y., CHOI, Y.C., BUSCH, H.: J. Biol. Chem. **245**, 1970 (1970).
RO-CHOI, T.S., REDDY, R., IIENNING, D., TAKANO, T., TAYLOR, C.W., BUSCH, H.: J. Biol. Chem. **247**, 3205 (1972).
ROIZMAN, B., BACHENHEIMER, S., WAGNER, E.K., SAVAGE, T.: Cold Spring Harbor Symp. Quant. Biol. **35**, 753 (1970).
RYSKOV, A.P., FARASHYAN, V.R., GEORGIEV, G.D.: FEBS Lett. **20**, 355 (1972).
SAMBROOK, J., WESTPHAL, H., SRINIVASAN, P.R., DULBECCO, R.: Proc. Nat. Acad. Sci. U.S. **60**, 1288 (1968).
SCHERRER, K., MARCAUD, L., ZAJDELA, F., LONDON, I.M., GROSS, F.: Proc. Nat. Acad. Sci. U.S. **56**, 1571 (1966).
SCHOCHETMAN, G., PERRY, R.P.: J. Mol. Biol. **63**, 591 (1972).
SHEARER, R., McCARTHY, B.J.: Biochemistry **6**, 283 (1967).
SHEARER, R.W., SMUCKLER, E.A.: Cancer Res. **32**, 339 (1972).
SHELDON, R., JURALE, C., KATES, J.: Proc. Nat. Acad. Sci. U.S. **69**, 417 (1972).
SHIH, T.Y., BONNER, J.: Biochim. Biophys. Acta **182**, 30 (1969).
SIBATANI, A.: Progr. Biophys. Mol. Biol. **16**, 15 (1966).
SIBATANI, A.: Personal communication (1972).
SIEBERT, G.: In Methods in cancer research (BUSCH, H., ed.), vol. 2, p.287. New York: Academic Press 1967.
SIVOLAP, Y.M., BONNER, J.: Proc. Nat. Acad. Sci. U.S. **68**, 387 (1971).
SOEIRO, R., AMOS, H.: Biochim. Biophys. Acta **129**, 406 (1966).
SOEIRO, R., DARNELL, J.E.: J. Cell Biol. **44**, 467 (1970).
SOEIRO, R., VAUGHAN, M.H., WARNER, J.R., DARNELL, J.E.: J. Cell Biol. **39**, 122 (1968).
SRINIVASAN, P.R., BOREK, E.: Science **145**, 548 (1964).
STEELE, W.J., BUSCH, H.: In Methods in cancer research (BUSCH, H., ed.), vol. 3, p.61. New York: Academic Press 1967.
STEPANOV, A.S., VORONINA, A.S., OVCHINNIKOV, L.P., SPIRIN, A.S.: FEBS Lett. **18**, 13 (1971).
STERN, H.: In Methods in enzymology (GROSSMAN, L., and MOLDAVE, K., eds.), vol. 12, part B, p.100. New York and London: Academic Press 1968.
STERN, R., FRIEDMAN, R.M.: Biochemistry **10**, 3635 (1971).
STOLLAR, V., STOLLAR, B.D.: Proc. Nat. Acad. Sci. U.S. **65**, 993 (1970).
SZESZAK, F., PIHL, A.: FEBS Lett. **20**, 177 (1972).
TONEGAWA, S., WALTER, G., BERNARDINI, A., DULBECCO, R.: Cold Spring Harbor Symp. Quant. Biol. **35**, 823 (1970).
WALL, R., DARNELL, J.E.: Nature New Biol. **232**, 73 (1971).
WANG, T.Y.: In Methods in enzymology (GROSSMAN, L., and MOLDAVE, K., eds.), vol. 12, part A, p.417. New York and London: Academic Press 1967.

Wang, T.Y.: In Methods in enzymology (Grossman, L., and Moldave, K., eds.), vol. 12, part B, p. 115. New York and London: Academic Press 1968.

Warner, J., Soeiro, R., Birnboim, C., Darnell, J.E.: J. Mol. Biol. **19**, 349 (1966).

Watts, J.W., Harris, H.: Biochem. J. **72**, 149 (1959).

Weinberg, R.A., Penman, S.: J. Mol. Biol. **38**, 289 (1968).

Weinberg, R.A., Penman, S.: Biochim. Biophys. Acta **190**, 10 (1969).

Whitelam, J.M., Naora, H.: Biochim. Biophys. Acta **272**, 425 (1972).

Whitely, A.H., McCarthy, B.J., Whiteley, H.R.: Proc. Nat. Acad. Sci. U.S. **55**, 519 (1966).

Willems, M., Wagner, E., Laing, R., Penman, S.: J. Mol. Biol. **32**, 221 (1968).

Yamashita, A., Naora, H.: Japanese J. Haemat., in press (1972).

Zapisek, W.F., Saponara, A.G., Enger, M.D.: Biochemistry **8**, 1170 (1969).

CHAPTER 4

Messenger RNA

A. J. HOWELLS

Introduction

Messenger RNA (mRNA) has been defined as "a polynucleotide which determines the sequence of amino acids in polypeptide chains. The mRNA is a carrier of the information of the genome" (SINGER and LEDER, 1966).

The approach taken to the subject of mRNA in this chapter will be a biochemical one and will be centred around information obtained from recent studies on the nucleotide sequences of mRNA. This approach has been adopted firstly, because much valuable information about the properties of mRNA has been obtained from such studies, and secondly, because other contributions to this series will deal with certain of the more biologically important aspects of mRNA.

The subject will be presented in two sections. In the first, the information gained from the sequencing studies with bacteriophage RNA will be discussed. In the second, progress which has been made on the purification and sequencing of other types of mRNA will be considered. Brief consideration will also be given to the stability of mRNA in various cell types in this section.

A. Nucleotide Sequencing Studies on Bacteriophage mRNA

1. The Single-Stranded RNA Viruses of *E. coli*

Nucleotide sequencing studies depend on the availability of RNA preparations of a high degree of purity. The purest mRNA preparations at present available are those from single-stranded RNA viruses of *E. coli*, since the RNA of the mature virus particle is not only the primary genetic material but is also a mRNA. It can be translated into protein both in the host cell and also in *E. coli* cell-free protein synthesizing systems; three proteins can be identified in both cases (VINUELA et al., 1967; EGGEN et al., 1967).

There are a number of single-stranded RNA viruses which infect *E. coli*. Most sequencing work has been done with RNA from the serologically related group—R17, MS2 and f2—although RNA from the serologically unrelated virus Qβ has also been extensively used. Since the studies with Qβ RNA have not provided any additional conceptual information to that found with RNA from the former group of viruses, they will not be discussed in this article.

a) The Viral Proteins. Information regarding the viral proteins has been summarized by Steitz (1969). Three virus-specific proteins have been found.

1) Coat protein—the major protein of the mature virus particle. Amino acid sequences are available for both f2 and R17; both consist of 129 amino acids and differ in sequence at only one position.

2) Maturation protein (A protein)—the minor protein of the mature virus particle. Although complete amino acid sequences are not available, the N-terminal dipeptide has been identified. Molecular weight data and the amino acid composition indicate that the protein contains approximately 350 amino acids.

3) RNA synthetase (replicase)—the enzyme which replicates the viral RNA within the host cell. This enzyme is not present in the mature virus particle. Again, although complete amino acid sequences are not available, the N-terminal dipeptide has been determined and data on the molecular weight and amino acid composition indicate that the protein contains 450 amino acids.

b) Size of the Genome. Genetic complementation analysis with R17 has shown that the genome contains three complementation groups and this correlates with the three virus-specific proteins described above (Gussin, 1966). Based on the sizes of these proteins, the genome codes for the condensation of 900 to 1,000 amino acids and thus should contain at least 2,700 to 3,000 nucleotides. In fact, sedimentation analysis indicates that both R17 and MS2 RNA have molecular weights of approximately 1.1×10^6 daltons (Gesteland and Boedtker, 1964). This indicates a polynucleotide approximately 3,400 nucleotides long and therefore suggests either that there may be untranslated sequences, or that short sequences may be translated to form products that are not recovered.

2. Techniques for Sequence Analysis

Sequence studies with these bacteriophage RNA molecules have involved the use of specific ribonucleases (T_1 which specifically cleaves on the 3'-side of guanylic acid residues and pancreatic, which cleaves on the 3'-side of pyrimidine nucleotides) and the oligonucleotide fingerprinting technique developed for use with ^{32}P-labelled RNA of high specific activity (Sanger and Brownlee, 1967). The application of these techniques to the sequencing of low molecular weight RNA molecules such as tRNA and 5S RNA are described by Goodman et al. (1968) and Brownlee et al. (1968). In the case of high molecular weight RNA molecules such as bacteriophage RNA, the oligonucleotide fingerprint patterns are too complex for effective analysis, and so several techniques have been developed to obtain oligonucleotide fragments of a size which can subsequently be handled by the sequencing technique.

a) Partial Ribonuclease Hydrolysis. The incubation of intact viral RNA molecules in the presence of low concentrations of ribonuclease T_1 at 0° results in reasonably controlled degradation, and yields large oligonucleotides which can be separated by electrophoresis on polyacrylamide gels (De Wachter and Fiers, 1971). Large oligonucleotides carrying either the 5'- or the 3'-terminus of the molecule can be easily identified by suitable labelling or modification techniques. Well separated oligonucleotides may also be selected and sequenced. Once a number of fragments (30–40 nucleotides long) have been sequenced and the fingerprint patterns for these have been established, longer oligonucleotides can be screened for these patterns. In this way oligonucleotides which contain the known shorter sequences can be detected.

b) Ribosome Binding Regions. Under appropriate conditions, ribosomes bind specifically to ribosome binding regions of mRNA *in vitro.* Such binding has been shown to protect the mRNA regions in contact with the ribosome surface from ribonuclease digestion. Thus, when ^{32}P-viral RNA is incubated with unlabelled ribosomes in the presence of magnesium ion, GTP, formylmethionyl-tRNA and initiation factors for protein synthesis (conditions which favour ribosome binding but do not permit polypeptide synthesis), and then treated with ribonuclease, intact fragments of labelled RNA can be dissociated from the particles (STEITZ, 1969). These fragments contain 35–40 nucleotides, and can be readily sequenced.

3. The Biochemical Map of R17 RNA

The best studied bacteriophage RNA is that from R17. On the basis of the overlapping oligonucleotide sequences which have been determined, it is possible to derive the biochemical map presented in Fig. 1 (JEPPESEN et al., 1970a).

a) Ribosome Binding Sites. The ribosome binding technique yields three different oligonucleotide fragments (STEITZ, 1969). Each sequence (A, C and S in Fig. 1) contains an AUG initiation codon. Since the N-terminal dipeptide has been identified for each of the three proteins of the virus, and each is different,

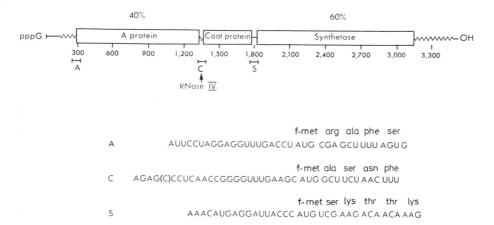

Fig. 1. The biochemical map of R17 RNA (drawn to scale). An RNA chain length of 3,500 nucleotides has been assumed and the best molecular weight estimates for the A protein and replicase (synthetase) were used. The ribosomal binding sites (sequences A, C, and S) and the point of ribonuclease IV cleavage, which initially divides the molecule into a 40% and a 60% fragment, have been located as accurately as possible along the RNA. Wavy lines indicate that the actual lengths of the corresponding regions are unknown. (From JEPPESEN et al., 1970a)

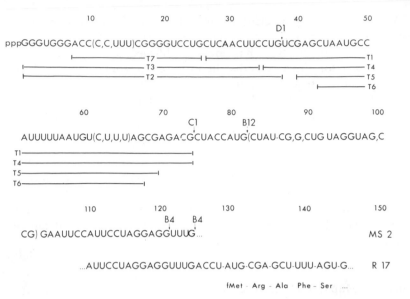

Fig. 2. The 5'-terminal sequence of MS2 RNA. The numbered arrows indicate the ends of the different 5'-terminal fragments that were isolated by electrophoresis on polyacrylamide gel slabs following limited T_1-ribonuclease digestion at 0° C. Products T1 to T7 were obtained by partial T_1-ribonuclease digestion of these fragments at 20° C. Mono- or oligonucleotides separated by commas within parentheses are of unknown sequence. The sequence of the A protein ribosome binding site of R17 RNA (STEITZ, 1969) has been written under the MS2 RNA sequence to show the overlapping region. (From DE WACHTER et al., 1971)

the three ribosome binding sites can be assigned to the three different genes. The following points can be deduced from the sequences of the ribosome binding sites.

1) Ribosomes bind to internal regions of polycistronic mRNA. Since there are three different sites (appropriate to each of the three genes) ribosomes do not bind only at 5'-ends of mRNA.

2) The ribosome binding sites are dissimilar in sequence. Other than the AUG triplet, other similarities seem to be minor. On the 5'-side, a pyrimidine is adjacent to AUG, A is common 4 nucleotides away, U is common 6 nucleotides away, G is common 9 nucleotides away, and purines are common 10 and 11 nucleotides away. It is clear, therefore, that ribosome binding does not involve recognition of a specific extended nucleotide sequence. It is interesting that the three genes are translated at different rates both *in vivo* and *in vitro;* the mole ratio of coat protein:replicase:A protein is approximately 20:6:1 (LODISH, 1968). This may mean that ribosomes exhibit different binding affinities for the different binding sites, although it is also possible that secondary and tertiary structure might be involved in differential rates of translation, as will be discussed later.

b) The Untranslated 5'-Terminal Sequence. The 5'-terminal sequence to position 74 has been determined for R17 RNA (ADAMS and CORY, 1970); this sequence does not contain the A protein ribosome binding region. The 5'-terminal sequence is also known for MS2 RNA (DE WACHTER et al., 1971) and this does overlap at its 3'-end with the sequence at the 5'-end of the A protein ribosome

(a)

(G)CA AAC UCC GGU AUC UAC UAA UAG AUG CCG GCC AUU CAA ACA UG

(b)

PyAA ACA UGA GGA UUA CCC AUG UCG AAG <u>ACA</u> <u>ACA</u> <u>AAG</u>

(c)

124   129   1   5
Ala Asn Ser Gly Ile Tyr   fMet Ser Lys Thr Thr Lys
(G)CA AAC UCC GGU AUC UAC UAA UAG AUG CCG GCC AUU CAA ACA UGA GGA UUA CCC AUG UCG AAG <u>ACA</u> <u>ACA</u> <u>AAG</u>

Coat protein cistron   Synthetase cistron

Fig. 3. (a) The nucleotide sequence of an R17 RNA fragment containing the chain terminating region of the coat protein cistron. The fragment was isolated as described in Fig. 2. (b) The sequence of the R17 RNA synthetase (replicase) ribosome binding site (STEITZ, 1969) written so as to show the overlapping region. The broken line shows the part of the sequence that is tentative. (c) Proposed nucleotide sequence of the region between the coat protein cistron and the synthetase cistron. (From NICHOLS, 1970)

binding region, although it does not contain the initiation codon (Fig. 2). It therefore seems highly likely that in the case of MS2 RNA, the initiation codon for the first translated region (A protein gene) is preceded on the 5'-side by 129 untranslated nucleotides. Since the MS2 5'-sequence is identical with the R17 5'-sequence to position 74, it seems possible that a similar situation exists in R17 RNA. It is particularly interesting that this untranslated "leader" sequence contains four potential initiation triplets, GUG at positions 3–5, and AUG at positions 46–48, 58–60 and 80–82, which are apparently not recognized by ribosomes.

c) Chain Termination Codons. An oligonucleotide containing 45 residues from T_1-digests of R17 RNA appears to contain the chain termination codons of the coat protein gene (Fig. 3) (NICHOLS, 1970). Since the amino acid sequence of the coat protein is known, possible nucleotide sequences for the coat protein gene can be written using the genetic code. The first seventeen positions (from the 5'-end) in the oligonucleotide correspond to a possible sequence for the last six amino acids of coat protein and are followed by two chain terminating codons in phase, UAA UAG. This finding raises the possibility that in natural mRNA, two nonsense codons may be used to terminate protein synthesis, the second ensuring termination in circumstances where the first is bypassed.

d) The Inter-Cistronic Region. The seven nucleotides at the 3'-end of the oligonucleotide sequenced by NICHOLS (1970) also overlap the sequence of the seven 5'-nucleotides of the replicase ribosome binding site, although the replicase initiation codon is not included (Fig. 3). It can be deduced, therefore, that there is an untranslated region 36 nucleotides long between the end of the coat protein gene and the start of the replicase gene. It is interesting that this apparently untranslated inter-cistronic divide contains an initiation triplet immediately after the two termination codons at the end of the coat protein gene and in phase with them. The fact that it is not recognised may mean that the ribosomes detach from the mRNA following termination of coat protein synthesis. It is also interesting to note however, that a chain terminating triplet (UGA) occurs in this untranslated sequence, in phase with the initiation triplet, perhaps ensuring that polypeptide synthesis which is initiated at this triplet is promptly aborted.

It is also certain that there is an untranslated inter-cistronic divide between the end of the A protein gene and the beginning of the coat protein gene. The length of this region is uncertain because the C-terminal amino acid sequence of the A protein is not known and hence potential chain terminating triplets in this region cannot be assessed. The nucleotide sequence available for this region comes partly from the coat protein ribosome binding site (STEITZ, 1969); this extends for 25 residues on the 5′-side of the initiation codon of the coat protein gene. A tentative sequence for an oligonucleotide 59 residues long, which contains

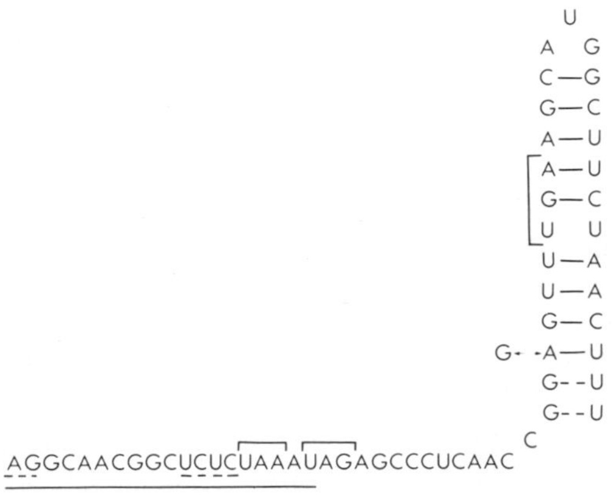

Fig. 4. The nucleotide sequence of an R17 RNA fragment containing the chain initiation region of the coat protein cistron. The fragment was isolated by electrophoresis on polyacrylamide gel slabs following partial ribonuclease IV digestion at 37° C. The dotted line indicates the parts of the sequence which are still tentative; the brackets above the sequence shows potential chain terminating triplets; the solid line underneath the sequence shows the residues which extend the coat protein ribosome binding sequence (STEITZ, 1969). A potential region of secondary structure within the ribosome binding sequence is also shown. (From CORY et al., 1970)

the coat protein ribosome binding sequence and extends for 19 residues on the 5′-side of this ribosome binding region, has also been presented (CORY et al., 1970) (Fig. 4). This tentative sequence of 44 nucleotides on the 5′-side of the coat protein initiation codon contains 3 potential chain terminating triplets; one, UGA is only 3 residues away from the initiation codon and the other two, UAA and UAG, are 27 and 23 residues away respectively. The UAA and UAG sequences are not adjacent but are separated by a single residue.

e) Untranslated Nucleotides at the 3′-End. An oligonucleotide of 11 residues from the 3′-end of R17 RNA has been sequenced (DAHLBERG, 1968). As this contains no chain terminating triplets, it is likely that there is an untranslated nucleotide region at the 3′-end. A tentative sequence of 51 nucleotides from the 3′-end (Fig. 5) is also available (CORY et al., 1970). This contains three potential chain terminating triplets; no two of these are adjacent, although two, UAA and UAG, are separated by a single nucleotide. It is not possible to say whether any are involved in the termination of the replicase because the C-terminal amino acid sequence of this protein is not known.

f) Gene Order in R17 mRNA. On the basis of the overlapping oligonucleotide sequences which have been discussed above, it is possible to derive the gene order given in Fig. 1 (JEPPESEN et al., 1970a). Experiments by these workers using the two specific fragments of R17 RNA produced by limited ribonuclease IV digestion provide additional evidence regarding this gene order. Ribonuclease IV initially cleaves R17 RNA at a position about 40% of the way into the molecule from the 5′-end; thus the 40% fragment has the 5′-end of the molecule and the 60% fragment the 3′-end (Fig. 1).

$$...ACCCG(GG)AUUCUCCCGAUUUGGUAACUAG...$$

$$...CUGCUUGGCUAGUUACCACCCA_{OH}$$

Fig. 5. The 3′-terminal sequence of R17 RNA. The fragment was isolated as described in Fig. 4. The brackets indicate potential chain terminating triplets. It is possible that the GG sequence shown at position *a* should be placed at position *b*. (From CORY et al., 1970)

Ribosome binding experiments were carried out with both the 40% and 60% fragments. An examination of the ribonuclease fingerprint patterns of the binding regions so obtained was not entirely conclusive, but did show that the A protein binding region is located in the 40% fragment, the coat protein binding region is mainly in the 40% fragment, while the replicase binding region occurs in the 60% fragment. Additionally, a number of shorter oligonucleotides (containing up to 20 nucleotides) were obtained from complete T_1 ribonuclease digestion and the sequences of a number of these could be assigned to positions within some of the longer known oligonucleotides sequences discussed earlier. Again it was possible to assign these either to the 40% or 60% fragments. These results confirm the gene order presented in Fig. 1.

4. Possible Secondary Structure

Many of the oligonucleotide sequences that have been obtained—from the 5′-leader sequence, from the coat protein gene, from the inter-cistronic divide between the coat protein and replicase genes, and from the 3′-terminal sequence— can be arranged in loops in which considerable base pairing is possible (Fig. 6). The frequency of these potential areas of secondary structure certainly seems too great to be explained by chance. Of particular interest are the looped structures presented for the coat protein gene; the chain initiation codon [Fig. 6(B)] and the first of the chain termination codons (UAA) [Fig. 6(D)] can be arranged at the closed ends of the loops, in non-hydrogen bonded regions. The three other oligonucleotides which correspond to amino acid residues 33–52, 57–75 and 81–99 can also be arranged in loops; the two major unknown portions (between residues 6 and 33, and 99 and 124) are long enough to give extra loops. If this proves to be the case, the coat protein gene could then be arranged in seven regularly spaced loops (Fig. 7). With regard to chain initiation and chain

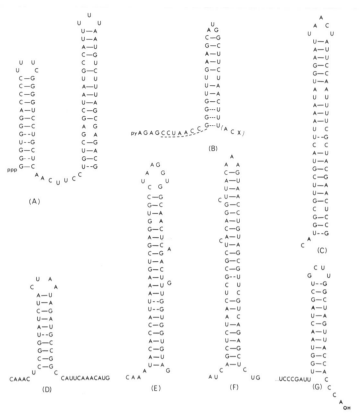

Fig. 6. Potential regions of secondary structure within the R17 RNA molecule. (A) the 5'-terminal sequence; (B)–(F) sequences from the coat protein cistron; (B) the ribosome binding site; (C) the sequence corresponding to amino acids 81–99; (D) the sequence corresponding to amino acids 124–129 (including the chain terminating region); (E) the sequence corresponding to amino acids 57–75; (F) the sequence corresponding to amino acids 33–52; (G) the 3'-terminal sequence. [(A) and (G) are from Cory et al. (1970); (B) from Steitz (1969); (C), (D), (E) and (F) from Jeppesen et al. (1970b)]

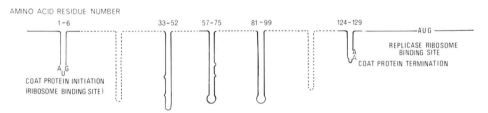

Fig. 7. Diagrammatic representation of the coat protein cistron in R17 RNA. Continuous lines represent portions whose sequences are known and show the five potentially hydrogen-bonded loops. The two loops in the unknown regions are speculative. (From Jeppesen et al., 1970b)

termination codons being located at the ends of loops, it should be pointed out that the chain initiation codons for the A protein and replicase genes (from the respective ribosome binding sequences) cannot easily be placed in similar positions.

In addition to this hypothetical arrangement of the nucleotide sequences into looped structures, there is considerable circumstantial evidence that bacterio-phage RNA does have secondary and tertiary structure in solution, and this can be summarized as follows.

1) Long oligonucleotides containing many guanylic acid residues are obtained following complete T_1 ribonuclease digestion. Since T_1 cleaves at guanylic acid residues, this suggests that many of these residues are inaccessible to nuclease attack, implying secondary or tertiary structure.

2) Ribonuclease IV cleaves intact R17 RNA initially in the region of the coat protein initiator. This observation, taken together with the fact that the coat protein gene both *in vivo* and *in vitro* is translated at a much greater frequency than the other genes, suggests that the ribosome binding site of the coat protein gene is readily accessible both to ribosomes and to ribonuclease IV. This also implies that the remainder of the molecule is not so readily accessible and thus that there is considerable secondary and tertiary structure.

3) Treatment of purified R17 RNA in ways likely to alter secondary and tertiary structure alters its translation characteristics *in vitro* (FUKAMI and IMAHORI, 1971). In a cell-free protein synthesizing system from *E. coli*, simultaneous initiation of both coat protein and replicase synthesis was obtained using untreated R17 RNA. Prior treatment for three minutes at 60° in 10 mM magnesium acetate resulted in a change in translation characteristics, such that coat protein synthesis was initiated before either of the other two proteins. After incubation for 3 minutes at 70° in the same buffer, replicase synthesis was preferentially initiated. Such results suggest that the ribosome binding sites may be particularly susceptible to changes in secondary or tertiary structure. A recent description of the nucleotide sequence of the coat protein gene of MS2 mRNA provides for a very substantial degree of structure stabilized by base pairing (MIN JOU et al., 1972).

5. Natural Degeneracy of the Genetic Code

The nucleotide sequences available for R17 RNA provide seven regions in which codon sequence and amino acid sequence can be compared. These are the two ribosome binding regions for the A protein and replicase genes, which give 9 codons (excluding the initiation codons); the nucleotide sequences which incorporate the ribosome binding region for coat protein (4 codons); the three internal sequences from the coat protein gene (58 codons); and the chain termina-tion sequence of the coat protein gene (6 codons). Eighteen amino acids are represented among these 77 codons (only histidine and aspartic acid are missing); of the 61 sense codons in the genetic code, 38 are represented (Table 1). The code is clearly degenerate in natural mRNA but the sample is too small for any assessment to be made as to whether certain alternative codons are preferred for any particular amino acid. Of great interest, however, is the finding that for serine all six alternative codons are represented UC(U, C, A, G) and AG(U, C); for valine all four alternative codons are represented GU(U, C, A, G); and similarly for threonine AC(U, C, A, G). Even on the basis of the wobble hypothesis of codon-anticodon pairing (CRICK, 1966), for each of these three amino acids

Table 1. The genetic code, indicating those codons found so far in the coat protein cistron, and the A protein and the replicase ribosome binding regions of R17 RNA. The figures in parentheses are the number of times that the particular codon occurs in R17 RNA. (From SANGER, 1971)

First letter	Second letter				Third letter
	U	C	A	G	
U	UUU(2) ⎫ Phe UUC(1) ⎬ UUA(2) ⎫ Leu UUG ⎭	UCU(4) ⎫ UCC(2) ⎬ Ser UCA(1) UCG(3) ⎭	UAU ⎫ Tyr UAC(4) ⎬ UAA(1) ⎫ C.T. UAG ⎭	UGU(1) ⎫ Cys UGC ⎬ UGA C.T. UGG(1) Tryp	U C A G
C	CUU ⎫ CUC ⎬ Leu CUA CUG ⎭	CCU(1) ⎫ CCC ⎬ Pro CCA(1) CCG ⎭	CAU ⎫ His CAC ⎬ CAA ⎫ GluN CAG(4) ⎭	CGU(2) ⎫ CGC(2) ⎬ Arg CGA(1) CGG ⎭	U C A G
A	AUU(3) ⎫ AUC(2) ⎬ Ileu AUA ⎭ AUG(1) Met	ACU(4) ⎫ ACC(2) ⎬ Thr ACA(2) ACG(1) ⎭	AAU(1) ⎫ AspN AAC(4) ⎬ AAA(3) ⎫ Lys AAG(3) ⎭	AGU(1) ⎫ Ser AGC(3) ⎬ AGA ⎫ Arg AGG ⎭	U C A G
G	GUU(2) ⎫ GUC(1) ⎬ Val GUA(2) GUG(2) ⎭	GCU(4) ⎫ GCC ⎬ Ala GCA(3) GCG ⎭	GAU ⎫ Asp GAC ⎬ GAA(1) ⎫ Glu GAG(1) ⎭	GGU(3) ⎫ GGC ⎬ Gly GGA GGG ⎭	U C A G

more than one iso-accepting species of tRNA (with different anticodons) would be necessary during protein synthesis for the translation of the range of codons found (for a discussion of the wobble hypothesis in relation to isoaccepting tRNA species see Chapter 5).

6. Summary of Studies of Bacteriophage mRNA

The nucleotide sequencing studies with bacteriophage RNA have provided the following information regarding the properties of mRNA.

1) Ribosome binding sites. All sites are internal in the polynucleotide chain, and have no fixed nucleotide sequence.

2) Initiation signals. All three genes are initiated with AUG.

3) Termination signals. Double nonsense codons are used in the case of the coat protein gene. Since the C-terminal amino acid sequences of the A protein and replicase enzyme are not known, no other chain terminating signals have yet been assigned. Studies with frame-shift mutants in the histidine operon of *Salmonella typhimurium* (RECHLER and MARTIN, 1970) suggest that a single nonsense codon terminates translation of *his* D (histidinol dehydrogenase) mRNA (for further discussion of nonsense codons and termination of translation see Chapter 7).

4) 5'- and 3'-ends of the molecule. There are untranslated sequences at both the 5'- and 3'-ends.

5) Inter-cistronic divides. There are untranslated regions between the end of the A protein and the start of the coat protein genes, and also between the end of the coat protein and the start of the replicase genes. Inter-cistronic divides also appear to exist in polycistronic bacterial mRNA (RECHLER and MARTIN, 1970).

6) Degeneracy of the genetic code. Many examples of degeneracy are demonstrated. All available alternative codons are translated for phenylalanine, valine, serine, threonine, asparagine, lysine and glutamic acid.

7) Secondary and tertiary structure. There is evidence for the existence of secondary and tertiary structure. With regard to such structure and also to the natural degeneracy of the code, bacteriophage mRNA may not be typical of all mRNA since it is both the primary genetic material and messenger. It seems likely that secondary and tertiary structure might be important in the packaging of the RNA within the mature virus particle. This raises an interesting question: is it possible, in an evolutionary sense, to select simultaneously for amino acid sequence of protein (compatible with biological activity) and nucleotide sequence (compatible with secondary and tertiary structure)? Perhaps two relevant points can be raised. Firstly, base pairing within each loop region is not perfect; apparently secondary structure can be adequately stabilized without maximum base homology. Secondly, there is the pattern of degeneracy in the genetic code. In most cases alternative codons for any amino acid differ in only the third (3') nucleotide of the triplet. As has been pointed out by SANGER (1970), within the limits of degeneracy it is possible to alter the nucleotides at the third position of codons, selecting for base pairing, without altering amino acid assignment.

B. Purification of Non-Viral mRNA Species and Preliminary Sequencing Studies

In view of the dual function of bacteriophage RNA and uncertainties concerning the importance of secondary structure, it is desirable to obtain information for other, perhaps more typical, species of mRNA. In this section, certain principles which are involved in the purification of particular species of non-viral mRNA will be considered, and the progress which has been made with the isolation and analysis of such mRNA species, particularly globin mRNA, will be discussed.

1. The Determination and Characterization of mRNA

In order to establish that an RNA fraction is a specific type of mRNA, it must be shown either that the RNA species has been synthesized on a genetically defined section of DNA, or that it is able to direct the synthesis of a specific protein. The technique of RNA-DNA hybridization may be used to demonstrate that an RNA preparation is complementary in base sequence to one strand of a specific section of DNA (and has, therefore, in all probability been synthesized on that section of DNA). The ability of an RNA preparation to direct the synthesis of a specific protein can be demonstrated using a cell-free protein synthesizing system.

a) Specific RNA-DNA Hybridization. While this technique has been used successfully to measure levels of specific mRNA of *lac* (lactose), *try* (tryptophan) and *gal* (galactose) operons in *E. coli* (ATTARDI et al., 1963; MORSE et al., 1969; GOSDEN et al., 1971), it has not yet been used in the isolation or purification of any specific species of mRNA. The major problem associated with the use of RNA-DNA hybridization in such procedures is that pure preparations of gene-specific DNA are required. It is now possible to obtain such DNA from *E. coli*. Episomes, such as the F factors and the temperate bacteriophages λ and ϕ80, may carry small specific regions of *E. coli* DNA within their own chromosomes. In the case of the *lac* operon, transducing particles of both λ and ϕ80 can be obtained, and genetic and biochemical methods have been developed for the isolation of pure *lac* operon DNA from these (SHAPIRO et al., 1969). Since techniques are available for obtaining transducing particles carrying any specific region of the *E. coli* genome (GOTTESMAN and BECKWITH, 1969) and because the methods used to obtain the *lac* operon DNA are probably generally applicable, it should therefore be possible to obtain specific DNA for any genetic region of *E. coli*.

The application of these techniques to a wide spectrum of organisms depends not only on the availability of viruses which can incorporate specific genetic regions, but also on detailed genetic information. Thus, while they may eventually be useful with other species of bacteria (as sufficient genetic information becomes available), application of these techniques to higher organisms seems unlikely in the near future.

b) Cell-Free Protein Synthesizing Systems. The unequivocal demonstration that an RNA preparation can direct the synthesis of a specific protein is possible in cell-free protein synthesizing systems in which polypeptide chain initiation occurs. Thus, with carefully prepared cell-free systems from *E. coli*, the addition of f2 RNA "informs" the ribosomes and brings about the synthesis of the three virus specific proteins (EGGEN et al., 1967).

Clearly the use of such a technique for characterizing mRNA is valid only in cases where the synthesis of a specific protein can be unequivocally demonstrated. Two principles related to the use of cell-free systems have therefore been established.

1) The protein being specified must be sufficiently well characterized so that it can be unambiguously distinguished from others. It must have either distinctive chromatographic or electrophoretic properties, or a characteristic peptide map so that its synthesis can be directly demonstrated. It is clearly not sufficient to show merely that an RNA preparation can stimulate amino acid incorporation in a cell-free protein synthesizing system. Thus DRACH and LINGREL (1966) have shown that although the addition of rabbit reticulocyte RNA stimulated amino acid incorporation in an *E. coli* system, no protein resembling globin either in chromatographic behaviour, peptide map or amino acid composition was produced. The peptide maps of the products obtained in the presence and absence of reticulocyte RNA were very similar, suggesting that the reticulocyte RNA

stimulated the translation of endogenous *E. coli* mRNA without "informing" the ribosomes in any way.

2) The cell-free protein synthesizing system must not be prepared from tissues which synthesize the particular protein being studied. Since added RNA can apparently stimulate the translation of endogenous mRNA, it is not valid for example, to use a rabbit reticulocyte system to assay for rabbit globin mRNA.

2. The Isolation of Globin mRNA

Rapid progress is at present being made towards the purification from eukaryotic organisms of a number of different types of specific mRNA. Globin mRNA was the first of these to be obtained and is the purest and best characterized non-viral mRNA preparation available. Its preparation and characterization will be dealt with in some detail in this section, since in general the techniques being used in the other cases are based on those developed for globin mRNA.

a) Advantages of Reticulocytes as a Source of Specific mRNA. There are two major problems associated with the isolation of specific mRNA from most tissues.

1) Homogeneity of cell types and of mRNA therein. Most tissues are composed of several different cell types, and even within one cell type many different proteins are synthesized. Thus mRNA, which is only a small percentage of the total cellular RNA, is a heterogenous collection of molecules.

2) Fragmentation during isolation. Since much mRNA exists in the cytoplasm as an unprotected, extended polynucleotide strand bound to ribosomes, it is particularly susceptible to physical shear and ribonuclease attack.

With reticulocytes these problems are less formidable than usual. Thus, the problem of homogeneity is largely overcome, firstly because reticulocytes are easy to obtain relatively uncontaminated by other cells active in protein synthesis, and secondly because they make one particular protein in very substantial quantities. Reticulocytes continue to make globin after losing their nuclei and at a stage when almost all other protein synthesis within the cell has ceased; globin synthesis can account for approximately 90% of total reticulocyte protein synthesis. The problem of fragmentation during isolation is also reduced since reticulocytes can be gently disrupted by osmotic shock, minimizing physical shear; in addition they contain extremely low levels of ribonuclease activity.

b) Methods Used for the Isolation of Globin mRNA. A discussion of the theory and practice of the methods used for the isolation of rabbit globin mRNA is presented by CHANTRENNE et al. (1967). Two different techniques have been used to separate mRNA from rRNA.

1) Sodium dodecyl sulphate (SDS) treatment. Polyribosome pellets prepared from reticulocyte lysates are dissolved in buffered 0.5% SDS; this treatment dissociates RNA from protein, freeing mRNA from the ribosome surface. When treated samples are analyzed by sucrose density gradient centrifugation, a small peak of RNA is found in the gradient profile between the 16S rRNA and 4S tRNA species. By selecting the peak fractions and by refractionating several times on sucrose gradients, a reasonably homogeneous peak of 9S RNA is obtained (Fig. 8).

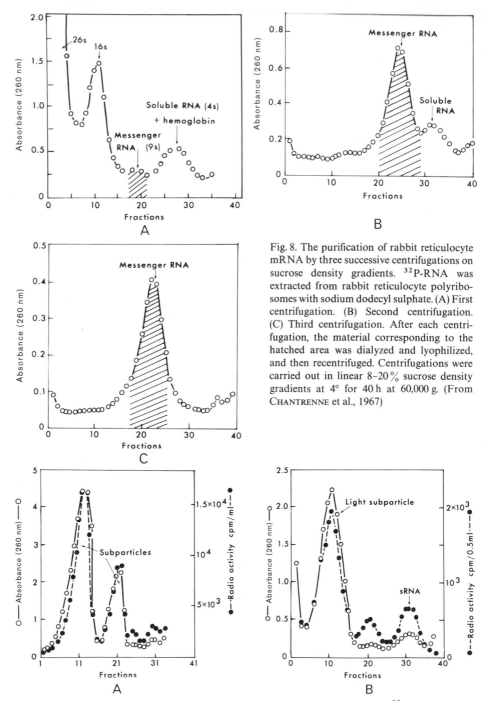

Fig. 8. The purification of rabbit reticulocyte mRNA by three successive centrifugations on sucrose density gradients. ^{32}P-RNA was extracted from rabbit reticulocyte polyribosomes with sodium dodecyl sulphate. (A) First centrifugation. (B) Second centrifugation. (C) Third centrifugation. After each centrifugation, the material corresponding to the hatched area was dialyzed and lyophilized, and then recentrifuged. Centrifugations were carried out in linear 8–20% sucrose density gradients at 4° for 40 h at 60,000 g. (From CHANTRENNE et al., 1967)

Fig. 9. The sedimentation pattern in linear 15–30% sucrose density gradients of ^{32}P-labelled ribosomal particles obtained by treating suspensions of rabbit reticulocyte polyribosomes with EDTA. (A) Centrifugation time 16 h. (B) Centrifugation time 40 h. Centrifugations were carried out at 60,000 g; temperature 4°

2) EDTA treatment. The treatment of polyribosome pellets with chelating agents such as EDTA dissociates the ribosomes into two subunits. Sucrose density gradient analysis of such treated samples from reticulocytes reveals a peak of RNA-containing material sedimenting between the smaller ribosomal subunit and 4S tRNA (Fig. 9). This material was also purified by several recentrifugation cycles and found to be a ribonucleoprotein complex. Following treatment with SDS or SDS plus phenol to dissociate these complexes, RNA of a similar size to that obtained by technique 1) was produced.

3. Characterization of Globin mRNA

In most of the studies which have been carried out with globin mRNA fractions, technique 2) (EDTA treatment) has been the preferred method of preparation.

a) Cell-Free Globin Synthesis. Conclusive evidence that such RNA preparations do contain globin mRNA comes from studies with cell-free protein synthesizing systems. Thus, mRNA from mouse reticulocytes was found to "inform" ribosomes in a rabbit reticulocyte system (LOCKARD and LINGREL, 1969, 1971). Since mouse β-chains can be separated chromatographically from rabbit α- and β-chains, it was possible to show, by co-chromatography and by tryptic peptide analysis, that mouse globin β-chain was produced. mRNA preparations from mouse reticulocytes have also been shown conclusively to stimulate the synthesis of mouse globin α- and β-chains in a system prepared from mouse Krebs II ascites cells (MATHEWS et al., 1971). Of great interest is the demonstration that RNA prepared from rabbit reticulocytes by differential salt precipitation can be translated in a defined *E. coli* system. In the presence of an initiator tRNA (*N*-acetylvalyl-tRNA, since valine is at the N-terminus of both α- and β-chains of rabbit globin), a product was formed which co-chromatographed with globin in a number of systems and which yielded typical globin tryptic peptides (LAYCOCK and HUNT, 1969). Thus eukaryotic mRNA can apparently be translated accurately by prokaryotic ribosomes.

mRNA prepared from rabbit reticulocytes has also been injected into frog oocytes, and in these cells programs the synthesis of rabbit haemoglobin. The identity of the apoprotein was established by co-chromatography with rabbit globin in several systems, co-electrophoresis with rabbit globin on polyacrylamide gels, and by tryptic peptide analysis (LANE et al., 1971). The living frog oocyte, as a general assay system for mRNA, may become very important because yields of translated product per molecule of injected mRNA appear to be much higher with these cells than in cell-free systems.

b) The Purity of the Preparations. All globin mRNA preparations are heterogeneous in that they contain separate mRNA species for α- and β-chains of globin, and it is also clear that they are impure. Early preparations isolated by the EDTA method were heavily contaminated with other types of RNA molecules; thus LABRIE (1969), using polyacrylamide gel electrophoresis, showed that a preparation from rabbit polyribosomes contained 28S, 18S, 7S and 5S rRNA species, 4S tRNA and uncharacterized RNA species of 14, 11 and 9S respectively. Globin mRNA, the major component, had a mobility equivalent to about 10S.

More recent preparations are less heavily contaminated with the well-characterized species of RNA (rRNA and tRNA) but do still contain appreciable levels of less clearly defined RNA species (GASKILL and KABAT, 1971).

c) Estimations of the Molecular Weight of Rabbit Globin mRNA. There is considerable variation in the estimates of molecular weight presented by different workers. Thus the S value of 9.0–9.3 obtained using the ultracentrifuge (CHANTRENNE et al., 1967) indicates a molecular weight of approx. 1.8×10^5 daltons. Based on mobilities in SDS-polyacrylamide gels, molecular weights varying between 1.92 and 2.35×10^5 daltons have been calculated (LABRIE, 1969; GASKILL and KABAT, 1971). Recently, formaldehyde-treated globin mRNA has been analysed in the ultracentrifuge and a molecular weight of $1.70 \pm 0.08 \times 10^5$ daltons was obtained (WILLIAMSON et al., 1971). Since formaldehyde treatment minimizes variations in sedimentation behaviour due to secondary structure, this is certainly the most accurate estimate so far available. The discrepancy between values obtained by electrophoresis and ultracentrifugation indicates that, perhaps due to some features of its structure, globin mRNA runs more slowly on polyacrylamide gels than would be predicted from comparisons with RNA molecules of similar size. Based on a molecular weight of 1.70×10^5, globin mRNA molecules are approximately 500 nucleotides long. The α-globin chain (141 amino acids) and the β-chain (146 amino acids) require mRNA molecules 429 and 444 nucleotides long respectively (allowing for single initiation and termination codons). It therefore seems that there may be about 60 untranslated nucleotides in each globin mRNA species.

d) High Adenylic Acid Content. A puzzling feature of the base ratio for globin mRNA estimated by CHANTRENNE et al. (1967) was a higher value for adenylic acid than would have been expected on the basis of the amino acid composition of globin. It now seems likely that adenylic acid rich regions are located in untranslated segments of the molecule. Thus when globin mRNA, extracted by EDTA treatment and further purified by gel electrophoresis, was digested with ribonuclease S, it yielded an oligonucleotide of 50–70 residues containing more than 70 % (mole basis) adenylic acid (LIM and CANELLAKIS, 1970). Further analysis indicated that it contained extended polyadenylic acid (poly A) sequences. Since AAA codes for lysine and there are no adjacent lysine residues in globin, this strongly suggests that the adenylic acid rich region lies in an untranslated region of the molecule. Additional studies by these workers indicated that the adenylic acid rich region may vary in length. Thus, mRNA extracted from regions of sucrose gradients containing larger polyribosome aggregates had longer adenylic acid rich regions than that extracted from the smaller polyribosomes.

Poly A sequences have been reported in both polyribosomal RNA and nuclear RNA extracted from cells of several different species of higher organisms (LEE et al., 1971; EDMONDS et al., 1971; see also Chapter 3). Two general proposals have been put forward to explain the presence of poly A in these types of RNA. Firstly, poly A regions might be involved in the processing of nuclear RNA into mRNA, and in its transport across the nuclear membrane (LEE et al., 1971; EDMONDS et al., 1971). Secondly, poly A sequences might exist at the 5'-end of certain mRNA molecules, limiting the rate of ribonuclease digestion, and so regulating the functional life of the molecules (LIM and CANELLAKIS, 1970). With regard to the latter proposal, recent studies indicate that poly A sequences in rabbit globin mRNA also exist at the 3'-end of the molecules (BURR and LINGREL, 1971).

A most useful practical development resulting from the discovery of poly A sequences in polyribosomal mRNA has been the use of poly-U impregnated filters and poly-U cellulose columns for separating cytoplasmic mRNA from rRNA and tRNA (LEE et al., 1971; SHELDON et al., 1972). These techniques will certainly play an important role in future studies on mRNA from eukaryotes.

4. Nucleotide Sequencing Studies with Globin mRNA

The two-dimensional oligonucleotide fingerprinting technique of SANGER and BROWNLEE (1967), which has been used successfully in the sequencing studies with bacteriophage RNA, requires radioactively labelled RNA of a very high specific activity, due to the low capacity of the ionophoresis step on cellulose acetate. Although mammalian reticulocytes are particularly suitable as a source of specific mRNA, such RNA cannot easily be radioactively labelled since these cells are enucleated and therefore inactive in RNA synthesis. Duck erythrocytes, however, which are nucleated cells and which can be labelled *in vitro* at the erythroblast stage, have recently been used as a source of radioactively labelled globin mRNA (PEMBERTON et al., 1972). This may prove suitable for sequencing studies.

Preliminary determinations of the nucleotide sequence at the 3'-end of globin mRNA have been made. By the use of periodate oxidation and ^{3}H-borohydride reduction to label the 3'-terminal nucleotide, followed by T_1 and pancreatic ribonuclease digestion, a 3'-terminal oligonucleotide approximately 6 nucleotides long and containing only adenylic acid residues, has been detected (BURR and LINGREL, 1971). It seems, therefore, that there is an untranslated region at the 3'-end of globin mRNA, perhaps included in which is the adenylic acid rich region previously discussed.

5. The Purification of Other Species of mRNA

a) Bacterial mRNA. The most obvious problem associated with the purification of specific species of bacterial mRNA is one of selection; bacterial cells synthesize a wide variety of proteins and therefore contain very heterogeneous populations of mRNA. Associated with this is the possibility that for short-lived mRNA species at least, intact mRNA molecules may not exist in the bacterial cell. Thus, not only is there evidence that transcription and translation are co-ordinated, the first ribosome attaching to nascent mRNA soon after its transcription has been initiated (MILLER et al., 1970), but it may also be that degradation of mRNA from the 5'-end begins before its synthesis at the 3'-end is complete. There is perhaps an alternative approach to the problem of isolating specific bacterial mRNA. Now that methods are available for the isolation of pure gene-specific *E. coli* DNA (SHAPIRO et al., 1969), it may be possible to synthesize specific species of mRNA *in vitro* using purified RNA polymerase. That faithful transcription of bacterial DNA into mRNA can be achieved *in vitro* is indicated by the experiments of ZUBAY and CHAMBERS (1969). In a combined transcription-translation system from cell-free extracts of *E. coli*, the addition of DNA from a transducing particle of ϕ80 carrying the *lac* operon of *E. coli* (ϕ80d*lac* DNA) brought about the synthesis of active β-galactosidase. It is also of interest to note that nucleotide sequencing studies with Qβ-bacteriophage RNA have involved the use of material which was synthesized *in vitro* (BILLETER et al., 1969).

b) Higher Organisms. In general, a biological approach has been taken to the problem of purifying other species of eukaryotic mRNA. Attempts have been made to identify other tissues which offer advantages similar to those associated with reticulocytes, particularly from the point of view of homogeneity of mRNA. The following species of mRNA are currently being investigated.

1) Histone mRNA. A 9–10S RNA fraction has been extracted from light polyribosomes prepared from sea urchin embryos (KEDES and GROSS, 1969). Although this preparation has not been rigorously characterized, circumstantial evidence indicates that it contains mRNA for histones. The light polyribosome fraction from which it has been prepared synthesizes protein with a high lysine and a low tryptophan content characteristic of histone. In addition, the amount of this RNA that can be extracted, and the rate of synthesis of nuclear proteins, are affected in parallel by normal changes in the rate of cleavage during embryogenesis or by drug-induced inhibition of cleavage and DNA replication.

2) Myosin mRNA. RNA from the large polyribosome fraction from chick embryonic leg muscle contains 26S material that may be the mRNA for myosin. Thus, it is the large polyribosome fraction that synthesizes myosin, and RNA extracted from small polyribosomes, which are not involved in myosin synthesis, does not contain such a labelled peak (HEYWOOD and NWAGWU, 1969). In addition, material isolated from the 25–27S region of the gradients, when added to a cell-free system prepared from chick reticulocytes, programs the synthesis of protein which co-electrophoreses with myosin on polyacrylamide gels (HEYWOOD, 1969). The purification of the 26S component will obviously be difficult because of its similarity in size to 28S rRNA.

3) Immunoglobulin (light chain) mRNA. A mouse myeloma which synthesizes and secretes large quantities of a kappa chain of immunoglobulin (K41) has been used in these studies. An RNA fraction which sediments in 9–13S regions of sucrose density gradients contains mRNA for the immunoglobulin. When added to a cell-free system prepared from rabbit reticulocytes, this RNA induces the synthesis of a protein which is specifically precipitated by antiserum prepared against K41 immunoglobulin and yields tryptic peptides characteristic of this protein. The RNA fraction is not homogeneous; polyacrylamide gel analysis has revealed that it contains 3 major components in the 9–13S range and that it is also contaminated with rRNA and tRNA species (STAVNEZER and HUANG, 1971).

4) Lens crystallin mRNA. Two RNA components of 10 and 14S have been obtained from polyribosome pellets isolated from calf lenses (BERNS et al., 1971). These preparations have been tested in cell-free systems isolated from both Krebs II ascites cells and rabbit reticulocytes and have been found to induce

the synthesis of protein which co-electrophoreses with lens crystallin protein in two different polyacrylamide gel systems (MATHEWS et al., 1972; BERNS et al., 1972).

5) Silk fibroin mRNA. Total ^{32}P-labelled RNA, extracted from posterior silkglands of *Bombyx mori*, has been fractionated on sucrose density gradients and an RNA fraction from the 45–65S region shown to contain fibroin mRNA by chemical analysis. Fibroin has a simple repetitive primary structure in which glycine comprises about 45 % of all residues and alternates in the sequence with alanine and serine. Based on codon assignments for these amino acids, predictions regarding the base composition of fibroin mRNA and its oligonucleotide frequencies following T_1 and pancreatic ribonuclease digestion can be made. The results of such analyses of silkgland 45–65S RNA correspond closely to predicted values (SUZUKI and BROWN, 1972). Additional information can also be derived from the oligonucleotide frequencies. Firstly, the fraction is at least 80 % pure fibroin mRNA and comprises between 0.8 and 1.4 % of the total RNA of the posterior silkgland. Secondly, glycine is coded for by two codons (GGU and GGA) but alanine (GGU) and serine (UCA) are coded for by single codons. Thirdly, poly A sequences longer than 10 residues are not present in the molecules.

It seems certain that progress in studies of eukaryotic mRNA will be rapid in the near future and detailed nucleotide sequence data for either globin mRNA or one of these other species will be available soon. Finally, although the number of examples is small, it is interesting to note that all eukaryotic mRNA species which have so far been examined are almost certainly monocistronic; poly-cistronic mRNA may be confined to prokaryotes and viruses.

6. The Stability of Non-Viral Species of mRNA

Most information on the stability of mRNA comes from indirect measurements. These involve for instance measuring changes in the amounts of specific proteins brought about by removal of an inducer, addition of co-repressor, or inhibition of RNA synthesis for example with actinomycin D. Such observations, which measure the translation product of mRNA, are certainly open to criticism since they ignore the possibility of other controls of protein synthesis. Nevertheless, certain features have emerged from these studies and they do seem to be reasonably well established.

a) Bacteria. The genetic regulation model of JACOB and MONOD (1961) predicts that the RNA templates for protein synthesis in bacteria will be short-lived. Indeed, many bacterial mRNAs seem to be degraded very rapidly. Thus the half-life for β-galactosidase mRNA of *E. coli* has been reported to be 1–2.5 minutes (KEPES, 1963; NAKADA and MAGASANIK, 1964); these estimates are indirect, being based on the declining rate of appearance of β-galactosidase activity following removal of inducer. In the case of the tryptophan operon, however, mRNA stability has been determined by direct measurement using specific RNA-DNA hybridization. Transcription, translation and mRNA degradation from the 5'-end appear to occur concurrently (MORSE et al., 1969). The degradative nucleases probably reach a particular region of the polycistronic messenger about 2 minutes after the first ribosome and immediately after the last. On the other hand, some bacterial mRNAs appear to exhibit much greater stability. Thus spore formation in *Bacillus cereus* seems to involve mRNAs which exist in the cell for several hours (ARONSON and ROSAS DEL VALLE, 1964). The translation of mRNA for penicillinase in *Bacillus cereus* can continue for at least one hour after all cellular RNA synthesis has been abolished with actinomycin D (HARRIS and SABATH, 1964).

b) Higher Organisms. It seems likely that most mRNA found in eukaryotes is of greater stability than that in bacteria. Studies with mouse liver indicate that total mRNA has an average half-life of about 2 hours (TRAKATELLIS et al., 1964). In the case of tissues specialized to produce predominantly a single protein, mRNA with much longer half-life seems to be involved. Thus cocoonase, an enzyme produced in the galea of developing silk moths (KAFATOS and REICH, 1968) and globin of mammalian reticulocytes (CHANTRENNE et al., 1967) are two specific proteins which are the products of RNA templates with lifespans of several days. In addition, in certain developmental situations, there is abundant evidence for the existence of mRNA of even greater stability. *Acetabularia* appears to contain mRNA species which can remain in the cytoplasm of enucleated cells for at least 40 days before being translated (HARRIS, 1968). In wheat embryos, protein synthesis commences immediately after imbibition and appears to be directed by mRNA preserved in the dry embryo (CHEN and OSBORNE, 1970). mRNA which is formed in the cytoplasm of sea urchin eggs during oogenesis is not translated until after fertilization. Evidence that such mRNA, which has been called "maternal" or "masked" messenger, can exist in the form of ribonucleoprotein particles ("informosomes") has been presented (NEYFAKH, 1971; GLISIN and SAVIC, 1971). There is evidence that some of these "masked" messengers code for microtubule proteins (RAFF et al., 1972). Control of the unmasking process may prove to be an important regulatory mechanism during development in eukaryotes.

The stability of mRNA is an important question since any model of genetic regulation must take it into account. There is clearly a lack of quantitative data on this subject, particularly in the case of eukaryotic organisms. An important outcome of the work on the specific species of eukaryotic mRNA which has been discussed in this article is that it should now be possible to measure directly the stabilities of these mRNA species.

References

Reviews

SANGER, F.: Biochem. J. **124**, 833 (1971).
SINGER, M. F., LEDER, P.: Ann. Rev. Biochem. **35**, 195 (1966).

Other References

ADAMS, J. M., CORY, S.: Nature **227**, 570 (1970).
ARONSON, A. I., ROSAS DEL VALLE, M.: Biochim. Biophys. Acta **87**, 267 (1964).

ATTARDI, G., NAONO, S., ROUVIERE, J., JACOB, F., GROS, F.: Cold Spring Harbor Symp. Quant. Biol. **28**, 363 (1963).
BERNS, A.J.M., DE ABREU, R.A., VAN KRAAIKAMP, M., BENEDITTI, E.L., BLOEMENDAL, H.: FEBS Lett. **18**, 159 (1971).
BERNS, A.J.M., STROUS, G.J.A.M., BLOEMENDAL, H.: Nature New Biol. **236**, 7 (1972).
BILLETER, M.A., DAHLBERG, J.E., GOODMAN, H.M., HINDLEY, J., WEISSMAN, C.: Nature **224**, 1083 (1969).
BROWNLEE, G.G., SANGER, F., BARRELL, B.G.: J. Mol. Biol. **34**, 379 (1968).
BURR, H., LINGREL, J.B.: Nature New Biol. **233**, 41 (1971).
CHANTRENNE, H., BURNY, A., MARBAIX, G.: Progr. Nucl. Acid Res. Mol. Biol. **7**, 173 (1967).
CHEN, D., OSBORNE, D.J.: Nature **226**, 1157 (1970).
CORY, S., SPAHR, P.F., ADAMS, J.M.: Cold Spring Harbor Symp. Quant. Biol. **35**, 1 (1970).
CRICK, F.H.C.: J. Mol. Biol. **19**, 548 (1966).
DAHLBERG, J.E.: Nature **220**, 548 (1968).
DE WACHTER, R., FIERS, W.: In Methods in enzymology (GROSSMAN, L., and MOLDAVE, K., eds.), vol. 21, p. 167. New York: Academic Press 1971.
DE WACHTER, R., VANDENBERGHE, A., MERREGAERT, J., CONTRERAS, R., FIERS, W.: Proc. Nat. Acad. Sci. U.S. **68**, 585 (1971).
DRACH, J.C., LINGREL, J.B.: Biochim. Biophys. Acta **129**, 128 (1966).
EDMONDS, M., VAUGHAN, M.H., NAKAZATO, H.: Proc. Nat. Acad. Sci. U.S. **68**, 1336 (1971).
EGGEN, K., OESCHGER, M.P., NATHANS, D.: Biochem. Biophys. Res. Commun. **28**, 587 (1967).
FUKAMI, H., IMAHORI, K.: Proc. Nat. Acad. Sci. U.S. **68**, 570 (1971).
GASKILL, P., KABAT, D.: Proc. Nat. Acad. Sci. U.S. **68**, 72 (1971).
GESTELAND, R.F., BOEDTKER, H.: J. Mol. Biol. **8**, 496 (1964).
GLISIN, V., SAVIC, A.: Progr. Biophys. Mol. Biol. **23**, 191 (1971).
GOODMAN, H.M., ABELSON, J., LANDY, A., BRENNER, S., SMITH, J.D.: Nature **217**, 1019 (1968).
GOSDEN, J.R., IRVING, M.I., BISHOP, J.O.: Biochem. J. **121**, 109 (1971).
GOTTESMAN, S., BECKWITH, J.R.: J. Mol. Biol. **44**, 117 (1969).
GUSSIN, G.N.: J. Mol. Biol. **21**, 435 (1966).
HARRIS, H.: Nucleus and cytoplasm, p. 1–15. Oxford: Clarendon Press 1968.
HARRIS, H., SABATH, L.D.: Nature **202**, 1078 (1964).
HEYWOOD, S.M.: Cold Spring Harbor Symp. Quant. Biol. **34**, 799 (1969).
HEYWOOD, S.M., NWAGWU, M.: Biochemistry **8**, 3839 (1969).
JACOB, F., MONOD, J.: J. Mol. Biol. **3**, 318 (1961).
JEPPESEN, P.G.N., STEITZ, J.A., GESTELAND, R.F., SPAHR, P.F.: Nature **226**, 230 (1970a).
JEPPESEN, P.G.N., NICHOLS, J.L., SANGER, F., BARRELL, B.G.: Cold Spring Harbor Symp. Quant. Biol. **35**, 13 (1970b).
KAFATOS, F., REICH, J.: Proc. Nat. Acad. Sci. U.S. **60**, 1458 (1968).
KEDES, L.H., GROSS, P.R.: Nature **223**, 1335 (1969).
KEPES, A.: Biochim. Biophys. Acta **76**, 293 (1963).
LABRIE, F.: Nature **221**, 1217 (1969).
LANE, C.D., MARBAIX, G., GURDON, J.B.: J. Mol. Biol. **61**, 73 (1971).
LAYCOCK, D.G., HUNT, J.A.: Nature **221**, 1118 (1969).
LEE, S.Y., MENDECKI, J., BRAWERMAN, G.: Proc. Nat. Acad. Sci. U.S. **68**, 1331 (1971).
LIM, L., CANELLAKIS, E.S.: Nature **227**, 710 (1970).
LOCKARD, R.E., LINGREL, J.B.: Biochem. Biophys. Res. Commun. **37**, 204 (1969).
LOCKARD, R.E., LINGREL, J.B.: Nature New Biol. **233**, 204 (1971).
LODISH, H.F.: Nature **220**, 345 (1968).
MATHEWS, M.B., OSBORN, M., BERNS, A.J.M., BLOEMENDAL, H.: Nature New Biol. **236**, 5 (1972).
MATHEWS, M.B., OSBORN, M., LINGREL, J.B.: Nature New Biol. **233**, 206 (1971).
MILLER, O.L., HAMKALO, B.A., THOMAS, C.A.: Science **169**, 392 (1970).
MIN JOU, W., HAEGEMAN, G., YSEBAERT, M., FIERS, W.: Nature **237**, 82 (1972).
MORSE, D.E., MOSTELLER, R.D., YANOFSKY, C.: Cold Spring Harbor Symp. Quant. Biol. **34**, 725 (1969).
NAKADA, D., MAGASANIK, B.: J. Mol. Biol. **8**, 105 (1964).
NEYFAKH, A.A.: Current Topics Devel. Biol. **6**, 45 (1971).
NICHOLS, J.L.: Nature **225**, 147 (1970).
PEMBERTON, R.E., HOUSMAN, D., LODISH, H.F., BAGLIONI, C.: Nature New Biol. **235**, 99 (1972).
RAFF, R.A., COLOT, H.V., SELVIG, S.E., GROSS, P.R.: Nature **235**, 211 (1972).

Rechler, M.M., Martin, R.G.: Nature **226**, 908 (1970).

Sanger, F., Brownlee, G.G.: In Methods in enzymology (Grossman, L., and Moldave, K., eds.), vol. 12A, p. 361. New York: Academic Press 1967.

Shapiro, J., Machattie, L., Eron, L., Ihler, G., Ippen, K., Beckwith, J.: Nature **224**, 768 (1969).

Sheldon, R., Jurale, C., Kates, J.: Proc. Nat. Acad. Sci. U.S. **69**, 417 (1972).

Stavnezer, J., Huang, R.C.C.: Nature New Biol. **230**, 172 (1971).

Steitz, J.A.: Nature **224**, 957 (1969).

Suzuki, Y., Brown, D.D.: J. Mol. Biol. **63**, 409 (1972).

Trakatellis, A.C., Axelrod, A.E., Montjar, M.: J. Biol. Chem. **239**, 4237 (1964).

Vinuela, E., Algranati, I.D., Ochoa, S.: Eur. J. Biochem. **1**, 3 (1967).

Williamson, R., Morrison, M., Lanyon, G., Eason, R., Paul, J.: Biochemistry **10**, 3014 (1971).

Zubay, G., Chambers, D.A.: Cold Spring Harbor Symp. Quant. Biol. **34**, 753 (1969).

CHAPTER 5

Transfer RNA and Cytokinins

D. S. LETHAM

Introduction

When it was observed that soluble RNA (sRNA) played a vital role in protein synthesis, this low-molecular-weight species of RNA attracted much attention. The typical sRNA molecule, with a molecular weight of about 25,000 and a sedimentation constant of 4S, is a polynucleotide chain containing about 80 nucleotides. Because sRNA was observed to accept amino acids and transfer them to growing polypeptide chains on ribosomes, sRNA is often referred to as transfer RNA (tRNA) and these terms are sometimes used interchangeably. This is not desirable. A small proportion of sRNA does not appear to be involved in transfer of amino acids. For example, *E. coli* contains a species of RNA (sedimentation constant 6S) which can be isolated from the supernatant fraction obtained by sedimentation of ribosomes. This polynucleotide, which has recently been completely sequenced, is composed of 184 nucleotides (BROWNLEE, 1971). Its function is unknown.

Each tRNA molecule is specific for a particular amino acid and is aminoacylated (charged) enzymically by an amino acid-tRNA ligase. In a particular tissue or micro-organism, amino acid-specific tRNA can exist in multiple forms, termed isoaccepting tRNA species. Thus in *E. coli* there are five leucine-accepting tRNAs and a normal unfractionated tRNA preparation probably contains about 60–80 molecular species. The development of effective methods for purifying amino acid-specific tRNAs from these extremely complex mixtures was a great achievement. These methods, which are briefly discussed in the first section of this chapter, were but the forerunner of an even greater achievement, the sequencing of tRNA molecules, which ranks among the truly great conquests of modern chemistry and molecular biology.

Some tRNA species in chloroplasts and mitochondria are organelle-specific. Chloroplast-specific species are aminoacylated exclusively by amino acid-tRNA ligases localized within the organelle. Mitochondrial and chloroplast tRNAs are discussed in Chapters 8 and 9 respectively.

A. Isolation of Amino Acid-Specific tRNA

In this section, a very brief description is presented of four types of procedure— countercurrent distribution, reversed-phase partition chromatography, methods based on chemical modification of uncharged tRNA, and procedures involving chemical modification of charged tRNA.

Those not familiar with countercurrent distribution are referred to CRAIG and CRAIG (1956) for an elegant discussion of this valuable purification method. Certain of the amino acid-specific tRNAs used for sequence studies were purified solely by this method utilizing differences in partition coefficients of tRNA species. Thus APGAR and co-workers (1962) investigating yeast tRNAs used a mixture of phosphate buffer, formamide and isopropanol to yield a two-phase solvent system for countercurrent fractionation of tRNA species. After redistribution using related solvent systems, these workers isolated alanine, valine and tyrosine tRNAs in a state of sufficient purity for structural investigation. To assess the distribution of tRNA species among the numerous fractions obtained by countercurrent distribution and other methods, specific amino acid-acceptance activity is determined with an amino acid-tRNA ligase preparation.

When compared with countercurrent distribution, reversed-phase partition chromatography affords much enhanced resolution of tRNA species. A quaternary ammonium extractant, e.g. tricaprylylmethylammonium chloride, as a Freon (tetrachlorotetrafluoropropane) solution is supported as a thin film on a column of hydrophobic diatomaceous earth. The chromatograms are developed with a sodium chloride gradient. The method, which gives excellent resolution of isoaccepting species, is of great value for both analytical and preparative experiments (WEISS and KELMERS, 1967; WEISS et al., 1968). It has recently been demonstrated that use of a polychlorotrifluoroethylene resin as the inert support in place of a diatomaceous earth enhances resolution. This eliminates the need to use Freon as a solvent (PEARSON et al., 1971).

Chemical modification of tRNA molecules can greatly facilitate separation of tRNA species. Several methods of separation are based on chemical modification of uncharged tRNA which facilitates separation of charged from uncharged species. In such methods, the tRNA mixture is first freed from amino acids by hydrolysis (pH 9) and the tRNA required is specifically aminoacylated. The amino acid is esterified to the 2' or 3' position on the terminal adenosine nucleoside while uncharged tRNA molecules possess vicinal hydroxyl groups at these positions and are subject to periodate oxidation to yield aldehydes. By reaction of the aldehyde groups with hydrazide or amino groups on polymers (e.g. polyacrylic acid hydrazide, or aminoethyl derivatives of cellulose or dextran) a separation of the oxidized tRNA from the aminoacyl tRNA can be achieved (ZACHAU et al., 1961; SAPONARA and BOCK, 1961; ZUBAY, 1962; MUENCH and BERG, 1964).

An unsatisfactory feature of the above method is degradation of all tRNAs except the charged species. In a more desirable procedure involving chemical modification, the unfractionated tRNA is freed from amino acids by hydrolysis (pH 9) and the required tRNA is charged enzymically. The aminoacyl moiety is then modified by reaction with the phenoxyacetyl ester of N-hydroxysuccinimide to yield a phenoxyacetyl derivative (GILLAM et al., 1968). The resulting derivatized aminoacyl tRNA is specifically retarded and separated from uncharged tRNA when the tRNA mixture is passed through a column of benzoylated DEAE-cellulose which shows a strong affinity for aromatic groups. Gentle hydrolysis of the purified tRNA derivative yields the tRNA species which is desired. It is noteworthy that coupling of aromatic amino acids to their corresponding tRNA species yields molecules which without modification bind strongly to benzoylated DEAE-cellulose and can be extensively purified by chromatography on this material (MAXWELL et al., 1968).

Methylated albumin-silicic acid has proved a useful matrix for chromatographic separation of tRNA species and is much superior to methylated albumin-kieselguhr. Chemical modification of an aminoacyl tRNA by conversion to an N-carbobenzyloxy derivative can greatly facilitate separation of the acyl tRNA from uncharged tRNA on such columns (STERN and LITTAUER, 1968).

The first successful complete purification of tRNA species was achieved by countercurrent distribution. Although such methods are still employed, there is now a preference for chromatographic procedures of the type outlined above because these offer greater speed and do not require sophisticated equipment. The various methods obviously do not exploit the same parameters of tRNA molecules. Combinations of methods can be employed to great effect. The resolution of tRNA preparations enabled the biochemist to enter a new but formidable domain of endeavour—the nucleotide sequence of pure RNA species.

B. The Structure of tRNA

In this section we will consider methods for determining nucleotide sequence, the primary and secondary structures of yeast alanine and other tRNAs, homology in primary and secondary structures of tRNAs, and finally tertiary structure.

1. Sequence Determination; Primary and Secondary Structures of Yeast Alanine tRNA and Certain Isoaccepting tRNAs

In 1965 HOLLEY and seven co-workers announced the nucleotide sequence of a tRNA molecule, yeast alanine tRNA. This truly great achievement required seven years of endeavour and the preparation of 1 g of pure alanine tRNA from 200 g of total RNA extracted from about 150 kg of yeast. The sequence was determined by enzymic dissection and similar enzymic procedures were later used to sequence other tRNA molecules. The principal enzymes used were pancreatic ribonuclease which cleaves the polyribonucleotide chain at the 3' side of pyrimidine nucleotides, takadiastase ribonuclease T_1 which cleaves at the 3' side of guanylic acid residues, and snake venom phosphodiesterase which cleaves nucleotides in a stepwise manner from the 3' end of small fragments. The first two enzymes yielded mononucleotides and short fragments usually not larger than hexanucleotides. The determination of the sequence within a hexanucleotide is now outlined to demonstrate the methods used to elucidate oligonucleotide structure. A hexanucleotide produced by T_1 ribonuclease was found by alkaline hydrolysis to contain one Ap, two Cp, one Gp and two Up moieties*. Because of the known specificity of the enzyme, Gp must occur at

---

* For the meaning of this symbolism, see the legend for Fig. 1.

the 3′ end. Since the terminal phosphate at this end inhibits the action of venom phosphodiesterase, this phosphate was removed with alkaline phosphatase. Complete hydrolysis with venom phosphodiesterase yielded the nucleoside adenosine plus the 5′-nucleotides pU, pC and pG*. Hence Ap was the 5′-terminal nucleotide. Digestion with pancreatic ribonuclease gave the dinucleotide ApUp and other products. Hence the hexanucleotide possessed the partial structure:

$$ApUp(CpUpCp)Gp$$

The sequence in parentheses remained to be determined. Partial hydrolysis with venom diesterase yielded a tetra- and a penta-nucleoside both of which yielded cytidine on alkaline hydrolysis. This nucleoside must have occurred at the 3′ end of both fragments. Hence the hexanucleotide possessed the following structure:

$$ApUpUpCpCpGp$$

The reactions involved are depicted in Fig. 1.

In order to relate the sequences determined for small fragments, Holley and co-workers now required large fragments to sequence. A brief treatment with T_1 at 0° in the presence of magnesium ions broke the alanine tRNA at only one point yielding two large fragments. Slightly more vigorous hydrolysis gave a number of additional large fragments. In sequencing these, minor bases sometimes served as "landmarks". From these results the complete nucleotide sequence (primary structure) of alanine tRNA was deduced (Holley et al., 1965).

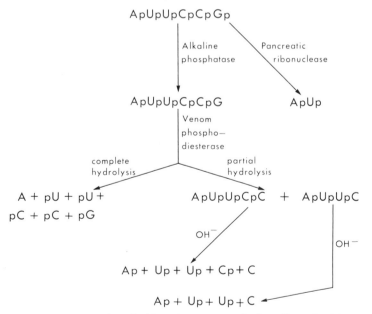

Fig. 1. Reactions used to elucidate the structure of a hexaribonucleotide. A, U, C and G represent adenosine, uridine, cytidine and guanosine respectively; p represents a phosphate group; Ap, Up, Cp and Gp represent 3′-nucleotides while pU, pC and pG denote 5′-nucleotides

---

* For the meaning of this symbolism, see the legend for Fig. 1.

By use of physical techniques, other workers had shown that tRNAs have an ordered secondary structure consisting of double helical regions with intra-molecular hydrogen bonding. Hence the Holley group proposed "folded" structures for alanine tRNA. The one which has become widely accepted is the clover-leaf structure; in Fig. 2 the nucleotide sequence of alanine tRNA is presented in this form. In this structure, 44% of the bases occur as G—C pairs. This is in accord with optical rotatory dispersion studies (VOURNAKIS and SCHERAGA, 1966). An interesting aspect of the sequence is the considerable

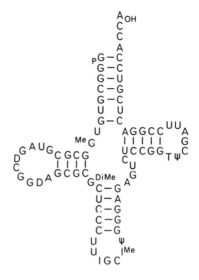

Fig. 2. The nucleotide sequence of yeast alanine tRNA presented in clover-leaf form. A, adenosine; G, guanosine; C, cytidine; U, uridine; I, inosine; T, thymine riboside; ψ, pseudouridine; D, dihydrouridine; I^Me, 1-methyl-inosine; ^MeG, 1-methylguanosine; G^DiMe, N_2N_2-dimethyl-guanosine

number of modified nucleosides; nine of the 77 nucleosides are modifications of the four common ribonucleosides. The anticodon, IGC, is located near the centre of the polynucleotide chain. Since the pairing between codon and anticodon is antiparallel, and because of "wobble" in the pairing of the third codon base (see under tertiary structure), the alanine tRNA anticodon would be expected to pair with the following codons: GCA, GCC, and GCU.

In 1966, ZACHAU and co-workers presented the nucleotide sequences for two species of serine tRNA isolated from yeast (ZACHAU et al., 1966). These tRNAs differ from yeast alanine tRNA in two important respects; they contain a pro-nounced extra arm and a modified adenine [6-(3-methylbut-2-enylamino)purine], a compound with plant hormone activity, adjacent to the 3′ end of the anticodon. The sequences presented in clover-leaf form are shown in Fig. 3; 14 of the 85 nucleo-sides are modifications of the four common ribonucleosides. The two species possess the same anticodon, IGA, and differ in sequence at only three positions. Because of the degeneracy of the genetic code, some isoaccepting tRNAs would be expected to differ in the first base of the anticodon. This prediction is in accord with the recently reported sequences for three E. coli tRNAs specific for valine

(YANIV and BARRELL, 1971; MURAO et al., 1970), which is coded for by the four triplets GUU, GUC, GUA, and GUG. Two species both possess the anticodon GAC, which corresponds to codons GUC and GUU, and differ in sequence at only six positions. The third isoacceptor, however, has uridine-5-oxyacetic acid·AC as anticodon which would be specific for codons GUA and GUG. At 52 positions the three sequences possess identical nucleosides. Isoaccepting tRNAs from a particular micro-organism therefore possess very closely related sequences.

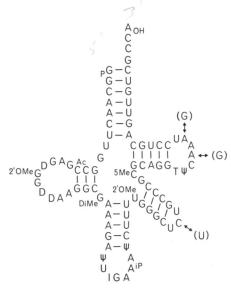

Fig. 3. The nucleotide sequence of yeast serine tRNA (species I and II) presented in cloverleaf form. The symbols listed in the caption for Fig. 2 also apply to Fig. 3. Additional symbols are:

C^{Ac}, N_4-acetylcytidine;

$^{2'OMe}G$, 2'-O-methylguanosine;

A^{iP}, N_6-(3-methylbut-2-enyl)-adenosine;

$^{2'OMe}U$, 2'-O-methyluridine;

^{5MeC},   5-methylcytidine.

The sequence shown is that of species I; substitution of the bracketed nucleosides yields the sequence for species II

The nucleotide sequences of over 20 tRNAs are now known. This rapid progress is due in part to the development by SANGER and co-workers of improved methods for separating fragments produced by enzymic dissection of nucleic acids. HOLLEY and other pioneers in the field had used DEAE-cellulose chromatography for such separations. To determine the nucleotide sequence of *E. coli* 5S ribosomal RNA, SANGER and collaborators first labelled the RNA with ^{32}P, and then separated the products of complete enzymic digestion by two-dimensional electrophoresis. Electrophoresis on cellulose acetate (first dimension) was followed by transfer of the partially resolved mixture to DEAE-cellulose paper for electrophoresis in the second dimension, the separated fragments being located by autoradiography (SANGER et al., 1965; BROWNLEE and SANGER, 1967). This procedure functioned satisfactorily for oligonucleotides up to about decanucleotides. For larger fragments, separation in the second dimension was achieved with a new elegant technique, homochromatography, in which a DEAE-cellulose paper chromatogram was developed with non-radioactive polynucleotides (BROWNLEE et al., 1968). Fragments containing up to 30 nucleotides were separated by this method. These procedures originally devised for sequencing 5S RNA have now been exploited to sequence several tRNAs, e.g. su_{III} tyrosine tRNA of *E. coli* (GOODMAN et al., 1968).

2. Homology in Primary and Secondary Structures

There are many processes in which all tRNAs from an organism must exhibit some degree of structural homology. These include interaction with ribosomes, interaction with common enzymes (e.g. methylases and enzymes which add CCA to the 3' end) and the recently observed co-crystallization (BLAKE et al., 1970). Hence tRNA molecules must possess many common structural features at the primary, secondary and tertiary levels. Consideration of the sequences

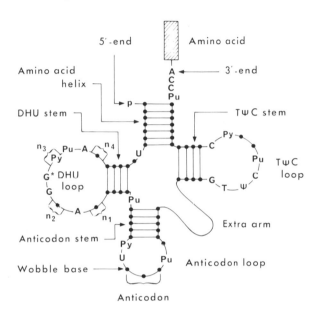

Fig. 4. Generalized clover-leaf structure for tRNA. Positions constantly occupied by a particular nucleoside are indicated. Pu = a purine nucleoside; Py = a pyrimidine nucleoside; G* represents either guanosine or 2'-O-methylguanosine; $n_1 = 0$–1, $n_2 = 1$–3, $n_3 = 1$–3 but usually 2, and $n_4 = 0$–2 nucleotides. ●—● indicates bases usually paired by hydrogen bonding. The DHU stem is shown with 4 base pairs; in some tRNAs only 3 base pairs are present

for over 20 tRNAs indicates that the primary and secondary structures do in fact exhibit much homology and it is possible to formulate a generalized clover-leaf structure (Fig. 4). tRNAs from micro-organisms, higher plants and animals have now been sequenced; all appear to conform to this generalized structure. The homology includes the general occurrence of identical nucleotides in particular positions and the requirement that pairs of nucleotides in other sites must be able to engage in Watson-Crick type (i.e. C—G, A—U) base pairing. Although tRNAs exhibit considerable homology, it should also be emphasized that marked differences in sequence occur both between tRNAs from the same organism and also between isoaccepting tRNAs from different species.

The generalized clover-leaf structure depicted in Fig. 4 can be divided into five folded regions—the amino acid arm, the dihydrouridine (DHU) arm, the anticodon arm, the TψC arm and the extra arm. The first of these arms has four unpaired bases (purine-CCA) as a terminal sequence to which the amino acid is attached at the 2' or 3' position of the adenosine moiety; the remainder of the arm is a double helical structure in which C—G base pairing usually predominates. The extra arm exhibits great variability in length. This arm when of sufficient length, and the DHU, TψC and anticodon arms can all be divided into two sections—a helical stem with paired bases and a loop with unpaired bases. The number of nucleotides present in, and significant features of, each section of tRNA are presented in Table 1. Non-complementation of bases in double helical regions is very infrequent and rarely exceeds one pair per molecule. The amino acid helical arm is frequently the site of mispairing which is usually the association of G with U, a combination which probably causes little distortion of the helix.

Table 1. Characteristics of regions of tRNA

Region	Number of nucleotides	Comments on structure
Amino acid helix	14 (7 base pairs)	The region where base mispairing occurs most frequently. Of the irregular pairings, G—U is the commonest; U—U, A—G and C—A occur rarely
DHU stem	6–8 (3–4 base pairs)	The first and last base pairs are always C—G
DHU loop	8–12	This region is usually rich in dihydrouridine; however it exhibits considerable variation in composition and *E. coli* tyrosine tRNA contains no DHU. A—purine nucleoside, G—G or $^{2'OMe}$G—G and A with spacing nucleotides are characteristic features of the sequence (see Fig. 4)
Anticodon stem	10 (5 base pairs)	The second base pair from the anticodon loop is usually C—G
Anticodon loop	7	Numbered from the 3′ end, the 3rd, 4th and 5th bases are the anticodon, the fifth being the wobble base. On the 5′ side of the anticodon is always uracil followed by a second pyrimidine; on the 3′ side is always a purine, often a modified purine, followed by the 1st base of the loop, frequently adenine but sometimes a pyrimidine
TΨC stem	10 (5 base pairs)	The base pair adjacent to the TΨC loop is always C—G. Imperfect pairing is found rarely in this region
TΨC loop	7	With only two exceptions (see text), all tRNAs contain the sequence T-Ψ-C-purine nucleoside at the same location in the loop. The purine is usually guanine
Extra arm	3–14	Extremely variable in structure and often lacks a helical stem

There are two common sequences in tRNA molecules. At the 3′ end, all tRNAs contain the following nucleoside sequence: purine nucleoside—C—C—A. With very rare exception, all tRNA molecules so far studied contain one thymine riboside residue. This is located in the TΨC loop and forms part of the sequence, G—T—ψ—C—purine nucleoside, which occurs in all these tRNAs. However isoleucine tRNA from *Mycoplasma* sp. (Kid) does not contain thymine but can be methylated by purified *E. coli* tRNA methylase to yield a tRNA containing thymine apparently in the normal position (JOHNSON et al., 1970). The tRNA of an *E. coli* mutant completely lacks thymine (SVENSSON et al., 1971). These unusual tRNAs lacking thymine function normally in protein synthesis. It has been suggested that the G—T—ψ—C sequence is involved in binding to ribosomes but there is no convincing evidence to support this.

In all tRNAs of known sequence, the base adjacent to the 3′ end of the anti-codon is a purine, often an unusual substituted purine. The known bases of this type are as follows (see Fig. 5): 6-(3-methylbut-2-enylamino)purine (I),

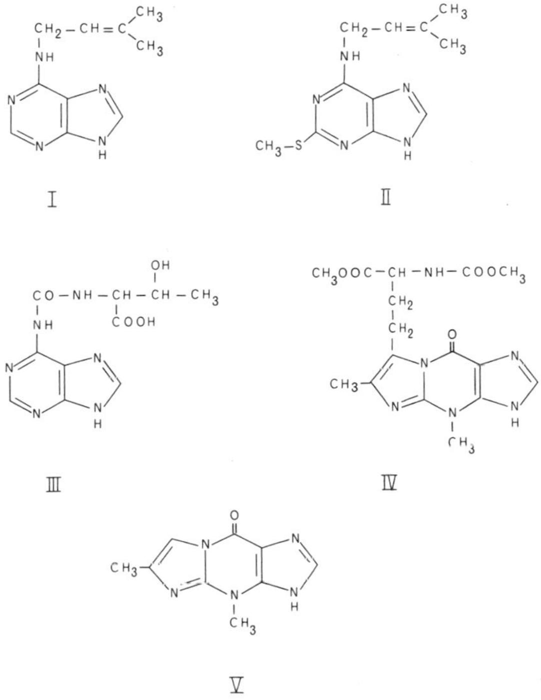

Fig. 5. Structures of unusual substituted purines thought to occur adjacent to the anticodon in tRNA molecules

6-(3-methylbut-2-enylamino)-2-methylthiopurine (II), *N*-(purin-6-ylcarbamoyl)-threonine (III), and a modified guanine previously termed "base Y" and shown very recently to possess structure (IV) (NAKANISHI et al., 1970; THIEBE et al., 1971). Phenylalanine tRNA from brewer's yeast probably contains the related base (V) (Fig. 5) at the 3′ end of the anticodon (KASAI et al., 1971). The two first-mentioned compounds exhibit pronounced plant hormone (cytokinin) activity and the possible significance of this will be discussed under regulatory mechanisms. It appears likely modified bases adjacent to the anticodon serve an important function. It has been suggested (FULLER and HODGSON, 1967) that the modifications to these bases inhibit codon base-pairing with the position adjacent to the anti-codon and also stabilize the single-stranded helix, a stacked conformation, in the anticodon loop (see under tertiary structure). Recent studies with synthetic model compounds support the latter suggestion (LEONARD et al., 1969).

Although the second and third positions (numbering from 5′ end) of the tRNA anticodons are almost invariably occupied by unmodified nucleosides, in the first or "wobble" (see under tertiary structure) position a modified nucleoside is often found. This is frequently inosine or 2′-O-methylguanosine, but the following nucleosides also occur in this position: N_4-acetylcytidine (OHASHI et al., 1972), 5-methylcytidine (CHANG et al., 1971), a modified guanosine of unknown structure termed base Q (HARADA and NISHIMURA, 1972), uridine-5-oxyacetic acid (MURAO et al., 1970) and probably the sulphur-containing nucleosides, 2-thiouridine-5-acetic acid methyl ester, 5-methylaminomethyl-2-thiouridine and 5-methyl-2-thiouridine (KIMURA-HARADA et al., 1971).

3. Tertiary Structure

tRNA molecules possess a specific tertiary structure. Several tRNAs may be stabilized in two states only one of which is biologically active. Magnesium ions influence the interconversion of forms which appear to differ in tertiary structure (LINDAHL et al., 1966; ADAMS et al., 1967). The exact tertiary structure of tRNA is not known at present although much information has recently been forthcoming regarding this important aspect of molecular biology. Numerous structures have been proposed, but the one represented schematically in Fig. 6 would appear to be the most probable. The amino acid arm is stacked on top of the TψC arm, while the DHU arm is stacked on the anticodon arm to give two double-helical structures with parallel axes (LEVITT, 1969). According to the proposed model, the anticodon and the aminoacyl acceptor site are at opposite ends of the molecule which is stabilized by interactions between certain of the nucleotides in the TψC, DHU and extra loops. Thus the three bases of the TψC sequence interact with adenine, guanine and guanine respectively in the DHU

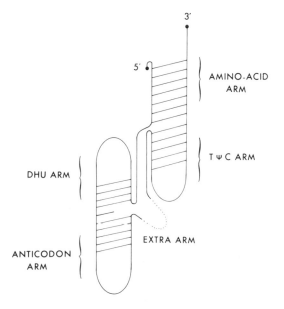

Fig. 6. A schematic diagram of the model proposed by Levitt for the tertiary structure of tRNA

loop. A diagram of tRNA tertiary structure drawn from photographs of an actual model is presented as Fig. 7. In the tRNA molecule shown, there are four base pairs in the DHU stem. The model is in accord with the physical properties of tRNA and with the differing reactivity of bases in different positions.

One very important aspect of tRNA structure is the conformation of the anticodon loop. Principally as a result of model building, FULLER and HODGSON (1967) proposed that the first five bases of the loop (the counting is from the

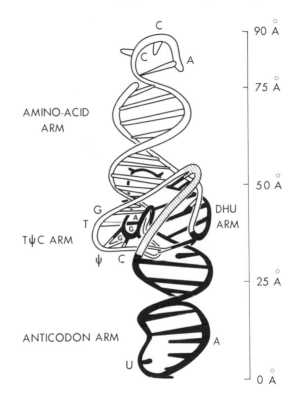

Fig. 7. A representation of tRNA tertiary structure derived from photographs of the actual model.
■■■■■■, DHU and anticodon arms
☐, TψC and amino-acid arms
▨▨▨▨▨, Extra arm
(From LEVITT, 1969)

3′ end; the third, fourth and fifth are the anticodon) are stacked on top of the helical stem as if continuing the helix. Although not base paired, these five bases are arranged in a helical manner. This model for the anticodon loop (see Fig. 8) permits slight distortion in the conformation of the anticodon such that certain abnormal pairings can occur with the third base of codon triplets. Hence the model provides a stereochemical basis for Crick's wobble hypothesis which states that "in the base-pairing of the third base of the codon there is a certain amount of play, or wobble, such that more than one position of pairing is possible" (CRICK, 1966). The alternative or "wobble" pairings predicted by CRICK can be accommodated in the Fuller and Hodgson model by distortion of the anticodon alone; no distortion of the codon is required. The anticodon base which engages the third codon base is appropriately termed the "wobble base"; it is the first base of the anticodon when this is written in the conventional 5′ to 3′ direction. As a wobble base, U can pair with A or G, G with U or C, and hypoxanthine with U, C or A. When an amino acid is coded for by all four bases in the third position of the codon, Crick's wobble theory predicts that there will be at least two isoacceptors differing in anticodon.

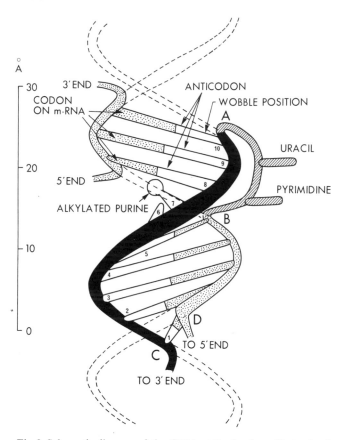

Fig. 8. Schematic diagram of the tRNA anticodon loop illustrating its relationship to the codon and the helical character of the structure. The bases of nucleotides 1 to 10 are stacked on one another and follow the regular helix which is shown black (bases numbered 6–10 are the first five bases of the anticodon loop). The complementary helix BD in the anticodon stem is shaded like the codon to indicate that they follow the same helix. The two nucleotides not in the standard conformation are represented by dark line shading. The representation of their conformation is very schematic because they lie behind nucleotides 8, 9 and 10 in the black chain. The dotted lines indicate the generic helix from which the structure can be imagined to be derived. (From FULLER and HODGSON, 1967)

FULLER and HODGSON (1967) also proposed that pairing between codon and anticodon can occur by hydrogen bonding with the stereochemistry which would exist if the interacting triplets were constituents of a regular RNA double helix (see Fig. 8). Thus when codon and anticodon are paired, the codon would be coincident with a hypothetical continuation of one strand of the helical anticodon stem.

Strong support for the concepts of FULLER and HODGSON regarding conformation of the anticodon loop has recently been provided by the work of EISINGER (1971) who studied interaction between tRNA molecules with complementary anticodons. Phenylalanine tRNA (anticodon GAA) and glutamic acid tRNA (anticodon UUC) were found to form a complex of high stability. Indeed it was so stable that it could be discerned in gel electrophoresis. The magnitude of the association constant and other characteristics of the interaction indicated that the paired anticodon triplets were both in a helical conformation.

In summary, tRNAs appear to be very compact elongated molecules about 90 Å in length with a rather rigid tertiary structure. The anticodon is conformationally fixed in a helical configuration and when hydrogen bonded to the codon probably becomes the equivalent of one strand of a double helical structure.

C. Biosynthesis of tRNA

In the synthesis of ribosomal RNA, polynucleotide precursors are modified by nucleotide cleavage. Similarly synthesis of tRNA appears to involve cleavage of the original transcription product of tRNA cistrons. Thus BERNHARDT and DARNELL (1969) detected in HeLa cells a precursor tRNA which migrated between 5S and 4S RNA during gel electrophoresis. This species of RNA appeared to possess a molecular weight greater than that of normal tRNA and was not merely tRNA with a modified conformation. From a phage-infected *E. coli* mutant, ALTMAN (1971) isolated a tyrosine tRNA precursor which has now been completely sequenced (ALTMAN and SMITH, 1971). Since the precursor has 41 additional nucleotides at the 5' end and three at the 3' end, cleavage at both ends is involved in synthesis of the mature tRNA. The precursor molecule lacks modified nucleosides but, surprisingly, contains the CCA sequence which constitutes the 3' end of the functional tRNA. Hence the CCA terminal triplet appears to be coded by the tRNA gene even though in the mature tRNA these nucleotides are subject to turnover as indicated by the following equation:

$$tRNA + 2CTP + ATP \rightleftharpoons tRNA—CCA + 3PP_i.$$

One enzyme appears to add this terminal sequence to all tRNAs in a particular tissue or micro-organism.

Nucleosides in precursor tRNA molecules are modified in many ways. The principal modifications are as follows:

1) Methylation. The main methylated nucleosides in tRNA are: 1-methyladenosine, 2-methyladenosine, N_6-methyladenosine, N_6,N_6-dimethyladenosine, 2'-O-methyladenosine, 1-methylguanosine, N_2-methylguanosine, N_2,N_2-dimethylguanosine, 7-methylguanosine, 2'-O-methylguanosine, 3-methylcytidine, 5-methylcytidine, 2'-O-methylcytidine, 2'-O-methyluridine, 2'-O-methylpseudouridine, thymine riboside and 1-methylinosine. Hence the site of methylation can be either in the ribose moiety or in the heterocyclic base.

2) Addition of the 3-methylbut-2-enyl group to the exocyclic nitrogen of adenine adjacent to the anticodon of some tRNAs. The adenine moiety may be further modified by addition of a methylthio group at position 2.

3) Conversion of adenine at the same position to N-(purin-6-ylcarbamoyl)-threonine.

4) Conversion of uridine to pseudouridine (5-β-D-ribofuranosyluracil) which contains a C—C glycosidic linkage instead of the usual N—C linkage.

5) Conversion of uracil to 4-thiouracil, derivatives of 2-thiouracil (see p. 90) and 5,6-dihydrouracil.

6) Deamination of adenine to yield hypoxanthine.

7) Conversion of cytosine to N_4-acetylcytosine and 2-thiocytosine.

8) Modification of guanine by ring closure to yield base Y [(IV) Fig. 5] and the related compound [(V) Fig. 5]. Formation of base Y is the most complex modification known.

Modifications 1) and 2) merit comment. For a brief period it was thought that tRNA methylases were localized in the nucleolus; however it now seems certain that the majority, if not all, activity is in the cytoplasm. A particular tissue appears to contain numerous methylases with considerable specificity (GAUSS et al., 1971). The methyl donor is S-adenosylmethionine and methylation can be achieved *in vitro* using methyl-deficient tRNA, that is tRNA from bacteria cultured in the absence of methionine. Nicotinamide (HALPERN et al., 1971) and certain cytokinins (WAINFAN and LANDSBERG, 1971) are inhibitors of tRNA methylation *in vitro*, while polyamines (e. g. spermidine) stimulate methylation (YOUNG and SRINIVASAN, 1971). Methylation may be subject to a variety of controls *in vivo*. The 3-methylbut-2-enyl group is derived from mevalonic acid. The addition of this group to tRNA has been performed *in vitro*. Normal tRNA is first gently oxidized to cleave the 3-methylbut-2-enyl groups present. In the presence of a suitable enzyme preparation and isopentenyl pyrophosphate, the oxidized tRNA will accept 3-methylbut-2-enyl groups (KLINE et al., 1969). There is good evidence that these groups are attached to adenine residues which normally carried a 3-methylbut-2-enyl group. In some tRNAs the adenine adjacent to the anticodon is further modified by the presence of a 2-methylthio group; the sulphur moiety can be labelled with ^{35}S-cysteine and the methyl is donated by S-adenosylmethionine (GEFTER, 1969). As will be seen later, the modifications briefly discussed in this paragraph profoundly influence the ability of tRNA to function in protein synthesis.

A recent notable achievement is the DNA-directed *in vitro* synthesis of tRNA. The tyrosine tRNA gene carried by bacteriophage ϕ80psu_{III} DNA has been transcribed *in vitro* by DNA-dependent RNA polymerase. The tRNA synthesized was larger than mature tRNA, exhibited a sedimentation coefficient of about 8S, and appeared to be the immediate transcription product of the tRNA gene (DANIEL et al., 1970). Maturation reactions would be required to yield normal functional tRNA. Very recently, ZUBAY et al. (1971) have achieved an *in vitro* synthesis of biologically competent su_{III} tyrosine tRNA using a cell-free system from *E. coli*. This tRNA suppressed amber mutations in the gene coding for β-galactosidase. Addition of isopentenyl pyrophosphate to the *in vitro* synthesizing system increased the activity of the tRNA four fold.

D. The Aminoacylation of tRNA; the Role of Aminoacyl-tRNA in Biosynthesis

In protein synthesis, an amino acid is transferred to a specific tRNA by a specific enzyme, an amino acid-tRNA ligase (these enzymes were formerly referred to as aminoacyl-tRNA synthetases). The amino acid carboxyl group esterifies the 2' or 3' hydroxyl group of the terminal adenosine present in all tRNAs; the 2'- and 3'-aminoacyl isomers probably exist in equilibrium. The amino acid-tRNA ligase (AL) catalyzes two reactions:

1) The formation of an aminoacyl adenylate, an "activated" amino acid, which possesses a high-energy bond and which remains firmly bound to the enzyme.

$$AL + ATP + AA \rightleftharpoons AL(AA \sim AMP) + PP_i$$

(where AA represents an amino acid and $\sim$ a high-energy bond).

2) The transfer of the amino acid to the tRNA molecule which is bound on an adjacent site on the enzyme.

$$AL(AA \sim AMP) + tRNA \rightleftharpoons AA - tRNA + AL + AMP.$$

The resulting aminoacyl tRNA then forms a complex with a protein factor and GTP for binding to the site on the ribosome. The subsequent role of tRNA in peptide bond formation is discussed in Chapter 7.

Aminoacyl-tRNA molecules readily undergo base-catalyzed hydrolysis at pH values over 7. Once attached to the tRNA, the amino acid can be altered without affecting the coding response of the tRNA. Thus the aminoacyl tRNA, cysteinyl-tRNA, can be hydrogenated using Raney nickel and the cysteine moiety converted into alanine, the —SH group being eliminated. When this modified aminoacyl-tRNA is added to an *in vitro* protein-synthesizing system, alanine is incorporated into polypeptide in response to poly UG which normally stimulates incorporation of cysteine but not alanine. Experiments of this type emphasize the adaptor role of tRNA in protein synthesis. Since mRNA translation is independent of the amino acid attached to the tRNA, esterification of the tRNA must be highly specific if incorrect insertions of amino acids into protein are to be avoided.

Each ligase is specific for one amino acid and the corresponding tRNA species. A particular ligase may charge several isoacceptors. Thus in *E. coli*, a single ligase appears to charge the five leucine-accepting tRNAs (KAN and SUEOKA, 1971). However leucine-tRNA ligase activity from soybean cotyledons has been fractionated into three components. One exclusively acylates two of the six leucine-accepting tRNAs found in the cotyledons; the remaining two enzymes both charge the other four species equally well (KANABUS and CHERRY, 1971). Two glycine ligases and two threonine ligases are known to occur in rat liver (FAVOROVA and KISELEV, 1970; ALLENDE et al., 1966).

In the aminoacylation of tRNA, the ligases may exhibit a low degree of species or organ specificity; that is enzyme from one organism can often amino-acylate tRNA from another. With rare exceptions, the aminoacylated tRNAs are identical to those produced by homologous acylation. For example, enzyme preparations from *Neurospora*, maize and mouse liver were used to aminoacylate tRNA from *E. coli*. The products were chromatographed. With one exception, the products from 18 combinations were indistinguishable from the amino-acylated tRNAs produced by homologous enzymes (Jacobson, 1971). The exception was a case of mischarging. When the enzyme was from *Neurospora*, the tRNA was from *E. coli* and the amino acid was phenylalanine, the products were phenylalanyl-tRNA(phenylalanine), phenylalanyl-tRNA(alanine) and phe-nylalanyl-tRNA(valine) (Barnett and Jacobson, 1964; Barnett, 1965; Barnett and Epler, 1966; Holten and Jacobson, 1969).

The structural features recognized in a tRNA molecule by an amino acid-tRNA ligase are not known precisely at present. The anticodon does not serve as the specific recognition site. The very recent sequencing of mutant tyrosine tRNAs of altered amino acid specificity provides strong evidence that the amino acid arm plays a very important role in the recognition of tRNA by amino acid-tRNA ligase (Shimura et al., 1972; Hooper et al., 1972). Attempts to elucidate the structural features involved in recognition are reviewed by Chambers (1971) and by Gauss et al. (1971).

Participation as an adaptor in ribosome-mediated *de novo* protein synthesis is the best understood and probably the principal biosynthetic role of tRNA. It should be emphasized however that aminoacyl tRNAs have other synthetic functions which tend to be overlooked. Thus the aminoacyl group is known to be transferred enzymically to the glycerol moiety of phosphatidyl-glycerol (Gould et al., 1968), and also to proteins (Leibowitz and Soffer, 1970). The enzymic transfer of an amino acid from tRNA into peptide linkage with the amino-terminal amino acid of specific acceptor proteins occurs in the absence of ribosomes and GTP. tRNA is thus involved in the specific modification of proteins, a reaction which may prove to be of considerable significance. Certain aminoacyl tRNAs also play an important role in the synthesis of bacterial cell walls. Glycine, serine, threonine and alanine are all transferred enzymically from tRNA to a lipid intermediate in the synthesis of peptidoglycan, a component of bacterial cell walls (Petit et al., 1968).

E. tRNA and Regulatory Mechanisms

Some evidence suggests that tRNA may play a role in the regulation of protein synthesis by directly influencing translational processes and by functioning in end-product repression of enzyme synthesis. Some recent elegant work shows that tRNA can also function as an enzyme inhibitor. The plant hormones, cyto-kinins, occur as bases adjacent to the anticodon in certain tRNA species. This occurrence may be related to mechanism of cytokinin action. Changes in relative amounts of isoaccepting tRNA species accompany differentiation and alteration of mode of growth, implicating tRNA in the regulation of these processes. The above concepts form the basis of the final part of this discussion.

1. Translation

a) Control of Translation by Modulating tRNA Species. Ames and Hartman (1963) observed that certain mutations in the histidine operon of *Salmonella typhimurium* not only prevented formation of the enzyme corresponding to the mutated gene, but also reduced production of other proteins coded by genes located on the side of the mutated gene distal to the operator. To explain these polar mutations, AMES and HARTMAN (1963) proposed a modulation mechanism for translation involving regulation by tRNA species. It was proposed that translation commenced at one end (now known to be the 5′ end) of the mRNA molecule and was limited by modulating triplets corresponding to modulator tRNA species. These species limited translation because of their scarcity or ability to dissociate the ribosome from the mRNA. Thus according to AMES and HARTMAN, a polar mutation was one in which a normal codon was replaced by a modulating triplet associated with a modulating species of tRNA. It is now almost certain that polar mutations, which have been observed in several operons, are in fact nonsense mutations (IMAMOTO et al., 1966; MARTIN et al., 1966) and the resulting nonsense codon (see under c, p. 99) causes premature termination of protein synthesis. The polar effect of the mutation may be a consequence of the susceptibility of the untranslated portion of the polycistronic messenger to enzymic degradation (MORSE and GUERTIN, 1971) or of the coupling of transcription to translation (IMAMOTO and KANO, 1971). These concepts are discussed more fully in Chapter 7. Hence polar mutations can now be adequately explained without invoking modulating tRNAs.

In the lactose operon of *E. coli* and the histidine operon of *S. typhimurium*, the cistron adjacent to the operator site is translated more frequently than non-adjacent cistrons (WHITFIELD et al., 1970). The hypothesis of AMES and HARTMAN has been invoked to account for such observations. However these results could now be explained equally well by:

1) possible variation in the coding of termination signals and the functional level of termination (release) factors which are codon specific (LU and RICH, 1971; BRETSCHER, 1968; CAPECCHI and KLEIN, 1970).

2) possible variation in spacer sequences (the intercistronic divide) between cistrons (RECHLER and MARTIN, 1970).

Despite these alternative explanations, modulation through tRNA remains a likely possibility which has recently received some experimental support. ANDERSON (1969) endeavoured to see if rate of protein synthesis *in vitro* can be regulated by concentrations of tRNA species as found *in vivo*. The *in vitro* protein-synthesizing system was based on an *E. coli* 30,000-g supernatant which contained tRNA species presumably in the proportions found *in vivo*. The effect of added tRNA species on arginine incorporation into protein in response to poly AG was studied. This polynucleotide directs incorporation of only four amino acids—arginine, glutamic acid, lysine and glycine. One minor species of arginine tRNA which recognizes codons AGA and AGG markedly stimulated protein synthesis while no other tRNA species was effective. ANDERSON proposed that AGA and AGG are regulatory codons in *E. coli;* they may be regulatory because the corresponding tRNA is present in suboptimal amounts which limits the rate of translation.

b) tRNA Modification and the Control of Translation. SUEOKA and SUEOKA (1964) observed a specific structural modification of leucine tRNA of *E. coli* following phage infection and proposed the "adaptor modification hypothesis".

This states (using the original wording) that "codon recognition of a particular adaptor out of a set of degenerate adaptors for an amino acid is changed by structural modification". It was suggested that by modification of a specific tRNA the functioning of some of the genes could be shut off. Convincing evidence that the above concept constitutes a regulatory mechanism *in vivo* is still lacking. However it is now known that certain structural modifications can greatly affect the functioning of tRNA molecules in translation. These are methylation and modification of the base at the 3' end of the anticodon.

tRNA, as we have already noted, contains a great number of methylated bases in addition to the four common bases. When *E. coli* is starved of methionine, it accumulates tRNA and this tRNA lacks the normal complement of methylated bases. Such methyl deficient tRNA has been compared with normal tRNA. Lack of methyl groups can markedly diminish the amino acid-acceptor capacity (Shugart et al., 1968), reduce binding to polynucleotide-ribosome complexes (Stern et al., 1970), and also alter coding properties. Thus normal *E. coli* leucine tRNA responds equally to poly UC and poly UG in ribosome binding assays; however methyl-deficient leucine tRNA responds preferentially to poly UC. Species of leucine tRNA which normally recognize poly UG appear to exhibit a changed codon response when methyl-deficient (Capra and Peterkofsky, 1968). Methylated bases are probably involved in correct codon recognition. Incomplete methylation and hypermethylation of tRNA is potentially capable of modifying protein synthesis. It is noteworthy that tumour cell extracts often possess elevated methylating activity when compared with extracts of normal tissue (see ref. listed by Gauss et al., 1971), and Mittelman et al. (1967) have observed that extracts of virus-induced tumour tissue hypermethylate homologous tRNA of normal tissue *in vitro*. Extracts of normal tissue do not.

Modification of the base at the 3' end of the anticodon can markedly affect the ability of tRNA to function in protein synthesis. Chemical modification of the 6-(3-methylbut-2-enylamino)purine base of yeast serine tRNA (Fittler and Hall, 1966) and the 6-(3-methylbut-2-enylamino)-2-methylthiopurine residue in *E. coli* phenylalanine tRNA (Faulkner and Uziel, 1971) does not affect amino acid-acceptor activity. However the modification to the serine tRNA markedly reduces binding to the mRNA-ribosome complex, while the modified phenylalanine tRNA does not function in polypeptide synthesis. The latter modification is particularly interesting because it is reversible and reversion restores ability to function in protein synthesis.

From *E. coli*, Gefter and Russel (1969) purified three species of suppressor tyrosine tRNA which differed only in the base adjacent to the 3' end of the anticodon. At this position, species A contained an unmodified adenine, B possessed 6-(3-methylbut-2-enylamino)purine and C had 6-(3-methylbut-2-enylamino)-2-methylthiopurine. These differences did not affect aminoacylation but did influence trinucleotide-dependent binding to ribosomes, the relative affinities being C > B > A. The three species also differed markedly in their ability to promote *in vitro* protein synthesis; A did not support synthesis, C was very effective, and B was intermediate in functional ability.

Adjacent to the anticodon of yeast phenylalanine tRNA, the modified guanine, base Y, is located. This unusual base can be selectively cleaved by very gentle acidic hydrolysis. Such cleavage alters coding properties (GHOSH and GHOSH, 1970). The observations presented in this section suggest that modification of the base adjacent to the anticodon is potentially an *in vivo* translational regulatory mechanism.

c) *Synthesis of New tRNA Species and the Control of Translation.* Synthesis of new species of tRNA with modified sequence is potentially capable of affecting translation. There is little to indicate that this occurs during normal regulation of growth. However synthesis of new tRNA is the key factor in the mechanism underlying suppressor mutations. Three codons UAA ("ochre"), UAG ("amber") and UGA do not specify an amino acid, but indicate to the translational mechanism that a genetic message has ended. These triplets, termed nonsense codons, bring to an end the synthesis of a protein molecule. Certain mutations in *E. coli* suppress the usual recognition of nonsense codons as peptide-chain termination signals. These mutations result in synthesis of new tRNA molecules which translate the nonsense codons as amino acids and termination no longer occurs. We will briefly consider two suppressor mutations—su_{III} (one of the several known UAG suppressor mutations), and a UGA suppression in which the triplet is read as tryptophan.

As a result of su_{III} mutation, the anticodon of a minor tyrosine tRNA species is changed from GUA to CUA. This enables the tRNA to recognize the amber codon (UAG) instead of the tyrosine codons UAU and UAC. Tyrosine is then inserted into protein by this tRNA in response to UAG (GOODMAN et al., 1968).

It is now known, however, that a change in anticodon of a tRNA is not necessary for suppression. UGA suppression is achieved by a tryptophan tRNA (anticodon CCA) whose codon recognition is altered by a structural change in the DHU stem. At position 24 of the suppressor tRNA, the base is adenine instead of guanine (HIRSH, 1971). Suppression involves A—C pairing in the wobble position which was not predicted by CRICK (1966). This important work by HIRSH emphasizes the sensitivity of codon-anticodon interaction at the wobble position to minor structural change elsewhere in the tRNA molecule.

2. End Product Repression; Enzyme Activity

A naturally occurring amino acid regulates its formation in two ways. It can inhibit the action of an enzyme of its biosynthetic pathway, a process termed end-product or feed-back inhibition; it can repress formation of the biosynthetic enzymes, which may be termed end-product or feed-back repression*. tRNA is strongly implicated in the latter control mechanism. Bacterial cells cultured on media containing the histidine analogue, α-methylhistidine, were found to contain derepressed levels of the histidine biosynthetic enzymes. The increased levels of these enzymes were shown to result not from an inhibition of histidine biosynthesis, but probably from an inhibition of transfer of histidine to tRNA. This indicated that the concentration of charged histidine tRNA, not free histidine, controlled repression of histidine biosynthetic enzymes (SCHLESINGER and MAGASANIK, 1964). This conclusion was supported by studies of mutants in which

---

* In this chapter, the term repression does not necessarily imply blocking of transcription.

there was a defect in the synthesis or aminoacylation of histidine tRNA (see references listed by BLASI et al., 1971). Other evidence suggests that the allosteric enzyme, phosphoribosyltransferase, which catalyzes the first step of the pathway for histidine biosynthesis, is intimately involved in the repression mechanism (KOVACH et al., 1969a; KOVACH et al., 1969b). This enzyme is of course sensitive to end-product inhibition by free histidine. BLASI et al. (1971) have now demonstrated that charged histidine-tRNA binds strongly to a site on phosphoribosyltransferase distinct from the end-product-sensitive site and the catalytic site. This histidyl-tRNA-enzyme complex may be involved in repression.

The enzymes of the *ilv* operon are repressed in the presence of all three branched-chain amino acids (isoleucine, leucine and valine) and are derepressed when any one of these becomes limiting. The first cistron of this operon specifies threonine deaminase. It has recently been demonstrated that charged leucine tRNA binds to an immature form of this enzyme and the resulting complex is possibly involved in repression (HATFIELD and BURNS, 1970). Uncharged leucine tRNA does not compete for the binding site. The aminoacyl-tRNA derivatives of isoleucine (SZENTERMAI et al., 1968), valine (WILLIAMS and FREUNDLICH, 1969), tryptophan (ITO et al., 1969) and methionine (CHEREST et al., 1971) have also been implicated in repression of the respective amino acid biosynthetic enzymes in bacteria. In the case of repression of the enzymes of methionine biosynthesis, it is a minor isoaccepting tRNA species which is implicated in the control process.

The intriguing question now is: How could an aminoacyl-tRNA-protein complex function in enzyme repression? Binding to DNA at a specific site to prevent transcription is probable. However the complex could regulate at the translational level utilizing the specificity of its anticodon. The actual complex formation *in vivo* might even occur *in situ* on the mRNA.

In end-product repression of amino-acid biosynthesis, an aminoacyl-tRNA may complex with the first enzyme of the biosynthetic pathway. However uncharged tRNA species also appear to bind to enzymes. JACOBSON (1971) observed that tryptophan pyrrolase (an enzyme which oxidizes tryptophan opening the pyrrole ring) activity was not detectable in crude enzyme extracts of a vermilion mutant of *Drosophila melanogaster*. However after treatment with ribonuclease, the crude enzyme preparations exhibited tryptophan pyrrolase activity. The activated enzyme was then purified and shown to be inhibited by one species of tyrosine-accepting tRNA. This finding was confirmed in the following manner. If the vermilion mutation was suppressed by the introduction of a second mutation, tryptophan pyrrolase activity was partially restored and the inhibitory tyrosine tRNA species disappeared. Hence an uncharged tyrosine tRNA species was responsible for the inactivation of tryptophan pyrrolase. This elegant work may have introduced a new concept in regulatory mechanisms, namely tRNA species as enzyme inhibitors.

3. Cytokinins and tRNA

There are three principal types of phytohormones—auxins, gibberellins and cytokinins. Each type induces a broad spectrum of biological responses and these spectra overlap to some degree. Cytokinins have the ability to induce cell division in certain plant tissue cultures (e.g. tobacco pith, soybean callus, and carrot phloem); this growth response distinguishes cytokinins from other phytohormones. Cytokinins evoke this response at extremely low concentrations. Zeatin, the most active known naturally-occurring cytokinin, induces growth of carrot phloem tissue (LETHAM, 1967) and soybean callus (MILLER, 1968) at concentrations less than 5×10^{-10} M, 0.1 µg/litre (see Fig. 9). Certain synthetic O-acyl zeatins exhibit slightly greater activity (SCHMITZ et al., 1971; LETHAM, 1972). Cytokinins are implicated in the control of many phases of plant growth and development

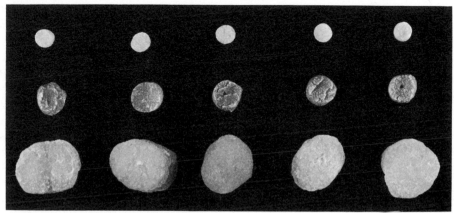

Fig. 9. Carrot secondary phloem explants. *Top row:* after excision; *middle row:* after growth on basal medium for 21 days; *bottom row:* after growth for 21 days on a basal medium containing zeatin at 5×10^{-10} M, 0.1 µg/litre. The explants after excision possessed a diameter of 2 mm

ranging from induction of seed germination to retardation of leaf senescence. For a comprehensive discussion of the role of cytokinins in control of plant growth, the reader is referred to the articles on cytokinins listed under Reviews in the References.

It should be emphasized that the growth regulatory activities of cytokinins are not confined to plant tissues. Exogenous cytokinins can very markedly promote the growth of one bacterium (QUINN et al., 1963) and stimulate other species to a lesser degree. These compounds also influence the growth of animal cells. Thus 6-(3-methylbut-2-enylamino)purine riboside strongly inhibits the growth of leukemic myeloblasts and Sarcoma-180 cells in culture (GRACE et al., 1967; FLEYSHER et al., 1968). Mitosis of human lymphocytes can be stimulated or inhibited by this riboside depending on the concentration (GALLO et al., 1969). The riboside also inhibits growth of Ehrlich carcinoma ascites in mice (SUK et al., 1970), and shows immunosuppressive properties *in vivo* (HACKER and FELDBUSH, 1969).

All naturally occurring cytokinins which have been isolated in a state of purity and which exhibit high activity are N_6-substituted adenines. Two compounds of this type, which occur adjacent to the anticodon in certain tRNAs, have been mentioned previously. A third compound termed zeatin, 6-(4-hydroxy-3-methylbut-*trans*-2-enylamino)purine (LETHAM et al., 1967), occurs in plants as the free base, the riboside and the riboside-5'-phosphate. The riboside of the *cis*-isomer of zeatin has been purified from hydrolysates of unfractionated sRNA of several plant tissues and recently zeatin riboside itself has been shown to occur in sRNA of pea epicotyls (HALL et al., 1967; PLAYTIS and LEONARD, 1971). Cytokinins are present in hydrolysates of sRNA from micro-organisms, animals and plants, but these compounds do not occur in purified ribosomal RNA (LETHAM and RALPH, 1967; SKOOG et al., 1966). In *E. coli*, cytokinin activity is associated with tRNA species corresponding to codons with the initial letter U, but not C, A or G (ARMSTRONG et al., 1969).

Cytokinin supplied exogenously to plant tissue also appears to occur in nucleotide form in the tissue sRNA. After the synthetic cytokinin, 6-benzyl-aminopurine, had been supplied to soybean tissue cultures, the cytokinin was detected chromatographically in hydrolysates of the sRNA. 6-Benzylaminopurine appeared to be present in only certain species of sRNA where it occurred as a nucleotide (Fox and CHEN, 1967). From the sRNA of tobacco-pith callus cultured in the presence of 6-benzylaminopurine, four cytokinin ribosides were purified (BURROWS et al., 1971). Three were the natural tRNA components 6-(4-hydroxy-3-methylbut-2-enylamino)purine riboside, 6-(4-hydroxy-3-methylbut-2-enylamino)-2-methylthiopurine riboside, and 6-(3-methylbut-2-enylamino)purine riboside. The fourth compound was 6-benzylaminopurine riboside, the riboside of the exogenous cytokinin. The benzylaminopurine moiety in sRNA could arise by incorporation of the purine *in toto* or by cleavage of the benzyl group from the supplied cytokinin and benzylation of adenosine residues in RNA perhaps by an active benzyl compound analogous to *S*-adenosylmethionine. Although the results of Fox and CHEN (1967) favour the first mechanism, there is insufficient evidence to reach a definite conclusion regarding formation of this unnatural component of sRNA.

Is the presence of cytokinins in tRNA related to the mechanism of action of these hormones? This question has been prominent in the thinking of workers in the hormonal domain for several years. From the material already presented in this chapter, it is clear that the presence or absence of a cytokinin in a tRNA molecule can markedly affect tRNA functional ability. Hence some held the view that cytokinin as a consequence of partial or *in toto* incorporation into tRNA affected protein synthesis and hence regulated growth. Unfortunately this hypothesis has not been substantiated by subsequent investigation. Firstly, 6-benzylamino-9-methylpurine is active as a cytokinin but neither the benzyl group nor the intact molecule is incorporated into tRNA in soybean callus tissue (KENDE and TAVARES, 1968). Secondly, although *E. coli* contains cytokinins in tRNA, no cytokinin auxotroph mutants could be found by KENDE and TAVARES (1968) who examined over 10,000 colonies. This suggests that cytokinins are not precursors in tRNA biosynthesis. Thirdly, cytokinins retard senescence of tobacco leaves but do not appear to be incorporated into tRNA in this tissue (RICHMOND et al., 1970). Fourthly, certain cytokinin analogues with a pyrazolo[4,3-*d*]-pyrimidine ring system are weakly active as cytokinins (HECHT et al., 1971). Since such 8-aza-9-deaza adenines contain a carbon atom in place of a nitrogen atom at the position normally engaged in nucleoside linkage, it is very unlikely that they form ribosides and are incorporated intact into tRNA. Fifthly, certain *N,N'*-disubstituted ureas exhibit cytokinin activity (BRUCE and ZWAR, 1966);

incorporation of such compounds into tRNA is highly improbable. When considered as a whole, the above results indicate that the mechanism of cytokinin action does not involve incorporation of cytokinins into tRNA.

The effects of exogenous cytokinins on synthesis of cytokinin-containing tRNA species and on the relative amounts of isoaccepting tRNAs have also been considered. Although tobacco-pith tissue requires a cytokinin for active growth, the tRNA of tissue cultured in the absence of cytokinin contains 6-(3-methylbut-2-enylamino)purine. The biosynthesis of this cytokinin in such tRNA appears to function normally and exogenously supplied cytokinin was not found to affect mevalonic acid incorporation into tRNA (CHEN and HALL, 1969). This suggests that exogenous cytokinins do not influence synthesis of cytokinin-containing tRNA species. However cytokinins appear to regulate the relative amounts of isoaccepting tRNAs possibly by controlling tRNA degradation. Although cytokinin application did not appear to cause any marked change in the level of leucine- or tyrosine-accepting tRNA in fenugreek cotyledons (RIJVEN and PARKASH, 1971), cytokinin treatment of soybean hypocotyls markedly altered the relative amounts of leucine-accepting tRNAs, the change being detectable after only three hours. The cytokinin also induced changes in serine and possibly tyrosine, but not valine or phenylalanine, isoacceptors. Cytokinin treatment appeared to result in changes in tRNA species which contain 6-(3-methylbut-2-enylamino)purine (ANDERSON and CHERRY, 1969; CHERRY and ANDERSON, 1971). These workers suggest that specific nucleases degrade cytokinin-containing tRNAs and that the enzymes bind to the cytokinin residues in the tRNA molecules. Exogenously supplied cytokinins bind to the nucleases and competitively inhibit their action. The tissue treated with cytokinin thus retains essential cytokinin-containing tRNA species which enable it to synthesize protein required for induction of growth. It should be emphasized that these suggestions are very speculative, but they certainly merit investigation.

Although the cytokinins and similar modified bases in tRNA undoubtedly play an important role in tRNA function probably by promoting binding to the mRNA-ribosome complex, and possibly by modifying codon recognition, it remains uncertain if the occurrence of cytokinins in tRNA is related to the mechanism of action of these hormones. However cytokinin-evoked changes in growth rate and differentiation may be the expression of changes in protein synthesis induced by regulation of the relative amounts of isoaccepting tRNAs.

Cytokinin-induced changes in patterns of isoaccepting tRNAs is but one instance of a general phenomenon. In a tissue or cell that is homeostatic, the proportions of isoacceptor tRNA species are relatively stable. However, when induction of growth or differentiation occurs, a change in the isoacceptor tRNA pattern usually results. Such changes have been shown to accompany differentiation of animal cells, viral infection, bacterial sporulation, change in growth conditions of bacterial cells (e. g. step-up, step-down) and conversion of a normal cell to a malignant cell. Appearance of new tRNA species occurs in some instances. All these changes have been discussed in the review by SUEOKA and SUEOKA (1970). It is particularly interesting that mouse plasma tumour cells which synthesize different proteins show differences in tRNA isoacceptor pattern (YANG and NOVELLI, 1968). It is tempting to assume that a species of tRNA which appears or greatly increases in amount has a special regulatory function. However this has yet to be demonstrated unequivocally. The study of the regulatory role of tRNA in growth and differentiation is really in its infancy. Today this regulatory function is little more than a hypothesis but one senses the concept has a promising future.

References

Reviews

Methods for purifying amino acid-specific tRNA:
(1) TANAKA, K.: In Procedures of nucleic acid research (CANTONI, G.L., and DAVIES, D.R., eds.), p. 466. New York and London: Harper & Row 1967.
(2) GILHAM, P.T.: Ann. Rev. Biochem. **39**, 227 (1970).

Methods for determination of nucleotide sequence:
(1) GILHAM, P.T.: Ann. Rev. Biochem. **39**, 227 (1970).

Structure of tRNA:
(1) ARNOTT, S.: Progr. Biophys. Mol. Biol. **22**, 181 (1971).
(2) MADISON, J.T.: Ann. Rev. Biochem. **37**, 131 (1968).
(3) STAEHELIN, M.: Experientia **27**, 1 (1971).
(4) WARING, M.J.: Ann. Rept. Progr. Chem. **65**(B), 551 (1968).
(5) GAUSS, D.H., HAAR, F., MAELICKE, A., CRAMER, F.: Ann. Rev. Biochem. **40**, 1045 (1971).

Aminoacylation of tRNA:
(1) CHAMBERS, R.W.: Progr. Nucl. Acid Res. Mol. Biol. **11**, 489 (1971).
(2) JACOBSON, K.B.: Progr. Nucl. Acid Res. Mol. Biol. **11**, 461 (1971).
(3) GAUSS, D.H., HAAR, F., MAELICKE, A., CRAMER, F.: Ann. Rev. Biochem. **40**, 1045 (1971).

Cytokinins:
(1) MILLER, C.O.: Ann. Rev. Plant Physiol. **12**, 395 (1961).
(2) LETHAM, D.S.: Ann. Rev. Plant Physiol. **18**, 349 (1967).
(3) LETHAM, D.S.: Bioscience **19**, 309 (1969).
(4) SKOOG, F., ARMSTRONG, D.J.: Ann. Rev. Plant Physiol. **21**, 359 (1970).

tRNA, protein synthesis, growth and differentiation:
(1) SUEOKA, N., KANO-SUEOKA, T.: Progr. Nucl. Acid Res. Mol. Biol. **10**, 23 (1970).

Other References

ADAMS, A., LINDAHL, T., FRESCO, J.R.: Proc. Nat. Acad. Sci. U.S. **57**, 1684 (1967).
ALLENDE, C.C., ALLENDE, J.E., GATICA, M., CELIS, J., MORA, G., MATAMALA, M.: J. Biol. Chem. **241**, 2245 (1966).
ALTMAN, S.: Nature New Biol. **229**, 19 (1971).
ALTMAN, S., SMITH, J.D.: Nature New Biol. **233**, 35 (1971).
AMES, B.N., HARTMAN, P.E.: Cold Spring Harbor Symp. Quant. Biol. **28**, 349 (1963).
ANDERSON, M.B., CHERRY, J.H.: Proc. Nat. Acad. Sci. U.S. **62**, 202 (1969).
ANDERSON, W.F.: Proc. Nat. Acad. Sci. U.S. **62**, 566 (1969).
APGAR, J., HOLLEY, R.W., MERRILL, S.H.: J. Biol. Chem. **237**, 796 (1962).
ARMSTRONG, D.J., BURROWS, W.J., SKOOG, F., ROY, K.L., SOLL, D.: Proc. Nat. Acad. Sci. U.S. **63**, 834 (1969).
BARNETT, W.E.: Proc. Nat. Acad. Sci. U.S. **53**, 1462 (1965).
BARNETT, W.E., EPLER, J.L.: Cold Spring Harbor Symp. Quant. Biol. **31**, 549 (1966).
BARNETT, W.E., JACOBSON, K.B.: Proc. Nat. Acad. Sci. U.S. **51**, 642 (1964).
BERNHARDT, D., DARNELL, J.E.: J. Mol. Biol. **42**, 43 (1969).
BLAKE, R.D., FRESCO, J.R., LANGRIDGE, R.: Nature **225**, 32 (1970).
BLASI, F., BARTON, R.W., KOVACH, J.S., GOLDBERGER, R.F.: J. Bacteriol. **106**, 508 (1971).
BRETSCHER, M.S.: J. Mol. Biol. **34**, 131 (1968).
BROWNLEE, G.G.: Nature New Biol. **229**, 147 (1971).
BROWNLEE, G.G., SANGER, F.: J. Mol. Biol. **23**, 337 (1967).
BROWNLEE, G.G., SANGER, F., BARRELL, B.G.: J. Mol. Biol. **34**, 379 (1968).

BRUCE, M. I., ZWAR, J. A.: Proc. Roy. Soc. Ser. B **165**, 245 (1966).

BURROWS, W. J., SKOOG, F., LEONARD, N. J.: Biochemistry **10**, 2189 (1971).

CAPECCHI, M. R., KLEIN, H. A.: Nature **226**, 1029 (1970).

CAPRA, J. D., PETERKOFSKY, A.: J. Mol. Biol. **33**, 591 (1968).

CHANG, S. H., MILLER, N. R., HARMON, C. W.: FEBS Lett. **17**, 265 (1971).

CHEN, C. M., HALL, R. H.: Phytochemistry **8**, 1687 (1969).

CHEREST, H., SURDIN-KERJAN, Y., ROBICHON-SZULMAJSTER, H.: J. Bacteriol. **106**, 758 (1971).

CHERRY, J. H., ANDERSON, M. B.: In plant growth substances 1970 (CARR, D. J., ed.), p. 181. Berlin-Heidelberg-New York: Springer 1971.

CRAIG, L. C., CRAIG, D.: In Technique of organic chemistry, 2nd edn. (WEISSBERGER, A., ed.), vol. 3 (part 1), p. 149. New York: Interscience 1956.

CRICK, F. H. C.: J. Mol. Biol. **19**, 548 (1966).

DANIEL, V., SARID, S., BECKMANN, J. S., LITTAUER, U. Z.: Proc. Nat. Acad. Sci. U.S. **66**, 1260 (1970).

EISINGER, J.: Biochem. Biophys. Res. Commun. **43**, 854 (1971).

FAULKNER, R. D., UZIEL, M.: Biochim. Biophys. Acta **238**, 464 (1971).

FAVOROVA, O., KISELEV, L.: FEBS Lett. **6**, 65 (1970).

FITTLER, F., HALL, R. H.: Biochem. Biophys. Res. Commun. **25**, 441 (1966).

FLEYSHER, M. H., HAKALA, M. T., BLOCH, A., HALL, R. H.: J. Med. Chem. **11**, 717 (1968).

FOX, J. E., CHEN, C. M.: J. Biol. Chem. **242**, 4490 (1967).

FULLER, W., HODGSON, A.: Nature **215**, 817 (1967).

GALLO, R. C., WHANG-PENG, J., PERRY, S.: Science **165**, 400 (1969).

GEFTER, M. L.: Biochem. Biophys. Res. Commun. **36**, 435 (1969).

GEFTER, M. L., RUSSELL, R. L.: J. Mol. Biol. **39**, 145 (1969).

GHOSH, K., GHOSH, H. P.: Biochem. Biophys. Res. Commun. **40**, 135 (1970).

GILLAM, I., BLEW, D., WARRINGTON, R. C., VON TIGERSTROM, M., TENER, G. M.: Biochemistry **7**, 3459 (1968).

GOODMAN, H. M., ABELSON, J., LANDY, A., BRENNER, S., SMITH, J. D.: Nature **217**, 1019 (1968).

GOULD, R. M., THORNTON, M. P., LIEPKALNS, V., LENNARZ, W. J.: J. Biol. Chem. **243**, 3096 (1968).

GRACE, J. T., HAKALA, M. T., HALL, R. H., BLAKESLEE, J.: Proc. Am. Assoc. Cancer Res. **8**, 23 (1967).

HACKER, B., FELDBUSH, T. R.: Biochem. Pharmacol. **18**, 847 (1969).

HALL, R. H., CSONKA, L., DAVID, H., McLENNAN, B.: Science **156**, 69 (1967).

HALPERN, R. M., CHANEY, S. Q., HALPERN, B. C., SMITH, R. A.: Biochem. Biophys. Res. Commun. **42**, 602 (1971).

HARADA, F., NISHIMURA, S.: Biochemistry **11**, 301 (1972).

HATFIELD, G. W., BURNS, R. O.: Proc. Nat. Acad. Sci. U.S. **66**, 1027 (1970).

HECHT, S. M., BOCK, R. M., SCHMITZ, R. Y., SKOOG, F., LEONARD, N. J., OCCOLOWITZ, J. L.: Biochemistry **10**, 4224 (1971).

HIRSH, D.: J. Mol. Biol. **58**, 439 (1971).

HOLLEY, R. W., APGAR, J., EVERETT, G. A., MADISON, J. T., MARQUISEE, M., MERRILL, S. H., PENSWICK, J. R., ZAMIR, A.: Science **147**, 1462 (1965).

HOLTEN, V. Z., JACOBSON, K. B.: Arch. Biochem. Biophys. **129**, 283 (1969).

HOOPER, M. L., RUSSELL, R. L., SMITH, J. D.: FEBS Lett. **22**, 149 (1972).

IMAMOTO, F., ITO, J., YANOFSKY, C.: Cold Spring Harbor Symp. Quant. Biol. **31**, 235 (1966).

IMAMOTO, F., KANO, Y.: Nature New Biol. **232**, 169 (1971).

ITO, K., HIRAGA, S., YURA, T.: Genetics **61**, 521 (1969).

JACOBSON, K. B.: Nature New Biol. **231**, 17 (1971).

JOHNSON, L., HAYASHI, H., SOLL, D.: Biochemistry **9**, 2823 (1970).

KAN, J., SUEOKA, N.: J. Biol. Chem. **246**, 2207 (1971).

KANABUS, J., CHERRY, J. H.: Proc. Nat. Acad. Sci. U.S. **68**, 873 (1971).

KASAI, H., GOTO, M., TAKEMURA, S., GOTO, T., MATSUURA, S.: Tetrahedron Letters 2725 (1971).

KENDE, H., TAVARES, J. E.: Plant Physiol. **43**, 1244 (1968).

KIMURA-HARADA, F., SANEYOSHI, M., NISHIMURA, S.: FEBS Lett. **13**, 335 (1971).

KLINE, L., FITTLER, F., HALL, R. H.: Biochemistry **8**, 4361 (1969).

KOVACH, J. S., BERBERICH, M. A., VENETIANER, P., GOLDBERGER, R. F.: J. Bacteriol. **97**, 1283 (1969a).

KOVACH, J. S., PHANG, J. M., FERENCE, M., GOLDBERGER, R. F.: Proc. Nat. Acad. Sci. U.S. **63**, 481 (1969b).

LEIBOWITZ, M. J., SOFFER, R. L.: J. Biol. Chem. **245**, 2066 (1970).

Leonard, N. J., Iwamura, H., Eisinger, J.: Proc. Nat. Acad. Sci. U. S. **64**, 352 (1969).
Letham, D. S.: Planta **74**, 228 (1967).
Letham, D. S.: Phytochemistry **11**, 1023 (1972).
Letham, D. S., Ralph, R. K.: Life Sci. **6**, 387 (1967).
Letham, D. S., Shannon, J. S., McDonald, I. R. C.: Tetrahedron **23**, 479 (1967).
Levitt, M.: Nature **224**, 759 (1969).
Lindahl, T., Adams, A., Fresco, J. R.: Proc. Nat. Acad. Sci. U. S. **55**, 941 (1966).
Lu, P., Rich, A.: J. Mol. Biol. **58**, 513 (1971).
Martin, R. G., Silbert, D. F., Smith, D. W. E., Whitfield, H. J.: J. Mol. Biol. **21**, 357 (1966).
Maxwell, I. H., Wimmer, E., Tener, G. M.: Biochemistry **7**, 2629 (1968).
Miller, C. O.: In Biochemistry and physiology of plant growth substances (Wightman, F., and Setterfield, G., eds.), p. 33. Ottawa: Runge Press 1968.
Mittelman, A., Hall, R. H., Yohn, D. S., Grace, J. T.: Cancer Res. **27**, 1409 (1967).
Morse, D. E., Guetin, M.: Nature New Biol. **232**, 165 (1971).
Muench, K. H., Berg, P.: Fed. Proc. **23**, 477 (1964).
Murao, K., Saneyoshi, M., Harada, F., Nishimura, S.: Biochem. Biophys. Res. Commun. **38**, 657 (1970).
Nakanishi, K., Furutachi, N., Funamizu, M., Grunberger, D., Weinstein, I. B.: J. Amer. Chem. Soc. **92**, 7617 (1970).
Ohashi, Z., Murao, K., Yahagi, T., von Minden, D. L., McCloskey, J. A., Nishimura, S.: Biochim. Biophys. Acta **262**, 209 (1972).
Pearson, R. L., Weiss, J. F., Kelmers, A. D.: Biochim. Biophys. Acta **228**, 770 (1971).
Petit, J. F., Strominger, J. L., Soll, D.: J. Biol. Chem. **243**, 757 (1968).
Playtis, A. J., Leonard, N. J.: Biochem. Biophys. Res. Commun. **45**, 1 (1971).
Quinn, L. Y., Oates, R. P., Beers, T. S.: J. Bacteriol. **86**, 1359 (1963).
Rechler, M. M., Martin, R. G.: Nature **226**, 908 (1970).
Richmond, A., Back, A., Sachs, B.: Planta **90**, 57 (1970).
Rijven, A. H. G. C., Parkash, V.: Plant Physiol. **47**, 59 (1971).
Sanger, F., Brownlee, G. G., Barrell, B. G.: J. Mol. Biol. **13**, 373 (1965).
Saponara, A., Bock, R. M.: Fed. Proc. **20**, 356 (1961).
Schlesinger, S., Magasanik, B.: J. Mol. Biol. **9**, 670 (1964).
Schmitz, R. Y., Skoog, F., Hecht, S. M., Leonard, N. J.: Phytochemistry **10**, 275 (1971).
Shimura, Y., Aono, H., Ozeki, H., Sarabhai, A., Lamfrom, H., Abelson, J.: FEBS Lett. **22**, 144 (1972).
Shugart, L., Novelli, G. D., Stulberg, M. P.: Biochim. Biophys. Acta **157**, 83 (1968).
Skoog, F., Armstrong, D. J., Cherayil, J. D., Hampel, A. E., Bock, R. M.: Science **154**, 1354 (1966).
Stern, R., Gonano, F., Fleissner, E., Littauer, U. Z.: Biochemistry **9**, 10 (1970).
Stern, R., Littauer, U. Z.: Biochemistry **7**, 3469 (1968).
Sueoka, N., Kano-Sueoka, T.: Proc. Nat. Acad. Sci. U. S. **52**, 1535 (1964).
Suk, D., Simpson, C. L., Mihich, E.: Cancer Res. **30**, 1429 (1970).
Svensson, I., Isaksson, L., Henningsson, A.: Biochim. Biophys. Acta **238**, 331 (1971).
Szentermai, A. M., Szentermai, M., Umbarger, H. E.: J. Bacteriol. **95**, 1672 (1968).
Thiebe, R., Zachau, H. G., Baczynskyj, L., Biemann, K., Sonnenbichler, J.: Biochim. Biophys. Acta **240**, 163 (1971).
Vournakis, J. N., Scheraga, H. A.: Biochemistry **5**, 2997 (1966).
Wainfan, E., Landsberg, B.: FEBS Lett. **19**, 144 (1971).
Weiss, J. F., Kelmers, A. D.: Biochemistry **6**, 2507 (1967).
Weiss, J. F., Pearson, R. L., Kelmers, A. D.: Biochemistry **7**, 3479 (1968).
Whitfield, H. J., Gutnick, D. L., Margolies, M. N., Martin, R. G., Rechler, M. M., Voll, M. J.: J. Mol. Biol. **49**, 245 (1970).
Williams, L., Freundlich, M.: Biochim. Biophys. Acta **186**, 305 (1969).
Yang, W. K., Novelli, G. D.: Biochem. Biophys. Res. Commun. **31**, 534 (1968).
Yaniv, M., Barrell, B. G.: Nature **233**, 113 (1971).
Young, D. V., Srinivasan, P. R.: Biochim. Biophys. Acta **238**, 447 (1971).
Zachau, H. G., Dutting, D., Feldmann, H.: Z. Physiol. Chem. **347**, 212 (1966).
Zachau, H. G., Tada, M., Lawson, W. B., Schweiger, M.: Biochim. Biophys. Acta **53**, 221 (1961).
Zubay, G.: J. Mol. Biol. **4**, 347 (1962).
Zubay, G., Cheong, L., Gefter, M.: Proc. Nat. Acad. Sci. U. S. **68**, 2195 (1971).

CHAPTER 6

Ribosomal RNA

L. DALGARNO and J. SHINE

Introduction

Ribosomes from all organisms are composed of two subunits which differ in size; each subunit contains an RNA species of high molecular weight and a variety of proteins. The larger of the two subunits contains, in addition, a low molecular weight RNA species, known as 5S RNA. In eukaryotes the cytoplasmic ribosome contains another small RNA component, lRNA, associated with the large subunit. This species appears to arise from cleavage of the high molecular weight ribosomal RNA precursor and, unlike 5S RNA, is hydrogen-bonded to the large ribosomal RNA in the ribosome. No such RNA has been found in the ribosomes of pro-karyotes.

In the first part of this chapter, the structure of these species of ribosomal RNA (rRNA) will be examined. The second section summarizes what is known of the synthesis of rRNA in prokaryotes and eukaryotes. The reader is referred to Chapters 8 and 9 respectively for a discussion of chloroplast and mitochondrial rRNA.

A. The Structure of Ribosomal RNA

Several recent reviews have appeared on the structure of rRNA and the reader is referred to ATTARDI and AMALDI (1970) for a detailed study of the size and base composition of rRNA, and to NOMURA (1970) for a more complete discussion of the structure of the bacterial ribosome.

1. Molecular Weight

The high molecular weight RNA from the large ribosome subunit has a sedimentation coefficient ranging from 23S in bacteria to 28S in mammals; the molecular weight ranges from 1.1 M (i.e. 1.1×10^6 daltons or about 3,000 nucleotides) to 1.8 M (5,000 nucleotides) respectively. The small ribosome subunit contains 16S RNA in bacteria and 18S RNA in mammals with molecular weights of 0.55 M (1,500 nucleotides) and 0.7 M (2,000 nucleotides) respectively.*

---

* Species of rRNA from the large and small subunits of eukaryote and prokaryote ribosomes will be referred to generically, where appropriate, as "large" and "small" RNAs.

When the major rRNA species from various organisms are compared, a sharp distinction in molecular weight is obvious between prokaryotes and eukaryotes. In prokaryotes both rRNA species have molecular weights nearly 20% less than those of rRNA from the simplest higher organisms (LOENING, 1968). During evolution within the eukaryotes the RNA from the small ribosome subunit has remained relatively constant in molecular weight (about 0.7 M), whereas the molecular weight of the large rRNA has increased from about 1.3 M in yeast to 1.4 M in sea-urchins and 1.8 M in mammals.

The molecular weight of 5S RNA appears to be very similar in prokaryotes and eukaryotes. Thus 5S RNA from *E. coli* and from a human cell line have identical chain lengths; both contain 120 nucleotides (FORGET and WEISSMANN, 1967; BROWNLEE et al., 1968).

The presence of lRNA associated with the large rRNA has been demonstrated in all eukaryotic cells so far examined. This low molecular weight rRNA has also been termed 7S RNA, 6S RNA, 5.8S RNA and $28S_A$ RNA, although the size of lRNA from various eukaryotes appears similar (130–160 nucleotides) when determined by polyacrylamide gel electrophoresis (KING and GOULD, 1970; PAYNE and DYER, 1972; SHINE and DALGARNO, 1972). Since lRNA is absent from the ribosomes of bacteria and blue-green algae, the presence of this rRNA species is probably characteristic of the cytoplasmic ribosomes of eukaryotes (PAYNE and DYER, 1972).

2. Base Composition

In rRNA there is no fixed relationship between the molar percentage of A and U or of G and C. Not all bases are involved in complementary pairing and the relative proportions of the four major ribonucleosides in rRNA varies considerably between different organisms. The G+C content of both high molecular weight rRNAs tends to increase from the lower to the more highly evolved eukaryotes (ATTARDI and AMALDI, 1970); e.g. *Neurospora* 25S and 17S RNA contain 51% and 49% G+C respectively, whilst the corresponding figures for man are 68% and 58%. The increase in G+C content with evolution is only a general trend however, and exceptions exist. Within the insect kingdom for example, the G+C content of rRNA from certain *Diptera* such as *Drosophila* appears to be unusually low (43%) compared with that of other insects, e.g. *Galleria* and *Hyalophora* (Order: *Lepidoptera*) which have G+C contents of approximately 56% (GREENBERG, 1969). The tendency of rRNA from higher organisms to contain higher proportions of G+C is probably accompanied by an increase in the extent and stability of secondary structure of these polynucleotides.

As well as the four common nucleosides, high molecular weight rRNA contains a small number of unusual nucleosides, including pseudouridine and various methylated derivatives of cytidine, guanosine, uridine and adenosine. Pseudouridine is present in higher proportion in eukaryotic rRNA (1–2 mole percent) than in bacterial rRNA (0.1–0.3 mole percent) and methylated nucleosides are also more abundant in eukaryotic rRNA (1.2–1.7 mole percent) than in *E. coli* rRNA (0.6–1 mole percent). Most methyl groups in bacterial rRNA are attached to bases; in mammalian rRNA they occur principally attached to the 2'-hydroxyl of the ribose moiety. The specific methylation of the small rRNA is up to 50% higher than that of the large rRNA in both bacterial and mammalian cells.

The base composition of 5S RNA does not reflect that of high molecular weight rRNA from the same source, although this has been examined in only a limited number of organisms. Thus the G+C content of *E. coli* 5S RNA is 64% whereas that from HeLa cells is 58%; the average G+C content of the high molecular weight rRNA from these two organisms is 54% and 63% respectively. In the bacterial, plant and animal cells so far examined, 5S RNA is characterized by the absence of methylated nucleosides (ATTARDI and AMALDI, 1970).

In mammalian lRNA the four major ribonucleosides are reported to be present in similar proportions to those found in the high molecular weight rRNA to which this small polynucleotide is hydrogen-bonded (PENE et al., 1968; KING and GOULD, 1970).

3. Base Sequence of High Molecular Weight rRNA

Using techniques available at present, sequence determination requires purification of the labelled RNA species and the specific cleavage of the polynucleotide into small oligonucleotides whose complete sequence can be determined. The specificity of the major ribonucleases used in sequence studies is summarized in Fig. 1. As separation of the individual oligonucleotides is achieved by electro-

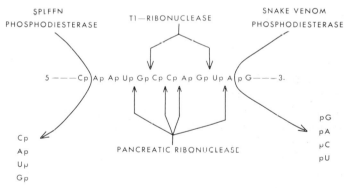

Fig. 1. Specificity of ribonucleases used in RNA sequence determination. Pancreatic and T_1-ribonuclease are the two major enzymes used for the specific cleavage of RNA. Both are termed "cyclizing endonucleases" because they act within a nucleotide chain and form intermediate cyclic phosphates. T_1-ribonuclease cleaves phosphodiester bonds adjacent to the 3'-phosphate attached to guanosine residues, giving rise to oligonucleotides terminating in guanosine-3'-phosphate. Pancreatic ribonuclease is pyrimidine-specific, cleaving bonds adjacent to the 3'-phosphate linked to cytidine and uridine and resulting in a mixture of oligonucleotides terminating in either cytidine-3'-phosphate or uridine-3'-phosphate. Oligonucleotides obtained by digestion with either nuclease can be sequenced with the aid of two exonucleases, spleen phosphodiesterase and snake venom phosphodiesterase, or with ribonuclease U_2 in combination with alkaline hydrolysis. Ribonuclease U_2 is purine-specific at low enzyme concentration, giving rise to oligonucleotides terminating mainly in guanosine-3'-phosphate or adenosine-3'-phosphate. Alkaline hydolysis results in the cleavage of nearly all phosphodiester bonds in rRNA producing a mixture of 2'- and 3'-mononucleotides; the only bonds not cleaved are those adjacent to 2'-O-methylated nucleosides

phoretic procedures which can accommodate only small amounts of RNA, the RNA must be labelled to a high specific activity with ^{32}P, and the specific oligo-nucleotides are then detected by autoradiography. Overlapping of these sequences is obtained by comparing larger oligonucleotides which are generated by more limited cleavage of the RNA molecule. Sequences derived from the large fragments are used to order the small oligonucleotides into an unambiguous total sequence. For a review of methods used in the sequence analysis of RNA the reader is referred to GILHAM (1970). The success of these procedures for the sequence analysis of high molecular weight rRNA depends on three factors:

1) The purity of the particular rRNA species. This provides little problem with high molecular weight rRNA which accounts for almost 80% of the cellular RNA; several simple methods are available for the fractionation and purification of rRNA (see Chapter 11);

2) labelling the rRNA to high specific activity with ^{32}P. Whilst this is relatively straightforward for bacteria it is not for eukaryotes and the sequencing procedures developed by BROWNLEE and SANGER (1967) have not yet been generally applied to high molecular weight rRNA from eukaryotic cells. Vertebrate cells grow slowly in the presence of levels of ^{32}P needed to label the rRNA to high specific activity; this is also partly due to the very low levels of total phosphate necessary in the labelling medium. As nucleosides can be readily incorporated into rRNA *in vivo*, sequence analysis of some eukaryotic rRNAs may be possible using fluorographic detection of oligonucleotides labelled with tritiated nucleosides. After coating the two-dimensional electropherogram with a scintillant, and exposing the coated electropherogram to X-ray film, the labelled oligonucleotides are seen as a blackening of the photographic emulsion by light emitted from the scintillant (RANDERATH, 1970), rather than directly by radioactive emission, as in the standard autoradiographic procedure;

3) the isolation of specific, large fragments of the rRNA molecule. Because of the relative absence of internal markers and of specifically susceptible cleavage points in high molecular weight rRNA, it has so far proved difficult to isolate very large oligonucleotides with which to overlap the sequences obtained from complete digestion with ribonucleases. This may be achieved by developing methods for limited digestion of isolated rRNA so that the secondary structure of the rRNA protects certain regions against attack by the nuclease. The specific orientation of the rRNA in the ribosome may also assist in obtaining limited cleavage of the molecule in certain exposed regions.

For a detailed discussion of sequence analysis of high molecular weight rRNA the reader is referred to studies on *E. coli* 16S RNA (FELLNER et al., 1970, 1972).

a) Internal Sequences. Using procedures developed by BROWNLEE and SANGER (1967), which are discussed in more detail in Chapter 5, it is possible in principle to isolate and sequence all oligonucleotides produced by specific ribonuclease digestion of rRNA labelled with ^{32}P. From knowledge of the arrangement of these oligonucleotides in the intact molecule, about 70% of the sequence of 16S RNA from *E. coli* has been determined (FELLNER et al., 1970, 1972). Apart from providing information on the heterogeneity and possible secondary structure of 16S RNA, as discussed below, the analysis has indicated some repetition of nucleotide sequences in this rRNA.

As mentioned above, these techniques are unfortunately limited by the requirement for RNA labelled to a high specific activity with ^{32}P, although some information, mainly of a comparative nature, can be obtained by digesting non-radioactive rRNA with pancreatic or T_1 ribonuclease and separating the resultant oligonucleotides by column or paper chromatography. The frequency of occurrence of certain oligonucleotides in the rRNA species can then be estimated. Using this method, ARONSON (1962) has shown a difference between the frequency of some short oligonucleotides (e.g. AU) in the 16S and 23S rRNA of *E. coli*. Similarly, pancreatic ribonuclease digestion of HeLa cell rRNA shows that the frequency of certain oligonucleotides (e.g. PyGC) differs considerably between 18S and 28S rRNA (AMALDI and ATTARDI, 1968).

b) 5'-Terminal Sequences. When rRNA is treated with a cyclizing endonuclease (T_1 or pancreatic), the interior of the molecule yields mono- and oligo-nucleotides with a 5'-hydroxyl and a 3'-phosphate. The fragment from the 5'-terminus has a phosphate at both the 5'- and 3'-ends, while the fragment from the 3'-terminus lacks an end phosphate. As alkaline phosphatase removes 3'- and 5'-terminal phosphates, all oligonucleotides from the interior of the RNA lose a single phosphate residue on treatment with this enzyme, whereas the 5'-terminal species lose two phosphates and the 3'-species none. 5'-Terminal fragments from nuclease digests of various rRNA species can then be identified by ion-exchange chromatography, since the loss of two phosphate groups after treatment with alkaline phosphatase is reflected in a marked change in mobility (TENER, 1967).

The 5'-terminus of rRNA has also been studied by a method involving the addition of ^{32}P to the 5'-end (TAKANAMI, 1967). Before the addition of ^{32}P any terminal phosphate must be removed with alkaline phosphatase which leaves a hydroxyl group at the 5'-terminus. The enzyme polynucleotide kinase is then used to catalyze the transfer of ^{32}P from γ-^{32}P-labelled ATP to the 5'-hydroxyl. The labelled RNA is digested with T_1-ribonuclease and yields a radioactive oligonucleotide derived from the 5'-terminus which can be isolated and sequenced.

Results for *E. coli* show that 16S and 23S rRNA have 5'-terminal pAAAUUG and pGGU respectively (TAKANAMI, 1967). In three species of bacilli, pU(X)$_4$G and pU(X)$_3$G have been found at the 5'-termini of the two major rRNA species, where X is A, C, or U (SUGIURA and TAKANAMI, 1967). From these limited results it appears that closely related bacteria have similar 5'-sequences, whereas these sequences differ between more distantly related species. Less is known of the 5'-terminal sequences of eukaryotic rRNAs. The main 5'-terminal nucleotides in L cell RNA are pU and pC for 18S and 28S rRNA respectively (LANE and TAMAOKI, 1967); pG is found at the 5'-end of 18S RNA from Novikoff hepatoma cells (EGAWA et al., 1971). The 5'-sequences of 18S and 25S rRNA from yeast are pUUG and pU(X)$_4$G respectively (SUGIURA and TAKANAMI, 1967). The finding of the same sequences at the 5'-end of rRNA also indicates a specific cleavage of the precursor rRNA during biosynthesis, and a lack of any gross heterogeneity in the terminal region; both of these points will be considered further below.

c) 3'-Terminal Sequences. Several methods commonly used to study 3'-terminal sequences in rRNA depend on oxidation by periodate of the 2',3'-diol group of the terminal ribose at the 3'-end of rRNA; this results in the formation of a dialdehyde (WHITFELD, 1954). Reagents such as sodium borohydride, isoniazid and dimedone can react with these oxidized nucleosides to form stable derivatives (Zamecnik et al., 1960; Hunt, 1965; DeWachter and Fiers, 1967; Glitz and Sigman, 1970). By labelling oxidized RNA with radioactive isoniazid and digesting with specific nucleases, the 3′-terminal oligonucleotide can be isolated and identified (Hunt, 1965). Alternatively, since the periodate-oxidized terminal nucleoside is susceptible to β-elimination in the presence of primary amines (Whitfeld, 1954; Khym and Uziel, 1968), it can be selectively removed by treatment with reagents such as aniline.

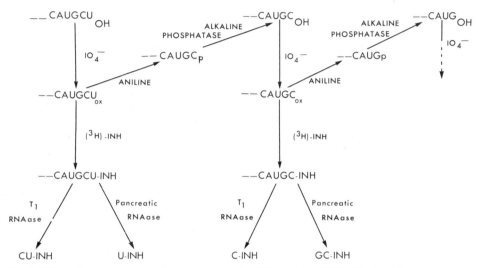

Fig. 2. Schematic representation of sequence analysis of 3′-termini of rRNA. Nucleosides are removed from the 3′-terminus by a combination of periodate oxidation and β-elimination. After removal of the resultant 3′-terminal phosphate with alkaline phosphatase, the procedure can be repeated in a step-wise manner. If oxidized RNA is labelled with ^{3}H-isoniazid (isonicotinic acid hydrazide, INH) after each periodate oxidation, then the 3′-terminal nucleoside or oligonucleotide arising from ribonuclease digestion of the labelled RNA can be readily identified

By using a combination of these two procedures (Fig. 2), Hunt (1970) has obtained the following sequences for the 3′-termini of rabbit reticulocyte rRNA: $GUUUGU_{OH}$ for 28S RNA, $GUCGCU_{OH}$ for lRNA, and $GAUCAUUA_{OH}$ for 18S rRNA.

d) Methylated Sequences. Another approach to the determination of base sequence in rRNA makes use of the small number of methylated nucleosides found in these molecules. Since these arise by transmethylation from S-adenosyl methionine, rRNA can be labelled in vivo with methyl-labelled methionine. The labelled RNA is digested and fractionated by the two-dimensional procedure mentioned earlier, to yield oligonucleotides containing methylated residues (Fig. 3). These can then be subjected to complete sequence analysis by standard methods. In 23S rRNA from E. coli, twelve unique T_1-oligonucleotides have been found which contain methylated nucleosides; each of these methylated oligonucleotides occurs twice per 23S RNA molecule (Fellner, 1969). In 16S rRNA, six unique species of methylated oligonucleotides are found after T_1-

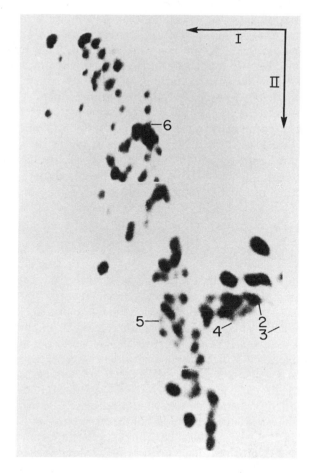

Fig. 3. Fingerprint of T_1-oligonucleotides from *E. coli* 16S RNA. Two-dimensional electropherogram or 'fingerprint' of a T_1-ribonuclease digest of *E. coli* 16S RNA generally labelled with ^{32}P and labelled in methyl positions with [^{3}H-methyl]-methionine. The position of the methylated oligonucleotides is determined by cutting out each spot and counting ^{3}H. Spot 1, N_2-methylguanylic acid (m^2 G), is not present on this fingerprint, having been lost into the anode buffer. The arrows refer to the direction of electrophoresis. Electrophoresis in the first dimension (I) is on cellulose acetate at pH 3.5 and in the second dimension (II) on DEAE-paper using 7% formic acid

Spot	Structure	Frequency (moles/mole of RNA)
1	m^{2}G	4
2	m^{4}Cm—C—m$_2$C—C—G	1
3	C—C—m^{7}G—C—G	2
4	m^{5}C—A—A—C—G	1
5	m$_2^6$A—m$_2^6$A—C—C—U—G	2
6	U—m^{5}C—A—C—A—C—C—U—A—G	1

m^{4}Cm, N_4-methyl-2'-*O*-methylcytidylic acid. m$_2$C, dimethylcytidylic acid. m^{7}G, 7-methylguanylic acid. m^{5}C, 5-methylcytidylic acid. m$_2^6$A, N_6, N_6-dimethyladenylic acid.
(After FELLNER, 1969)

ribonuclease digestion; these all have frequencies of occurrence which are close to integral numbers (Fig. 3), suggesting the lack of any major heterogeneity within the rRNA in these specific areas. Thus methylation is specific to a small number of sequences which appear to be duplicated in the 23S rRNA. Of the nine moles of pseudouridine present per mole of 23S rRNA, six are found in the methylated sequences which comprise only 5% of the molecule. This implies a clustering of modified nucleosides in the rRNA which is also confirmed by the presence of more than one methylated nucleoside in many of these short sequences. It is likely that there is a special structural significance to such a non-random distribution of modified nucleosides. In fact, in those parts of *E. coli* 16S RNA which have been sequenced, clusters of two or more methyl groups are only found in loops at certain proposed points of folding of this polynucleotide (Ehresmann et al., 1970; Fellner et al., 1972).

4. Base Sequence of Low Molecular Weight rRNA

The complete base sequence of 5S RNA from both *E. coli* (Brownlee et al., 1968) and human KB cells (Forget and Weissmann, 1967) has been determined. Both have a 5′-terminal sequence which is probably involved in base pairing with a complementary sequence found at the 3′-end of the molecule. Certain sequences are repeated twice in *E. coli* 5S RNA and the molecule can be divided into two "halves" which display considerable sequence homology. In 5S RNA from KB cells there is also a tendency for sequences to be duplicated, although the repeated sequences themselves are different to those in *E. coli* 5S RNA. The oligonucleotide pattern of ribonuclease digests of HeLa cell 5S RNA indicates that the sequences present in this RNA are similar to those derived from KB cell 5S RNA, although some differences are apparent (Hatlen et al., 1969). Similarly, the oligonucleotide "fingerprint" of marsupial 5S RNA (Averner and Pace, 1972) is very similar to that of KB cell and mouse 5S RNA (Williamson and Brownlee, 1969). In HeLa cells, both ppG and pppG as well as pG have been found at the 5′-terminus of 5S RNA. The significance of this finding in relation to the biosynthesis of 5S RNA is discussed below.

The only sequence information available for lRNA is that for the 3′-terminus ($GUCGCU_{OH}$) mentioned above (Hunt, 1970). Interestingly, an identical 3′-tri-nucleotide (GCU_{OH}) is found in insect lRNA (Shine and Dalgarno, unpublished results) and may indicate a similar specificity of precursor cleavage for the two cell types.

5. Heterogeneity in the Base Sequence of rRNA

Despite the multiplicity or redundancy of ribosomal DNA cistrons in both prokaryotes and eukaryotes (see below), the termini and methylated regions of rRNA show no evidence of any gross heterogeneity of base sequence within a population of rRNA molecules. On the other hand, limited heterogeneity in other areas of the molecule has now been demonstrated. Thus, in an extensive analysis of *E. coli* 16S rRNA, some heterogeneity in the primary sequence was encountered (FELLNER et al., 1972). In a certain region, the sequence AACUG is found in approximately 70% of the RNA molecules, whilst in the remaining 30% it is replaced by AACCUG. Similarly, the immediately adjacent sequence GCAUCUG, found in 70% of the molecules, is replaced in the remainder by GAUCUG. It is of interest that in sections where variation is observed, more than one change is normally found. This could arise from "hot-spots" in the rDNA cistrons, or indicates that a change in one position requires a compensatory change in a nearby sequence to retain the secondary structure necessary for a functional molecule.

Some sequence heterogeneity is also found in 5S RNA. In a particular strain of E. coli, two forms of 5S RNA have been found in about equal amounts; these differ by only one nucleotide in a single position (BROWNLEE et al., 1968). There may also be more than one form of 5S RNA in HeLa cells since some of the larger oligonucleotides in ribonuclease digests are found in less than molar amounts (HATLEN et al., 1969). Two 3'-sequences, CUU_{OH} and $CUUU_{OH}$ have been found in KB cell 5S RNA, each in nearly half molar amounts (FORGET and WEISSMAN, 1967).

6. The Existence of 'Hidden Breaks' in rRNA

The presence of breaks in specific regions of rRNA has been shown to occur in the rRNA of an amoeba (STEVENS and PACHLER, 1972), in sea urchins (NEMER and INFANTE, 1967) and in chloroplast 23S rRNA (see Chapter 9). A break also exists close to the centre of the large (26S) rRNA isolated from a wide variety of insects (SHINE and DALGARNO, 1973). Such breaks are normally masked in these rRNAs by the presence of intra-molecular hydrogen bonding and are only expressed after treatments which disrupt these bonds, e.g. heat or exposure to dimethyl sulphoxide. The terminal labelling procedure previously illustrated (Fig. 2) may be used to determine the number of polynucleotide chains in a particular RNA species. This is illustrated in Fig. 4 where insect 26S RNA is shown to contain three polynucleotide chains; two of these arise from the presence of a central break in the RNA molecule and lRNA accounts for the other. A terminal labelling procedure such as this is particularly valuable for the identi-fication of a small RNA species such as lRNA since, unlike general radioactive labelling, it leads to equivalent incorporation into polynucleotides of markedly different chain length.

Unlike scissions in insect rRNA, the breaks seen in the rRNA of sea urchins and amoeba do not occur in the centre of the RNA molecule, and consequently dissociation results in polynucleotides of unequal length. It is not clear whether these scissions occur at specific nucleotides or are limited merely to a particular region of the RNA. It is likely however that they are at nuclease-sensitive regions in the RNA, perhaps those areas of the polynucleotide which are exposed on the surface of the ribosome or ribosome precursor. This view is supported by the demonstration that when E. coli ribosomal subunits are subjected to limited nuclease digestion large fragments of rRNA are produced (ALLET and SPAHR, 1971).

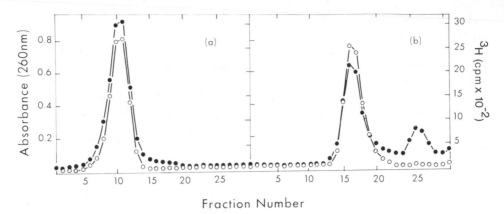

Fig. 4. Use of a 3'-end labelling procedure to determine the number of polynucleotide chains in a species of rRNA. Isolated 26S rRNA from cultured *Aedes aegypti* cells was oxidized with periodate before and after heating, and the 3'-termini labelled with ³H-isoniazid as described in Fig. 2. The labelled RNA was centrifuged on a sucrose gradient (from left to right).
　　　　(a) Unheated RNA
　　　　(b) Heated RNA (60° for 5 minutes).
From the specific activity of unheated 26S RNA, it can be calculated that there are three 3'-ends per RNA of aggregate molecular weight 1.5 M. After heating, one-third of the radioactivity (i.e. one of the 3'-ends) is associated with material in the 7S region of the gradient; the other two-thirds of the radioactivity sediments with the optical density peak at 18S. The two 18S 'halves' of 26S RNA sediment coincidentally indicating that the break in the polynucleotide chain occurs close to the centre of the molecule. (From SHINE and DALGARNO, 1973). ‑●‑●‑ radioactivity; ‑o‑o‑ absorbance

　　　　The apparent lability of the central region of *E. coli* 23S rRNA (MIDGLEY, 1965a, 1965b; MIDGLEY and MCILREAVY, 1967), the duplication of methylated sequences in 23S rRNA (FELLNER, 1969), and the existence of a central scission in insect 26S rRNA may indicate a limited two-fold symmetry of folding in the RNA from the large ribosome subunit. In fact MOLLER (1969) has proposed a structure for the large ribosome subunit in which two equal halves are arranged cylindrically around a central axis. The preferential breakage of the nucleotide chain linking the two 'halves' may explain why the large ribosome subunit from *E. coli* occasionally yields a mixture of 16S and 23S RNA (KURLAND, 1960). However, the question of a possible two-fold symmetry in the larger rRNA is unlikely to be resolved until more complete sequence data are available.

7. Secondary Structure of rRNA

a) High Molecular Weight rRNA. The secondary structure of rRNA was first studied in detail by DOTY et al. (1959). By measuring the hyperchromicity of rRNA at increasing temperatures it was concluded that at moderate ionic strength the high molecular weight rRNA has a secondary structure consisting of small base-paired helical regions involving about 60% of the polynucleotide chain. The stability of these helices is dependent upon their ionic environment; their heterogeneity is reflected in a typical melting curve for rRNA characterized by a broad transition from the partly double-helical to the completely single-stranded form (Fig. 5). If the temperature of a solution of denatured rRNA is lowered, it regains its original absorbance, indicating extensive reformation of base-pairs.

The temperature at which one half of the absorbance increase at 260 nm is seen, is known as the T_m for the particular RNA and is a measure of the stability of its secondary structure. As G—C interactions are more stable than A—U interactions, RNA which contains a large proportion of G—C pairs will exhibit a higher T_m than RNA containing more A—U pairs. It follows that the T_m of rRNA reflects its base composition. Thus, in 0.1 M salt *Drosophila melanogaster* 26S RNA (42% G+C) has a T_m of 49° (Fig. 5; see SHINE and DALGARNO, 1973),

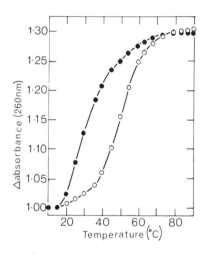

Fig. 5. Effect of ionic strength on the thermal denaturation of rRNA. *Drosophila melanogaster* 26S rRNA (G+C content 42%) was heated to 90° in 5° increments. At each temperature the optical density at 260 nm was recorded and plotted relative to that at 10°. -o-o-, RNA heated in 0.09 M NaCl, 0.01 M phosphate (pH 7.4). T_m: 49° -●-●-, RNA heated in 0.01 M phosphate (pH 7.4) T_m: 32°. (From SHINE and DALGARNO, 1973)

whereas the T_m of 28S RNA from rabbit reticulocytes (67% G+C) is 58° (Cox, 1970). Apart from its influence on T_m, the base composition of the helical regions also affects the hyperchromicity at 260 nm, since the melting of G—C pairs contributes less to the increase in absorbance at this wavelength than does that of A—U pairs. Since an increase in absorbance at 280 nm is also seen when rRNA is heated, and as this is due predominantly to the melting of G—C pairs, it is also possible to obtain an estimate of the base composition of helices melting over a certain temperature range by comparing the change in absorbance at 260 nm with that at 280 nm.

The increase in T_m of RNA with increase in ionic strength (Fig. 5) is consistent with a structure in which helices are stabilized by an increase in the number of cations available for shielding negative phosphate groups. The contribution of phosphate repulsion to helix instability is thus markedly enhanced at low ionic strength.

The large temperature range over which the melting of rRNA occurs is due to the denaturation of short base-paired regions and can be thought of as the summation of the individual melting curves for a number of helices differing both in length and composition. Under conditions of moderate ionic strength the length of double-helical regions has been estimated as 4–17 base pairs (FRESCO et al., 1960; Cox, 1970). Unpaired residues probably occur in such helices and it has been suggested that up to one third of the nucleotides in a "paired" helix can loop-out without disrupting the helix (FRESCO et al., 1960; FINK and CROTHERS, 1972).

A detailed examination of the secondary structure of mammalian rRNA and the length and composition of the helical regions is given by COX (1970). Studies such as these have led to the view that isolated rRNA under conditions of moderate ionic strength contains many regions in which the single-stranded chain doubles back upon itself forming double-stranded hairpin loops which alternate with flexible single-stranded regions as shown in Fig. 6 (FRESCO et al., 1960; GOULD and SIMPKINS, 1969; COX, 1970). This type of structure is consistent with that recently proposed for extensive tracts in *E. coli* 16S rRNA (FELLNER et al., 1972).

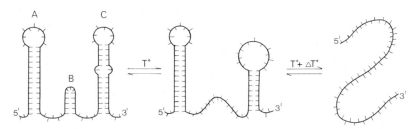

Fig. 6. Possible conformations of rRNA. (A) and (B), hairpin loops containing only one double-helical segment which melt to single-stranded forms at different temperatures due to different lengths or nucleotide composition. (C), hairpin loop with two double-helical segments which melt independently because they are separated by residues which are not base-paired. The base residues are indicated by dashes. (From COX, 1970)

The proposed secondary structure of these fragments was derived from knowledge of the primary sequence, taking into account the particular susceptibility of certain regions of these fragments to ribonuclease. It is notable in making comparisons of this sort that the melting curves of rRNA from various organisms show an increased stability of secondary structure with evolutionary progression; this is paralleled by an increased $G+C$ content (see above), and by increased resistance to RNAase.

To determine whether isolated rRNA and native rRNA *in situ* have similar secondary structures, studies of hyperchromicity (SCHLESSINGER, 1960), optical rotatory dispersion (GOULD and SIMPKINS, 1969) and X-ray diffraction (ZUBAY and WILKINS, 1960) have been carried out on whole ribosomes. These show that the degree of secondary structure of the rRNA *in situ* is similar to that of isolated rRNA in solutions of moderate ionic strength (see ATTARDI and AMALDI, 1970; NOMURA, 1970).

Little is known about the relative arrangement of individual helical regions within the rRNA molecule. It is certain however that the tertiary structure of rRNA in the ribosome is different from that of isolated rRNA, since the effective hydrodynamic volumes of isolated 23S and 16S rRNA are much larger than those of the 50S and 30S ribosome subunits respectively (MIALL and WALKER, 1969). Interaction with ribosomal proteins thus appears necessary for the compact conformation of rRNA in the ribosome. Although the accessibility of rRNA in the ribosome to certain chemicals has indicated that a large proportion of the rRNA is exposed (COTTER et al., 1967), more detailed knowledge of the tertiary folding of rRNA within the ribosome depends on determining the mutual arrangement of RNA and protein. The interested reader is referred to a review of ribosome structure by NOMURA (1970).

b) Low Molecular Weight rRNA. Several models for the structure of 5S RNA have been proposed since the sequence of this molecule was determined (BROWNLEE et al., 1968; CANTOR, 1968; JORDAN, 1971). Most envisage 5S RNA as a ring-like structure containing three short, helical regions, although other base-pairing arrangements are possible. All models appear to agree that base-pairing exists between the two ends of the molecule. These models have arisen by considering the following types of experimental data:

1) Information from optical measurements showing that 5S RNA contains 35–40 base pairs, about two-thirds of which are G—C pairs;

2) information from limited enzyme digestion indicating the pairing state of a particular base or region;

3) the accessibility of certain bases to specific chemical modification (LEE and INGRAM, 1969). Recently the binding of oligonucleotides to 5S RNA has been used to locate the single-stranded regions (LEWIS and DOTY, 1970). JORDAN (1971) has used limited hydrolysis of *E. coli* 5S RNA to determine the most exposed point on the molecule, and has found that this contains a sequence complementary to a region found in all transfer RNAs.

The secondary structure of lRNA appears to be similar to that of the associated high molecular weight rRNA (KING and GOULD, 1970).

c) RNA-Protein Interaction. Although it is now possible to reconstitute functional ribosomes from rRNA and ribosomal proteins (NOMURA, 1970), little is known about how a protein recognizes a particular region of rRNA, or about the nature of the interactions which stabilize this association. For these reasons the isolation and characterization of sites on rRNA which bind ribosomal proteins is now attracting attention. By limited ribonuclease digestion of *E. coli* 16S RNA, large fragments have been isolated which bind particular ribosomal proteins (ZIMMERMANN et al., 1972). Of the six 'core' proteins which bind directly to 16S RNA, five appear to bind to a large fragment which comprises almost the entire 5′-terminal half of the 16S molecule. Only one binds to a similar fragment from the 3′-end. Since the 5′-end of 16S RNA is released first during transcription of precursor RNA, as discussed below, the binding of ribosomal proteins to the nascent rRNA may aid release from the DNA template and permit a concomitant initiation of assembly of the ribosome subunit.

The specificity of binding sites for the individual proteins could be due to either the local conformation of particular nucleotide sequences, or to the interaction of two or more separated sequences maintained in the proper configuration by the three-dimensional folding of the RNA chain. A choice between these two alternatives cannot yet be made, although SCHAUP et al. (1972) found that the binding of a particular ribosomal protein to *E. coli* 16S RNA protects a large region of the rRNA from ribonuclease. The protected RNA consists of several fragments all of which can reassociate with the protein, and this suggests that the protein interacts with more than one portion of the RNA chain.

B. The Synthesis of Ribosomal RNA

1. Eukaryotes

a) The Nucleolus. The major processing steps in the synthesis of mature rRNA and the essential events in ribosome maturation occur in the nucleolus of eukaryotic cells. The DNA template on which transcription occurs (rDNA) is associated with the *nucleolar organizer* region. This region is associated with a limited, fixed number of chromosomes in any tissue and in some eukaryotes it can be identified cytologically as a secondary constriction in the metaphase chromosomes of somatic cells.

The number of nucleoli per somatic cell varies greatly, even within a species, but generally there are between one and ten. The number per somatic cell is probably the same as the number of nucleolar organizer regions within the genome.

Table 1. Multiplicity of genes for 23S–28S and 16S–18S rRNA in various organisms

Tissue or cell type	Multiplicity per genome
HeLa	1,100 (heteroploid)
Rat liver	750 (ploidy not determined)
Xenopus	900 (diploid)
Xenopus	1,600 (diploid)
Drosophila melanogaster	260 (diploid)
Tobacco leaves [a]	1,500–2,000 (diploid)
Saccharomyces cerevisiae	140 (haploid)
Escherichia coli	5–6
Bacillus subtilis	9–10
Bacillus megaterium	35–45

From MADEN (1971).
[a] MATSUDA and SIEGEL (1967); TEWARI and WILDMAN (1968).

b) The Multiplicity of rRNA Genes. Within a somatic cell each nucleolar organizer region contains a number of apparently identical genes for rRNA. Table 1 shows the extent of this multiplicity or redundancy in several representative eukaryotes and also includes comparative data for a number of prokaryotes. This information has been obtained from saturation hybridization experiments between purified rRNA and cellular DNA. Such experiments show that eukaryotes may contain several hundred rRNA gene copies in the nucleolar organizer region. Prokaryotes have roughly an order of magnitude fewer rRNA gene copies. In eukaryotes the multiplicity of genes for 5S RNA, which is coded by extra-nucleolar DNA, is even higher than that for high molecular weight rRNA (BROWN and WEBER, 1968). The presence of multiple copies of genes for rRNA is presumably related to the requirement, in a somatic cell, for rapid production of rRNA molecules.

In the amphibian *Xenopus laevis*, about half of the nucleotide sequence of the coding strand of rDNA is complementary to the two high molecular weight species of rRNA (BIRNSTIEL et al., 1968), which are represented in equimolar proportions. The coding unit has a molecular weight of about 9×10^6 daltons and is repeated about 450 times in each nucleolar organizer region; about 50% of this repeated unit is "spacer" DNA (DAWID et al., 1970). Part is transcribed and is found in the primary product of transcription; the major portion is not transcribed and separates the tandem cistrons from each other. Both transcribed and non-transcribed spacer regions have a very high (approximately 70%) $G+C$ content (DAWID et al., 1970).

The extent of repetition of the coding unit for rRNA varies widely even in a single natural population of organisms such as the amphibian toad *Bufo marinus* (MILLER and BROWN, 1969). This appears to be due to a particular susceptibility of the nucleolar organizer region to spontaneous deletions and additions (RITOSSA, 1968; MILLER and BROWN, 1969). In wild-type *Drosophila* females there are approximately 250 rRNA genes in each nucleolar organizer region of the two X-chromosomes. When this same nucleolar organizer is present in only a single dose, the number of rRNA genes increases in a controlled fashion to approximately 400. This results from a process which has been termed "disproportionate replication"; it occurs progressively in the somatic cell during the normal development of larva to pupa to adult fly and appears to be analogous to gene amplification [see g) below, and TARTOF, 1971].

c) The Primary Transcription Product. In all eukaryotic tissues examined to date, transcription of rDNA by nucleolar RNA polymerases involves the polymerization of ribonucleotides into a precursor polynucleotide of higher molecular weight than the sum of the molecular weights of the two mature rRNA species (see reviews by DARNELL, 1968; ATTARDI and AMALDI, 1970; MADEN, 1971). This precursor molecule has been termed "r-pre-RNA" and "precursor RNA"; it has also been referred to by sedimentation coefficient (*e.g.* "45S RNA" in mammalian cells).

The transcription of precursor RNA has been visualized directly by electron autoradiography in the elegant studies of MILLER and BEATTY (1969). At any particular time, transcription of *Xenopus* rDNA is performed by about 100 polymerase molecules in linear array along the rDNA: these cover about one third of the total length of the gene. The precursor RNA appears to be coated with protein (possibly ribosomal) as it is transcribed. The product of transcription is therefore probably a ribonucleoprotein. Between the rRNA genes there is, according to the micrographs of MILLER and BEATTY, a region which is free of associated polymerase molecules and of nascent RNA; this probably represents the non-transcribed spacer DNA (see below, however, and CASTON and JONES, 1972). These regions are homogeneous within a species but differ greatly in sequence between species (BROWN et al., 1972). The size of the primary transcription product varies between organisms. In general it appears that the more primitive eukaryotes have precursor molecules of lower molecular weight than do the higher eukaryotes such as birds and mammals (PERRY et al., 1970; see also Fig. 9). In addition to the differences in size between organisms, it has also been reported that even within the tissue of a single species, several different size-classes of the presumed nucleolar, high molecular weight precursor RNA can exist (TIOLLAIS et al., 1971; GRIERSON and LOENING, 1972).

From kinetic studies it has been estimated that in mammalian cells the precursor molecule (approximately 4.1 M in molecular weight) is synthesized in about 2.5 minutes (GREENBERG and PENMAN, 1966). Interestingly, in prokaryotes the rate of polymerization of ribonucleotides into both rRNA and messenger RNA is not markedly different from this (BREMER and YUAN, 1968; BREMER and BERRY, 1971). Thus the rate-limiting steps in transcription may be the same in eukaryotes and prokaryotes; this could conceivably be the rate of strand separation of template DNA (BROWN and HASELKORN, 1971).

Methylation of rRNA occurs at the level of the precursor molecule, close to the point of polymerization of ribonucleotides by the RNA polymerase, *i.e.* during, and not after transcription of the precursor. The significance of methylation of rRNA is poorly understood, but it is possible that, occurring as it does in specific nucleotide sequences, it serves to disrupt hydrogen bonding in these areas and thus permits a certain RNA species to adopt a particular overall secondary and tertiary structure. This may be important as a determinant of the specific recognition and processing by nucleases.

d) Processing of Precursor RNA. The precursor RNA is cleaved to form various intermediates of defined size which ultimately give mature rRNA molecules. As already indicated, the precursor RNA interacts with proteins during transcription, and other studies, beyond the scope of this article, indicate that all steps in the cleavage of the precursor probably occur within ribosome precursors which are found primarily in the nucleolus (see MADEN, 1971). The processing of these intermediates during ribosome biosynthesis and that of the RNA molecules which they contain is illustrated schematically in Fig. 7.

The most extensive studies of the processing of precursor molecules have been with free-living cultured cells whose RNA can be radioactively labelled in a relatively simple medium of defined composition. The development of methods

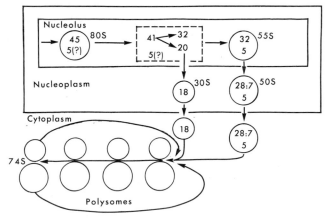

Fig. 7. Schematic representation of ribosome formation in HeLa cells. The numbers outside the particles represent the sedimentation constants of the ribonucleoprotein particles. The numbers inside represent the sedimentation constants of the RNA species which they contain. Only trace amounts of 30S ribonucleoprotein are found in the nucleoplasm. Note (1) that 5S RNA is synthesized in the nucleoplasm but is also found in the nucleolus, associated with the precursor of the large subunit, and (2) that 7S RNA is probably generated during the 32S → 28S conversion. (From MADEN, 1971)

for disrupting cells of this type and of fractionating their subcellular constituents into relatively pure fractions has been an important contribution to these studies. PENMAN's procedures have been commonly adopted in studies with HeLa cells (PENMAN, 1966; WEINBERG et al., 1967) which have provided an important model system in this area of research.

The detailed steps in the processing of precursor RNA have been deduced using a variety of experimental approaches. These include:

1) Treating pulse-labelled cells with actinomycin which stops further RNA synthesis but allows the processing of most species of prelabelled RNA to continue unaffected (cf. however PENMAN, 1966);

2) labelling with methyl-labelled methionine which does not label hetero-geneous nuclear RNA (Chapter 3) and hence reduces the background against which synthesis and processing of precursor RNA is followed;

3) infecting HeLa cells with poliovirus which leads to the accumulation of intermediates which are seen in only trace amounts in normal cells (WILLEMS et al., 1968; WEINBERG and PENMAN, 1970);

4) comparing "fingerprints" of possible precursors and mature RNA species (MADEN et al., 1972);

5) competitive hybridization studies which serve a similar purpose to 4).

When RNA extracted from HeLa cell nucleoli is examined by acrylamide gel electrophoresis, three major species are commonly seen (WEINBERG et al., 1967; WEINBERG and PENMAN, 1968; WILLEMS et al., 1968; WEINBERG and PENMAN, 1970). They have molecular weights of about 4.1 M (45S precursor RNA), 2.1 M (32S RNA) and 1.7 M (28S rRNA). The 45S and 32S species are invariably present in nucleolar preparations but 28S RNA is often present in only trace amounts suggesting that it has only a transient life in the nucleolus and that when present in large amounts this represents cytoplasmic contamination (compare Fig. 7 and Fig. 8). The 3.1 M RNA (seen in Fig. 8) is generally regarded as an intermediate

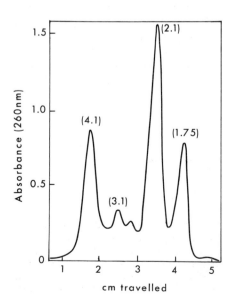

Fig. 8. Gel electrophoresis of HeLa cell nucleolar RNA. Numbers in parentheses indicate molec-ular weights in million daltons. Electrophoresis is from left to right. (From WEINBERG et al., 1967)

in normal processing, but trace amounts of 36S RNA (evident between the 2.1 M and 3.1 M species in Fig. 8) and 24S RNA (see WEINBERG and PENMAN, 1970) are thought to represent products of abnormal cleavages which appear to occur to a limited extent in normal cells, but whose occurrence is stimulated by stresses such as polio infection [see WEINBERG and PENMAN, 1970; Fig. 2(b) of their paper illustrates this proposed alteration in processing in polio-infected cells]. Little if any RNA characteristic of the small ribosomal subunit (0.65 M, 18S) is found in RNA from nucleolar fractions or from pure nuclei.

When HeLa cells are labelled with ^{3}H-uridine for 10 minutes at 37°, a large proportion of incorporation is into 45S (4.1 M) precursor which is synthesized on the rDNA template in the nucleolus. After labelling for 20 minutes this precursor eventually gives rise to 2.1 M RNA and mature 0.65 M rRNA (PENMAN, 1966). It is probable that an RNA intermediate of 3.1 M is an early product of cleaved 4.1 M RNA [see Fig. 9(a)]. The 0.65 M RNA passes relatively quickly (within 30 minutes) to the cytoplasm as a ribonucleoprotein complex of about 40S (the small ribosome subunit); it is therefore found in only trace amounts in the nucleus. The rapid maturation of 0.65 M RNA contrasts with the relatively slow conversion of 2.1 M RNA to 1.75 M RNA; the latter species appears only after a longer period of labelling (PENMAN, 1966).

The essential steps in the maturation pathway generally accepted for HeLa cells are shown in Fig. 9(a). In other eukaryotes the underlying features of processing are similar although quantitative differences exist. Fig. 9(b), (c) and (d) illustrate proposed pathways of rRNA synthesis in insects, *Xenopus* and mung beans respectively. The major points of dissimilarity between different eukaryotes are in 1) the size of the precursor RNA, and in 2) the proportion of the precursor which is lost during processing and is not present in the mature rRNA molecules. From Fig. 9 it can be seen that the precursors have estimated molecular weights of 4.1 M (HeLa), 3.8 M (insect), 2.5–2.6 M (*Xenopus*) and 2.9 M (mung bean leaves), and that the proportion of precursor lost during processing to mature rRNA molecules (of different molecular weights) differs markedly from HeLa

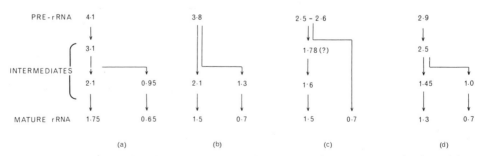

Fig. 9. Steps in the processing of rRNA in various eukaryotes. Numbers represent molecular weights in million daltons.

    (a) HeLa cells (WEINBERG and PENMAN, 1970)

    (b) *Aedes aegypti* cells (DALGARNO et al., 1972)

    (c) *Xenopus* eggs (LOENING et al., 1969)

    (d) Mung bean leaves (GRIERSON and LOENING, 1972)

cells to *Xenopus*. It has been suggested that there is a degree of correlation between the size of the primary transcription unit and the proportion lost during processing (LOENING et al., 1969). Thus the larger HeLa precursor loses about 40% of its weight during processing whereas the smaller 2.6 M molecule in *Xenopus* loses far less. However, since it is possible that cleavage of the precursor occurs during its synthesis (GRIERSON and LOENING, 1972), "small precursors" may, in some instances, represent intermediates rather than primary transcription products. This view is supported by studies of RNA processing in nuclei isolated from the amphibian *Rana pipiens* (CASTON and JONES, 1972). As in *Xenopus*, they find the expected 2.6 M RNA in relative abundance; however a 4.4 M RNA with the properties of a precursor species is also found in limited amounts. Their results suggest that the primary transcription product is cleaved very rapidly to give more stable molecules which are intermediates in further processing, suggesting an analogy with processing of prokaryotic rRNA (see below).

It is probable that lRNA also derives from the large precursor. The kinetics of its synthesis parallel those of 28S RNA in HeLa cells and 26S RNA in yeast (KNIGHT and DARNELL, 1967; UDEM and WARNER, 1972). This suggests that when the immediate precursor RNA is cleaved to give 26S or 28S RNA, there is a simultaneous scission introduced which generates a small RNA molecule which is retained in association with the larger one (see Fig. 10). Synthesis according to a scheme of this type seems inherently more likely than by a mechanism in which a small species is synthesized independently of 28S RNA and, subsequent to its synthesis, associates by hydrogen-bonding with the larger molecule.

As discussed earlier, the cytoplasmic ribosomes of certain eukaryotes (insects, amoeba) may contain rRNA harbouring "hidden breaks". In cultured *Aedes aegypti* cells this break is introduced after cleavage of 1.5 M (26S) RNA from its immediate precursor (2.1 M). Thus newly synthesized, pulse-labelled 26S RNA is free of any central scission but when labelled for a longer time effectively all 26S RNA has an introduced break (DALGARNO, unpublished results). The scission in insect 26S RNA results apparently from the action of a specific ribonuclease, since when 26S RNA is freed of lRNA, and the two remaining 3'-ends are reacted with labelled isoniazid, the T_1-ribonuclease digestion products contain two identical terminal sequences (SHINE, HUNT, and DALGARNO, unpublished results).

e) Non-Ribosomal Sequences in Precursor RNA. Although the sum of the molecular weights of the two mature rRNA molecules is significantly less than the molecular weight of the precursor, the sequences removed during processing have not been detected as discrete RNA species. This suggests either or both of the following possibilities: 1) that the enzymes which process the precursor are exonucleases; 2) that any large fragments produced from endonuclease action are rapidly degraded.

Although the non-ribosomal sequences have not been isolated, some information about their properties can be deduced. The base composition and extent of methylation of precursor RNA can be compared with that of mature rRNA species and the properties of the non-ribosomal sections determined by difference. Using this approach it is clear that these sequences differ from rRNA in that:

1) They show a far lower degree of methylation (WEINBERG et al., 1967; WEINBERG and PENMAN, 1970; VAUGHAN et al., 1967);

2) they have a considerably higher G+C content (AMALDI and ATTARDI, 1968; JEANTEUR et al., 1968; WILLEMS et al., 1968);

3) they are very low or lacking in pseudouridine (AMALDI and ATTARDI, 1968; JEANTEUR et al., 1968).

It seems likely that the properties of these regions are important determinants in processing (see PERRY and KELLEY, 1972).

f) The Arrangement of rRNA Species within the Precursor Molecule. Although the mechanism of processing precursor RNA has received considerable attention, the relative disposition of the large and small rRNA molecules with respect to the 5′- and 3′-ends of the precursor is not known with certainty. Kinetic data on the synthesis of cytoplasmic rRNA in *Euglena gracilis* (BROWN and HAZELKORN, 1971) and studies in HeLa cells with the inhibitor cordycepin (SIEV et al., 1969) suggest that the small rRNA is found at the 5′-end of the precursor. REEDER and BROWN (1970) have purified rDNA from *Xenopus* and demonstrated that *E. coli* RNA polymerase preferentially initiates transcription at the normal initiation site within the sequence coding for 18S rRNA. The kinetics of appearance of label in 18S and 28S RNA suggest that 18S RNA is at the 5′-terminus of the precursor. On the other hand, kinetic studies of RNA synthesis in isolated *Rana pipiens* nuclei (CASTON and JONES, 1972), and sequence studies on rRNA and its precursors in Novikoff hepatoma cells (CHOI and BUSCH, 1970; EGAWA et al., 1971; SEEBER and BUSCH, 1971) favour the view that 18S RNA is at the 3′-end.

As with the rRNA species, the position of the spacer RNA is not known. It is not clear whether the immediate precursors of 18S and 28S RNA are degraded at the 3′-end, the 5′-end, or both. However, the results of PERRY and KELLEY (1972) suggest that a nucleolar 3′-exonuclease may play a critical role in processing

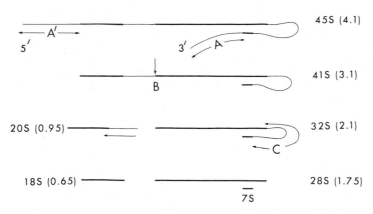

Fig. 10. Possible arrangement of rRNA and spacer RNA within the HeLa precursor molecule. Numbers represent molecular weights in million daltons. The mature ribosomal RNA sequences are depicted as heavy lines. Removal of non-ribosomal sequences A and A′ would generate 41S RNA containing both 28S and 18S sequences. Cleavage at B would then generate 32S and 20S RNA. Cleavage and removal of C would generate 28S RNA with attached 7S RNA. Cleavage at B before removal of A and A′ could generate 36S and 24S molecules; the latter may be capable of giving rise to mature rRNA (from MADEN, 1971). Slightly different models for the HeLa precursor have also been proposed (WEINBERG and PENMAN, 1970)

L-cell precursor RNA. This enzyme preferentially degrades methyl-deficient sequences removing approximately 600 nucleotides from the 3′-ends of both 45S and 32S precursors. Interestingly there is a great decelleration in rate when the enzyme reaches certain internal points in the molecule.

Although substantial gaps in our knowledge exist, this has not discouraged the construction of 'topographical' models of the precursor molecule (see MADEN, 1971; PERRY and KELLEY, 1972). Fig. 10 represents one such model for HeLa cell 45S RNA which demonstrates schematically how processing could give rise to the intermediates observed by cleavage of the precursor at certain nuclease-sensitive sites (MADEN, 1971).

g) *Amplification of rRNA Genes.* In the maternal germ cells of certain amphibia and insects, the number of nucleoli is greatly increased to permit a rapid synthesis of ribosomes required for utilization by the early embryo (MILLER, 1966; BROWN and DAWID, 1968). Thus the *Xenopus* oocyte contains more than 1,000 free-floating or extra-chromosomal nucleoli which permit the accumulation of approximately 10^{12} ribosomes in each mature egg. In a somatic cell this would require 10^5 days of normal rRNA synthesis, assuming no selective replication of rDNA. Reduction of this time by a factor of more than 1,000 (the number of nucleoli present), brings the required time down to 3–6 months (see DAVIDSON, 1968).

Although they are apparently identical in all other ways, chromosomal and extrachromosomal nucleoli differ in the base composition of the constituent DNA. Chromosomal rDNA contains 4.5% 5-methyl deoxycytidine which is found on both coding and non-coding DNA chains; amplified DNA does not contain this methylated nucleoside (DAWID et al., 1970). The significance of this difference in base composition is not known.

In *Xenopus* it is likely that the template for amplification is the chromosomal nucleolar organizer region of the oogonium (BROWN and BLACKLER, 1972). The stage at which amplification is thought to occur has retreated further and further into the early stages of oogonium development as methods of detecting redundant sequences have improved. It is now thought that amplification occurs in two phases: 1) very early in the oogonial stage, and 2) at meiotic prophase when there is a sudden and more substantial synthesis of rDNA (see MACGREGOR, 1972).

Amplification in *Xenopus* results in the formation of thousands of closed DNA rings which vary in size; some are longer, and some shorter than nucleolar organizer DNA. Synthesis is probably by a "cascade" mechanism since the instantaneous rate of formation of extra-chromosomal nucleoli is dependent on the number present at that particular time (MACGREGOR, 1972). It has been suggested that chromosomal rDNA may cyclize, be excised, and then replicate in similar fashion to that of some DNA bacteriophages (MACGREGOR, 1972). Alternatively, it has been proposed that amplification occurs by reverse transcription of rDNA from rRNA by an RNA-dependent DNA polymerase. An enzyme of the appropriate specificity has been found in *Xenopus* oocytes (CRIPPA and TOCCHINI-VALENTINI, 1971; BROWN and TOCCHINI-VALENTINI, 1972; MAHDAVI and CRIPPA, 1972). In addition, DNA synthesis in the cap region of *Xenopus* oocytes is inhibited by a derivative of rifampicin which inhibits "reverse transcriptase" (FICQ and BRACHET, 1971). Whatever mechanism is involved, the

circular DNA molecules can apparently reattach to the nucleolar organizer
region of *Xenopus*, suggesting that "recombination" may occur between replicated
extranucleolar DNA and chromosomal rDNA.

h) The Biosynthesis of 5S RNA. 5S genes are located in the chromosomal
DNA in the extra-nucleolar region of the nucleus. 5S DNA has been isolated
from *Xenopus laevis*; on partial denaturation it appears in the electron microscope
as a long repeated sequence in which high GC and low GC regions alternate
(Brown et al., 1971). About 17% of the length of each repeating sequence
(probably within the high GC region) codes for 5S RNA, *i.e.* about 83% of the
length is apparently spacer DNA. Hence the spacer represents a larger proportion
of 5S DNA than it does of the ribosomal RNA genes. Brown et al. (1971) estimate
that about 0.7% of the total nuclear DNA codes for 5S RNA; this represents
about 24,000 5S genes for each haploid cell. Although 5S RNA genes are highly
clustered, in HeLa cells these clusters occur on a number of chromosomes (Aloni
et al., 1971) whereas in *Drosophila* they are found on a single chromosome
(Wimber and Stephenson, 1970).

In HeLa cells, mature 5S RNA has 5'-terminal pppGp indicating that transcription of each 5S molecule is from a separate cistron and that no post-transcriptional cleavage occurs at the 5'-terminus (Hatlen et al., 1969). The absence
of a precursor for 5S RNA in eukaryotes contrasts with the existence of a precursor
of slightly larger size than 5S RNA in prokaryotes, as described below.

After synthesis in the nucleoplasm, 5S RNA migrates to the nucleolus, where
it is found associated with the precursor of the large ribosome subunit in a 1:1
ratio with 28S rRNA (Knight and Darnell, 1967); the same 1:1 ratio is found
in the mature ribosome subunit and in monomeric ribosomes. Some 5S RNA
accumulates in the nucleus as a pool which contains 20–25% of the total cellular
5S RNA. This contrasts with 28S RNA; only 1.5–2% of the total cellular 28S
RNA is found in the nucleus either as 28S RNA *per se*, or as its precursor
(Vaughan et al., 1967).

It has already been pointed out that in eukaryotes, the degree of redundancy
of 5S genes is considerably higher than that of 28S and 18S genes. This is perhaps
surprising in view of the fact that these three species are found in an equimolar
ratio in ribosomes; it may indicate that under normal conditions most 5S DNA
sites are not transcribed, or that they are transcribed slowly. During oogenesis,
however, additional 5S genes may be transcribed, or the rate of transcription
may increase to match the transient gene amplification of the 28S and 18S genes.
5S RNA synthesis is co-ordinated with that of the high molecular weight species.
Thus in the anucleolate mutant of *Xenopus* (which lacks the 28S and 18S genes
but retains the chromosomal 5S genes), no 5S RNA synthesis occurs (Brown
and Weber, 1968). During normal development of the *Xenopus* oocyte there is
a coordinate acceleration of synthesis of all three species although there is no
amplification of 5S genes (Brown and Dawid, 1968). This coordinate synthesis
may break down however: on treating L-cells with low concentrations of actinomycin, synthesis of 28S and 18S RNA can be inhibited without a corresponding
decrease in 5S RNA which accumulates in the nucleoplasm (Perry and Kelley,
1968). The mechanism which permits the coordinate synthesis of high molecular
weight rRNA and 5S RNA is not known.

The full significance of these studies on 5S RNA will only be apparent when its role is known. At present it is only clear that 5S RNA, as part of a ribonucleoprotein complex, is essential in some way for the biological activity of reconstituted bacterial ribosomes; no comparable information exists for eukaryotic cells.

i) RNA Polymerases and rRNA Synthesis. Ribosomal RNA and messenger RNA have different roles, are synthesized at different overall rates, and are made in separate regions of the nucleus by enzymes which differ in their properties. RNA polymerases present in the nucleoplasm and the nucleolus are discussed in Chapter 2.

2. Prokaryotes

a) Redundancy and Linkage of RNA Genes. The bacterial chromosome, a single duplex of circular DNA, is not enclosed within an organelle and thus, in contrast to eukaryotes, there is no morphologically recognizable structure which represents a site of rRNA synthesis and ribosome maturation. There are 5–10 copies of each rDNA species per genome in *Bacillus subtilis;* this is roughly an order of magnitude less redundancy than exists for genes for rRNA in eukaryotes (Table 1) and represents physically about 0.5% of the coding DNA of the bacterial chromosome. The 16S and 23S genes are closely linked (COLLI et al., 1971) and probably alternate with each other as in eukaryotes.

In at least two types of bacteria, the tandem cistrons of rDNA are not clustered in a single region of the bacterial chromosome (SMITH et al., 1968; PURDOM et al., 1970). In *Proteus mirabilis* for example, those regions which contain rDNA are separated from each other by fairly extensive regions of "chromosomal" DNA; these "spacer" regions have molecular weights of $1—3 \times 10^7$ daltons, and presumably contain genes for other species of RNA (PURDOM et al., 1970).

In *Bacillus subtilis,* 5S genes make up part of each tandem set of rDNA cistrons, which is in contrast to eukaryotes. The 5S gene is more closely linked to the 23S gene than to the 16S gene, suggesting the following arrangement: 16S—23S—5S—space—16S—23S—5S—space (COLLI et al., 1971).

b) The Existence of a Single Transcription Unit for Prokaryotic rRNAs. Until recently it was generally believed that the transcription of 16S and 23S RNAs occurred independently of each other. It is now almost certain that 16S and 23S RNAs are initially transcribed from a single cistron with only one site of attachment for RNA polymerase (PETTIJOHN et al., 1970; DOOLITTLE and PACE, 1971; KOSSMAN et al., 1971; GUPTA and SINGH, 1972). In arriving at this conclusion workers have, in general, considered three possible arrangements of the genes for 16S and 23S rRNA (Fig. 11) and attempted to design experiments to distinguish between them. Model A has separate transcriptional units for 16S and 23S RNA, while B and C are tandem models, with a single initiation site, in which 16S is transcribed before (B) or after (C) transcription of 23S RNA. To test these models STAMATO and PETTIJOHN (1971) developed a technique which permits the synchronous re-initiation of new RNA chains at promoter sites. Under these conditions label appears in 16S RNA before it appears in 23S RNA so that either the genes are independently transcribed, or they are present as

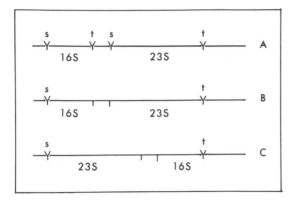

Fig. 11. Possible arrangement of rRNA genes in the prokaryotic cell. The letters *s* and *t* respectively represent starting and terminating sites for the RNA polymerase. It is assumed that transcription is from left to right. Genes for 5S RNA have not been included.
A. Separate transcriptional units for 16S and 23S RNA. B and C. Combined or tandem transcriptional units for 16S and 23S RNA. 16S RNA is transcribed first in B and last in C.
(After KOSSMAN et al., 1971)

one unit in which the 16S gene is transcribed first. Thus model C is eliminated. Models A and B differ in their predictions. Model A predicts that soon after synchronous re-initiation of RNA synthesis, nascent RNA will contain equal amounts of the nucleotide sequences of 16S and 23S RNA; on the other hand B predicts that such chains will have mainly 16S RNA sequences. Competitive hybridization experiments designed to test these predictions (KOSSMAN et al., 1971) support the view that in *E. coli* the two rRNAs are transcribed as a single transcriptional unit, that the 16S gene is at the 5′-end of this unit, and that the 16S RNA is cleaved from the nascent large precursor before the polymerase has completed transcription of 23S rRNA at the 3′-end. Direct evidence for the existence of a large precursor molecule exists. PETTIJOHN et al. (1970) have found an RNA species of the appropriate size which is generated during *in vitro* transcription of isolated rDNA from *E. coli*. In addition, a large RNA species is found in *E. coli* grown in the presence of 8-azaguanine, a purine analogue which is incorporated into RNA and presumably affects normal processing. Experiments with rifampicin, a specific inhibitor of RNA chain initiation, but not of chain elongation, also support the view that 16S and 23S rRNA are transcribed as a single unit (BREMER and BERRY, 1971). On the other hand, experiments in which the pattern of inhibition of 16S and 23S synthesis by actinomycin D has been examined, have led some workers to conclude that the two species are transcribed independently. In spite of this the accumulated evidence indicates that in bacteria, as in eukaryotes, the two rRNA genes are transcribed as part of a single large unit.

c) *Processing the Precursor.* It is now evident that 16S rRNA is released from the primary transcription unit as a molecule of slightly higher molecular weight (denoted p16) than the mature 16S RNA (denoted m16); a similar situation probably holds for 23S RNA. As already mentioned above, cleavage to form p16 probably occurs before the polymerase has reached the 3′-end of the precursor molecule; this explains the difficulty in obtaining direct evidence for the existence of a high molecular weight primary transcription unit.

The first indications for the existence of p16 and p23 came from the work of HECHT and WOESE (1968) with *Bacillus subtilis* and ADESNIK and LEVINTHAL (1969) with *E. coli*. They demonstrated the existence, in pulse-labelled bacterial

RNAs, of two rapidly-labelled species of slightly higher molecular weight than the respective mature 16S and 23S molecules. When *E. coli* was pulse-labelled with ^{3}H-uridine and then treated with actinomycin, p16 was converted to an RNA corresponding to m16; similarly p23 appeared to be converted to m23. This result implied a precursor-product relationship between p16 and m16 and between p23 and m23.

In *E. coli*, potassium-dependent ribonuclease II (one of six ribonucleases characterized in this bacterium) has been implicated in the conversion of p16 to m16 (YUKI, 1971; CORTE et al., 1971); this conversion probably occurs within a ribonucleoprotein precursor of the small ribosome subunit. Thus when *E. coli* with a temperature-sensitive mutation in the gene for ribonuclease II is in-cubated at the restrictive temperature (43°), ribonucleoproteins accumulate which are similar to the natural ribosome precursors. *In vitro*, purified ribo-nuclease II converts both the ribonucleoprotein precursor of the 30S subunit and p16 RNA itself to the respective mature form; a small oligonucleotide (5–6S) is released in both instances (CORTE et al., 1971). Sequence data show that the maturation of p16 involves the removal of about 10% of the polynucleotide chain; this occurs at both the 5'- and 3'-termini (see SOGIN et al., 1971; BROWNLEE and CARTWRIGHT, 1971; LOWRY and DAHLBERG, 1971; HAYES et al., 1971). No strong evidence exists to implicate ribonuclease II in the maturation of 23S RNA even though, at the restrictive temperature, precursors of 50S subunits accumulate. These contain m23 RNA, suggesting that ribonuclease II is not required for the processing of p23 RNA. The accumulation of 50S subunits may mean that the maturation of the 50S precursor depends on a parallel maturation of the 30S precursor (see NOMURA, 1970).

The significance of the fact that mature rRNA is made, in both eukaryotes and prokaryotes, from high molecular weight precursors is unknown; it has been speculated however (NOMURA, 1970) that it relates to the control of ribosome maturation.

d) 5S RNA. Several lines of evidence indicate that in prokaryotes, 5S RNA is cleaved from a slightly larger precursor molecule (HECHT et al., 1968; JORDAN et al., 1970). When RNA is extracted from pulse-labelled *E. coli*, a species of 5S RNA is obtained which is 1–3 nucleotides longer at the 5'-end than the 5S molecule isolated from ribosomes (JORDAN et al., 1970); this species is apparently the same as the 5S RNA which accumulates in chloramphenicol-treated *E. coli* (JORDAN et al., 1971) and, from its labelling kinetics and subcellular localization, behaves as a precursor of mature 5S RNA. None of the precursors bear a di- or triphosphate group at the 5'-end (FEUNTEUN et al., 1972) contrasting with the situation in eukaryotes. It is possible that p23 RNA gives rise to the precursor of 5S RNA during the conversion of p23 to m23 RNA (see PACE et al., 1970). The maturation of 5S RNA takes place on the ribosome by the stepwise removal of nucleotides from the 5'-end of the 5S precursor (JORDAN et al., 1970; FEUNTEUN et al., 1972). Cleavage may therefore require a specific, ribosome-bound, 5'-exo-nuclease, as yet unidentified, or could be due to ribonuclease II which, although mainly acting in the $3' \rightarrow 5'$ direction, does have some endonuclease activity (SPAHR, 1964).

References

Reviews

Attardi, G., Amaldi, F.: Ann. Rev. Biochem. **39**, 183 (1970).
Darnell, J.E.: Bact. Rev. **32**, 262 (1968).
Davidson, E.H.: Gene activity in early development. New York: Academic Press 1968.
Gilham, P.T.: Ann. Rev. Biochem. **39**, 227 (1970).
Macgregor, H.C.: Biol. Rev. **47**, 177 (1972).
Maden, B.E.H.: Progr. Biophys. Mol. Biol. **22**, 127 (1971).
Nomura, M.: Bact. Rev. **34**, 228 (1970).

Other References

Adesnik, M., Levinthal, C.: J. Mol. Biol. **46**, 281 (1969).
Allet, B., Spahr, P.F.: Eur. J. Biochem. **19**, 250 (1971).
Aloni, Y., Hatlen, L.E., Attardi, G.: J. Mol. Biol. **56**, 555 (1971).
Amaldi, F., Attardi, G.: J. Mol. Biol. **33**, 737 (1968).
Aronson, A.I.: J. Mol. Biol. **5**, 453 (1962).
Averner, M.J., Pace, N.R.: J. Biol. Chem. **247**, 4491 (1972).
Birnsteil, M., Spiers, J., Purdom, I., Jones, K.: Nature **219**, 454 (1968).
Bremer, H., Berry, L.: Nature New Biol. **234**, 81 (1971).
Bremer, H., Yuan, D.: J. Mol. Biol. **38**, 163 (1968).
Brown, D.D., Blackler, A.W.: J. Mol. Biol. **63**, 75 (1972).
Brown, D.D., Dawid, I.B.: Science **160**, 272 (1968).
Brown, D.D., Weber, C.S.: J. Mol. Biol. **34**, 661 (1968).
Brown, D.D., Wensink, P.C., Jordan, E.: Proc. Nat. Acad. Sci. U.S. **68**, 3175 (1971).
Brown, D.D., Wensink, P.C., Jordan, E.: J. Mol. Biol. **63**, 57 (1972).
Brown, R.D., Haselkorn, R.: J. Mol. Biol. **59**, 491 (1971).
Brown, R.D., Tocchini-Valentini, G.P.: Proc. Nat. Acad. Sci. U.S. **69**, 1746 (1972).
Brownlee, G.G., Cartwright, E.: Nature New Biol. **232**, 50 (1971).
Brownlee, G.G., Sanger, F.: J. Mol. Biol. **23**, 337 (1967).
Brownlee, G.G., Sanger, F., Barrell, B.G.: J. Mol. Biol. **34**, 379 (1968).
Cantor, C.R.: Proc. Nat. Acad. Sci. U.S. **59**, 478 (1968).
Caston, J.D., Jones, P.H.: J. Mol. Biol. **69**, 19 (1972).
Choi, Y.C., Busch, H.: J. Biol. Chem. **245**, 1954 (1970).
Colli, W., Smith, I., Oishi, M.: J. Mol. Biol. **56**, 117 (1971).
Corte, G., Schlessinger, D., Longo, D., Venkov, P.: J. Mol. Biol. **60**, 325 (1971).
Cotter, R.I., McPhie, P., Gratzer, W.B.: Nature **216**, 864 (1967).
Cox, R.A.: Biochem. J. **117**, 101 (1970).
Crippa, M., Tocchini-Valentini, G.P.: Proc. Nat. Acad. Sci. U.S. **68**, 2769 (1971).
Dalgarno, L., Hosking, D., Shen, C.: Eur. J. Biochem. **24**, 498 (1972).
Dawid, I.B., Brown, D.D., Reeder, R.H.: J. Mol. Biol. **51**, 341 (1970).
DeWachter, R., Fiers, W.: J. Mol. Biol. **30**, 507 (1967).
Doolittle, W.F., Pace, N.R.: Proc. Nat. Acad. Sci. U.S. **68**, 1786 (1971).
Doty, P., Boedtker, H., Fresco, J.R., Haselkorn, R., Litt, M.: Proc. Nat. Acad. Sci. U.S. **45**, 482 (1959).
Egawa, K., Choi, Y.C., Busch, H.: J. Mol. Biol. **56**, 565 (1971).
Ehresmann, C., Fellner, P., Ebel, J.P.: Nature **227**, 1321 (1970).
Fellner, P.: Eur. J. Biochem. **11**, 12 (1969).
Fellner, P., Ehresmann, C., Ebel, J.P.: Nature **225**, 26 (1970).
Fellner, P., Ehresmann, C., Stiegler, P., Ebel, J.P.: Nature New Biol. **239**, 1 (1972).
Feunteun, J., Jordan, B.R., Monier, R.: J. Mol. Biol. **70**, 465 (1972).
Ficq, A., Brachet, J.: Proc. Nat. Acad. Sci. U.S. **68**, 2774 (1971).
Fink, T.R., Crothers, D.M.: J. Mol. Biol. **66**, 1 (1972).
Forget, B.G., Weissmann, S.H.: Science **158**, 1695 (1967).
Fresco, J.R., Alberts, B.M., Doty, P.: Nature **188**, 98 (1960).
Glitz, D.G., Sigman, D.S.: Biochemistry **9**, 3433 (1970).

GOULD, H.J., SIMPKINS, M.: Biopolymers **7**, 223 (1969).
GREENBERG, H., PENMAN, S.: J. Mol. Biol. **21**, 527 (1966).
GREENBERG, J.R.: J. Mol. Biol. **46**, 85 (1969).
GRIERSON, D., LOENING, U.E.: Nature New Biol. **235**, 80 (1972).
GUPTA, R.S., SINGH, U.N.: Biochim. Biophys. Acta **277**, 567 (1972).
HATLEN, L., AMALDI, F., ATTARDI, G.: Biochemistry **8**, 4989 (1969).
HAYES, F., HAYES, D., FELLNER, P., EHRESMANN, C.: Nature New Biol. **232**, 54 (1971).
HECHT, N.B., BLEYMAN, M., WOESE, C.: Proc. Nat. Acad. Sci. U.S. **59**, 1278 (1968).
HECHT, N.B., WOESE, C.: J. Bacteriol. **95**, 986 (1968).
HUNT, J.A.: Biochem. J. **95**, 541 (1965).
HUNT, J.A.: Biochem. J. **120**, 353 (1970).
JEANTEUR, P., AMALDI, F., ATTARDI, G.: J. Mol. Biol. **33**, 757 (1968).
JORDAN, B.R.: J. Mol. Biol. **55**, 423 (1971).
JORDAN, B.R., FEUNTEUN, J., MONIER, R.: J. Mol. Biol. **50**, 605 (1970).
JORDAN, B.R., FORGET, B.G., MONIER, R.: J. Mol. Biol. **55**, 407 (1971).
KHYM, J.X., UZIEL, M.: Biochemistry **7**, 422 (1968).
KING, H.W.S., GOULD, H.: J. Mol. Biol. **51**, 687 (1970).
KNIGHT, E., Jr., DARNELL, J.E.: J. Mol. Biol. **28**, 491 (1967).
KOSSMAN, C.R., STAMATO, T.D., PETTIJOHN, D.E.: Nature New Biol. **234**, 102 (1971).
KURLAND, C.G.: J. Mol. Biol. **2**, 83 (1960).
LANE, B.G., TAMAOKI, T.: J. Mol. Biol. **27**, 335 (1967).
LEE, J.C., INGRAM, V.M.: J. Mol. Biol. **41**, 431 (1969).
LEWIS, J.B., DOTY, P.: Nature **225**, 5110 (1970).
LOENING, U.E.: J. Mol. Biol. **38**, 355 (1968).
LOENING, U.E., JONES, K.W., BIRNSTEIL, M.L.: J. Mol. Biol. **45**, 353 (1969).
LOWRY, C.V., DAHLBERG, J.E.: Nature New Biol. **232**, 52 (1971).
MADEN, B.E.H., SALIM, M., SUMMERS, D.F.: Nature New Biol. **237**, 5 (1972).
MAHDAVI, V., CRIPPA, M.: Proc. Nat. Acad. Sci. U.S. **69**, 1749 (1972).
MATSUDA, K., SIEGEL, A.: Proc. Nat. Acad. Sci. U.S. **58**, 673 (1967).
MIALL, S.H., WALKER, F.O.: Biochim. Biophys. Acta **174**, 551 (1969).
MIDGLEY, J.E.M.: Biochim. Biophys. Acta **95**, 232 (1965a).
MIDGLEY, J.E.M.: Biochim. Biophys. Acta **108**, 348 (1965b).
MIDGLEY, J.E.M., McILREAVY, D.J.: Biochim. Biophys. Acta **145**, 5 (1967).
MILLER, O.L.: Nat. Cancer Inst. Monograph **23**, 53 (1966).
MILLER, O.L., BEATTY, B.R.: Science **164**, 955 (1969).
MILLER, O.L., BROWN, D.D.: Chromosoma **28**, 430 (1969).
MOLLER, W.: Nature **222**, 979 (1969).
NEMER, M., INFANTE, A.A.: J. Mol. Biol. **27**, 73 (1967).
PACE, B., PETERSON, R.L., PACE, N.R.: Proc. Nat. Acad. Sci. U.S. **65**, 1097 (1970).
PAYNE, P.I., DYER, T.A.: Nature New Biol. **235**, 145 (1972).
PENE, J.J., KNIGHT, E., DARNELL, J.E.: J. Mol. Biol. **33**, 609 (1968).
PENMAN, S.: J. Mol. Biol. **17**, 117 (1966).
PERRY, R.P., CHENG, T.Y., FREED, J.J., GREENBERG, J.R., KELLEY, D.E., TARTOF, K.D.: Proc. Nat. Acad. Sci. U.S. **65**, 609 (1970).
PERRY, R.P., KELLEY, D.E.: J. Cell. Physiol. **72**, 235 (1968).
PERRY, R.P., KELLEY, D.E.: J. Mol. Biol. **70**, 265 (1972).
PETTIJOHN, D.E., STONINGTON, G.O., KOSSMAN, C.R.: Nature **228**, 235 (1970).
PURDOM, I., BISHOP, J.O., BIRNSTEIL, M.L.: Nature **227**, 239 (1970).
RANDERATH, K.: Anal. Biochem. **34**, 188 (1970).
REEDER, R.H., BROWN, D.D.: J. Mol. Biol. **51**, 361 (1970).
RITOSSA, F.M.: Proc. Nat. Acad. Sci. U.S. **60**, 509 (1968).
SCHAUP, H.W., SOGIN, M., WOESE, C., KURLAND, C.G.: Mol. Gen. Genetics **114**, 1 (1972).
SCHLESSINGER, D.: J. Mol. Biol. **2**, 92 (1960).
SEEBER, S., BUSCH, H.: J. Biol. Chem. **246**, 7144 (1971).
SHINE, J., DALGARNO, L.: J. Mol. Biol. **75**, 57 (1973).
SIEV, M., WEINBERG, R., PENMAN, S.: J. Cell. Biol. **41**, 510 (1969).
SMITH, I., DUBNAU, D., MORELL, P., MARMUR, J.: J. Mol. Biol. **33**, 123 (1968).

SOGIN, M., PACE, B., PACE, N. R., WOESE, C. R.: Nature New Biol. **232**, 48 (1971).
SPAHR, P. F.: J. Biol. Chem. **239**, 3716 (1964).
STAMATO, T. D., PETTIJOHN, D. E.: Nature New Biol. **234**, 99 (1971).
STEVENS, A. R., PACHLER, P. F.: J. Mol. Biol. **66**, 225 (1972).
SUGIURA, M., TAKANAMI, M.: Proc. Nat. Acad. Sci. U. S. **58**, 1595 (1967).
TAKANAMI, M.: J. Mol. Biol. **23**, 135 (1967).
TARTOF, K. D.: Science **171**, 294 (1971).
TENER, G. M.: In Methods in enzymology, vol. 12, part A, p. 398. New York: Academic Press 1967.
TEWARI, K. K., WILDMAN, S. G.: Proc. Nat. Acad. Sci. U. S. **59**, 569 (1968).
TIOLLAIS, P., GALIBERT, F., BOIRON, M.: Proc. Nat. Acad. Sci. U. S. **68**, 1117 (1971).
UDEM, S. A., WARNER, J. R.: J. Mol. Biol. **65**, 227 (1972).
VAUGHAN, M. H., SOEIRO, R., WARNER, J. R., DARNELL, J. E.: Proc. Nat. Acad. Sci. U. S. **58**, 1527 (1967).
WEINBERG, R. A., LOENING, U. E., WILLEMS, M., PENMAN, S.: Proc. Nat. Acad. Sci. U. S. **58**, 1088 (1967).
WEINBERG, R. A., PENMAN, S.: J. Mol. Biol. **38**, 289 (1968).
WEINBERG, R. A., PENMAN, S.: J. Mol. Biol. **47**, 169 (1970).
WHITFELD, P. R.: Biochem. J. **58**, 390 (1954).
WILLEMS, M., WAGNER, E., LAING, R., PENMAN, S.: J. Mol. Biol. **32**, 211 (1968).
WILLIAMSON, R., BROWNLEE, G. G.: FEBS Lett. **3**, 306 (1969).
WIMBER, D. E., STEPHENSON, D. M.: Science **170**, 639 (1970).
YUKI, A.: J. Mol. Biol. **56**, 439 (1971).
ZAMECNIK, P. C., STEPHENSON, M. L., SCOTT, J. F.: Proc. Nat. Acad. Sci. U. S. **46**, 811 (1960).
ZIMMERMAN, R. A., MUTO, A., FELLNER, P., EHRESMANN, C., BRANLANT, C.: Proc. Nat. Acad. Sci. U. S. **69**, 1282 (1972).
ZUBAY, G., WILKINS, M. H. F.: J. Mol. Biol. **2**, 105 (1960).

CHAPTER 7

Translation of Messenger RNA

G. D. CLARK-WALKER

Introduction

The term translation is used herein to mean the biosynthesis of a protein by assembly of amino acids into a sequence specified by a messenger RNA (mRNA) template. Messenger RNA, produced from a DNA template with the aid of DNA-dependent RNA polymerase, and carrying the genetic information of that DNA, is often a transient intermediate between the stable genetic material and the final gene product, protein. However the final product from some genes is not protein but other RNA. Two types of this RNA are transfer RNA (tRNA) and ribosomal RNA (rRNA) and both are involved in the translation of mRNA.

The translation process can conveniently be separated into three steps: initiation, elongation and termination. Both the initiation and termination steps are unique events in the synthesis of any protein and occur at particular sites on the mRNA molecule determined by specific triplet codons. Elongation involves the repetitive synthesis of a peptide bond by the attachment of an amino acid to the growing peptide chain and the shift, by the distance of three nucleotide bases, of the ribosomal complex towards the 3′ end of the mRNA. These three steps in the translation of mRNA will be treated separately below.

Translation of mRNA in prokaryotic organisms occurs on ribosomes which have a sedimentation coefficient of about 70S (PETERMAN, 1964). In eukaryotes, the majority of proteins are synthesized in the cytosol on 80S ribosomes, but it is now recognized that in these organisms mitochondria and chloroplasts have their own machinery for protein synthesis. This was initially established by both *in vivo* and *in vitro* studies showing that the translation process in these organelles differs from the cytoplasmic process. For instance in growing yeast it has been shown that synthesis of a mitochondrial protein, cytochrome aa_3, is inhibited by chloramphenicol and other antibiotics which inhibit prokaryotic translation (CLARK-WALKER and LINNANE, 1966) and this can be correlated with an *in vitro* inhibition of amino acid incorporation by isolated mitochondria (LAMB et al., 1968). This and similar observations with chloroplasts and mitochondria from other organisms implies that the translation process in these organelles and in prokaryotes differs from that in the eukaryote cytoplasm. Although this supposition has proved to be true in certain details, in general the translation mechanisms are similar (VAZQUEZ et al., 1969; CIFERRI and PARISI, 1970). Accordingly the detailed events in translation will be discussed for the bacterium *E. coli*, and attention will be focussed on the mechanism in eukaryotes where the processes appear to differ.

Several recent reviews of translation are listed in the references.

A. Initiation

Initiation of protein synthesis in *E. coli* and other prokaryotic organisms requires the formation of an initiation complex consisting of a 70S ribosome, mRNA, the protein initiation factors IF1, IF2 and IF3, GTP and formylmethionyl-tRNA (fmet-tRNA) (LUCAS-LENARD and LIPMANN, 1971).

1. Formylmethionyl-tRNA

Two species of methionyl-tRNA exist in *E. coli* (MARCKER and SANGER, 1964; CLARK and MARCKER, 1966). One which can be formylated by the enzyme met-tRNA transformylase (MARCKER, 1965) is used for initiation of protein synthesis; the other is not a substrate for this enzyme, and is used in peptide chain elongation. Fmet-tRNA is required for initiation of synthesis of all bacterial proteins and is specified by the AUG codon in messenger RNA. Subsequently the formyl group, and in some cases the methionine residue, is cleaved from the N-terminus of the growing polypeptide chain (TAKEDA and WEBSTER, 1968; HOUSMAN et al., 1972).

2. Protein Initiation Factors

Three factors termed IF1, IF2 and IF3 (previously f1, F1 and A; f2, FIII and C; f3, FII and B respectively) (CASKEY et al., 1972) were isolated from ribosomes by extraction with ammonium chloride (STANLEY et al., 1966) and have molecular weights of 9,000 (IF1) (HERSHEY et al., 1969), 65,000–80,000 (IF2) (CHAE et al., 1969; KOLAKOFSKY et al., 1969), and 21,000 (IF3) (WAHBA et al., 1969). These three protein initiation factors are present in the 30S subunit isolated from *E. coli* but they are not present in the 70S particle from fragmented polysomes (PARENTI-ROSINA et al., 1969). They are therefore not constant components of the ribosome and appear to dissociate from the 30S particle once initiation has occurred.

3. Steps in Initiation

The 70S initiation complex in the *E. coli* system is formed through the association of a 50S subunit with a 30S complex. The 30S complex at this stage is composed of the 30S subunit, fmet-tRNA, the three protein initiation factors, GTP and mRNA. The formation of this 30S complex is thought to arise by the prior association of IF2, fmet-tRNA and GTP to form a ternary complex (Fig. 1) (LOCKWOOD et al., 1971; RUDLAND et al., 1971). This ternary complex together with IF1 then binds to a 30S subunit complex composed of a 30S subunit, IF3 and mRNA (DAVIS, 1971). However, these proposals for the formation of a 30S initiation complex must remain tentative as evidence has been presented that the two inititiation factors IF2 and IF3 associate to form a complex which then binds fmet-tRNA and GTP (GRONER and REVEL, 1971).

Association of the 50S subunit with a 30S initiation complex to form a 70S initiation complex involves the hydrolysis of GTP catalyzed by IF2 (KOLAKOFSKY et al., 1969). The initiation factors are thought to be released from the 30S subunit at this stage. Evidence has also accumulated that the binding of fmet-tRNA occurs directly at the peptidyl-tRNA binding site on the ribosome (P site; Fig. 1) (THACH and THACH, 1971; DE GROOT et al., 1971) and is not translocated there via the aminoacyl-tRNA binding site (A site) as previously proposed (KOLAKOFSKY et al., 1968).

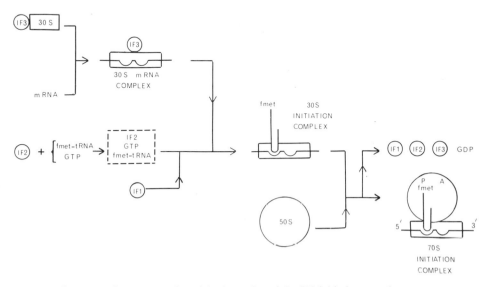

Fig. 1. A diagrammatic representation of the formation of the 70S initiation complex

In addition to binding fmet-tRNA and catalyzing GTP hydrolysis, the initiation factor IF2 also promotes the attachment of ribosomes to mRNA. However in template competition experiments using synthetic polynucleotides, it was shown that IF2 promotes the binding of ribosomes to any site on the mRNA (REVEL et al., 1970), whereas IF3 is concerned with the recognition of specific sites on the natural messenger.

4. Regulation of Translation by Initiation Factors

Evidence that differences in ribosomes are significant in the initiation of translation has been obtained with bacteriophage RNA as a natural messenger. It was found that ribosomes from *Bacillus stearothermophilus* initiate translation of only one cistron on the f2 phage RNA molecule whereas ribosomes from *E. coli* initiate at all three cistrons (LODISH, 1970). To explain this effect, work has now centered on the initiation factor IF3 as previously it was observed that a crude preparation of initiation factor IF3 from phage T4-infected *E. coli* cells promoted the formation of an initiation complex with T4 messenger RNA.

Furthermore, ribosomes from T4-infected cells are restricted in their ability to translate mRNA from the host or RNA from bacteriophages MS2 or f2 (Hsu and Weiss, 1969). It therefore seemed that a phage specified or modified initiation factor was responsible and further investigations have confirmed this view. In experiments with MS2-infected *E. coli*, an initiation factor IF3-B2 has been isolated which is required for the recognition, by the 30S subunit, of the three initiation sites on MS2 RNA (Berissi et al., 1971). Similarly, two IF3 factors have been isolated from infected *E. coli* which promote the recognition of phage MS2 initiation sites and T4 messenger RNA sites respectively (Lee-Huang and Ochoa, 1971).

It now seems highly probable that further specific initiation factors will be found for different RNA phage systems. What would be interesting however, from the point of view of translational regulation of gene expression, would be the isolation of specific initiation factors from uninfected cells promoting initiation with specific messengers. However, the investigation of such a system must await the purification of specific mRNA molecules from bacteria.

5. Initiation in Eukaryotic Organisms

Eukaryotic organisms appear to lack the transformylase activity needed to formylate met-tRNA (Caskey et al., 1967). However, two types of met-tRNA have been found in yeast (Takeishi et al., 1968) and mouse liver (Smith and Marcker, 1970) and three types in rat liver (Burgess and Mach, 1971). In each case, one of these types of met-tRNA (met-tRNA$_f$) can be formylated *in vitro* by *E. coli* transformylase and this species is the one capable of forming an initiation complex. Supporting evidence that met-tRNA$_f$ is used in initiation in eukaryotes has been obtained with a rabbit reticulocyte system synthesizing globin, where it was shown that small nascent peptides contained N-terminal methionine (Wilson and Dintzis, 1970; Hunter and Jackson, 1971).

In general outline, the mechanism of initiation in eukaryotic systems seems to be analogous to that found in bacteria (Crystal and Anderson, 1972) with soluble initiation factors being involved (Heywood, 1970; Crystal et al., 1971). In the brine shrimp an IF2 like factor (IFM2) has been shown to promote the binding of *E. coli* fmet-tRNA to 40S ribosomal subunits in the presence of the AUG codon (Zasloff and Ochoa, 1971), and a preparation giving similar results has been reported from rat liver (Moldave and Gasior, 1971). The formation of an initiation complex involving 40S subunits from mouse tumor cells, fmet-tRNA, GTP, the AUG codon and initiation factors has also been reported (Burgess and Mach, 1971), and hydrolysis of GTP mediated by initiation factor has been found to occur when fmet-tRNA binds to reticulocyte ribosomes (Shafritz et al., 1971).

The specificity of initiation with various purified mRNAs *in vitro* is claimed to reside in soluble initiation factors; mRNA from one tissue or organ may not be translated by ribosomes from a different part of the organism (Heywood, 1969; Naora and Kodaira, 1970). However, this is an open question as *in vivo* protein synthesis experiments with frog oocytes injected with various purified mRNAs have shown little species specificity with respect to the type of mRNA which they can translate (GURDON et al., 1971). Similarly it has been shown that globin mRNA can be translated by a nonerythropoietic ascites cell system in the absence of reticulocyte components (MATHEWS et al., 1971). It was suggested by these authors that this may be a consequence of the undifferentiated tumour cell retaining the apparatus to make a wide variety of proteins. Further aspects of the specificity of ribosomes are considered in Chapter 3. The study of initiation of translation in eukaryotic organisms undoubtedly will receive increasing attention in the next few years because it may be an important selective point in determining which messenger RNA species are translated.

B. Elongation

After the formation of the 70S initiation complex in the *E. coli* system, fmet-tRNA is located at the P site (peptidyl or puromycin reactive site, also called the D or donor site) on the 50S subunit and the A site (aminoacyl or acceptor site) is free for occupation by a charged tRNA. The ribosome is now ready to commence peptide bond synthesis and polypeptide chain elongation.

The process of elongation, which is a repetitive process, can be divided into three stages: 1) binding of the charged tRNA to the ribosome complex; 2) linkage of the incoming amino acid to the nascent peptide chain by peptide bond formation; 3) translocation of the newly extended peptidyl tRNA from the A site to the P site (Fig. 2). During this cycle, protein elongation factors play an important role. These factors termed EF G, EF Tu and EF Ts (previously factors G, Tu and Ts of LIPMANN (1969), and equivalent to S_2 and F_{II}, S_3 and F_{Iu}, and S_1 and F_{IS} respectively) occur in the soluble fraction of the cell wherein factors EF Tu and EF Ts are associated in a complex termed EF T. The elongation factors have been purified and molecular weights of 39,000–42,000 for EF Tu, 19,000 for EF Ts and 72,000–84,000 for EF G have been determined (LUCAS-LENARD and LIPMANN, 1971). Interestingly, the two elongation factors EF Tu and EF Ts have been found to be two of the four subunits of the replicase of the RNA bacteriophage $Q\beta$ (BLUMENTHAL et al., 1972).

1. Binding

The binding of charged tRNA requires an association of EF Tu, GTP and charged tRNA in a complex which is analogous to the ternary initiation complex between IF2, GTP and fmet-tRNA (Fig. 2). This ternary aminoacyl-tRNA binding complex is formed by the association of GTP and aminoacyl-tRNA with EF T which, as noted above, is a complex of EF Tu and EF Ts. EF Ts is displaced during formation of the ternary binding complex, leaving EF Tu associated with aminoacyl-tRNA and GTP. GTP hydrolysis accompanies the binding of the complex to the ribosome, and the energy released is thought to be used to orient the charged tRNA at the A site (LUCAS-LENARD and LIPMANN,

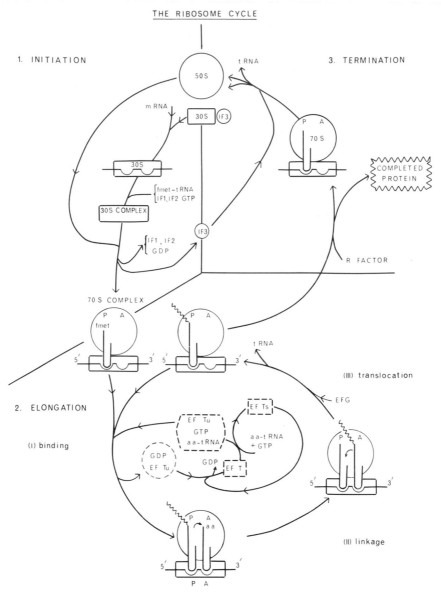

Fig. 2. The ribosome cycle, a diagrammatic representation of the three steps in the translation process—initiation, elongation and termination

1971). This binding of aminoacyl tRNA to the A site can be inhibited by tetracyclines and certain other antibiotics (HIEROWSKI, 1965; CERNA et al., 1969). Following hydrolysis of GTP, the resulting GDP and EF Tu are released from the ribosome as a stable complex (Fig. 2). The factor EF Ts then binds to EF Tu displacing GDP and regenerating EF T which may then interact with more GTP and charged tRNA (MILLER and WEISSBACH, 1970; WATERSON et al., 1970; WEISSBACH et al., 1970).

Ribosome configuration appears to influence the binding reaction as it has been found that binding of charged tRNA at the A site occurs only if the P site is occupied (DE GROOT et al., 1971). That is, the A site binds charged tRNA only as a result of induced conformational changes brought about by the presence of a charged tRNA at the P site. Conformational change in the ribosome is also implicated in conferring a selective or recognition function on the ribosome so that only the specific charged tRNA appropriate to the codon in mRNA can be bound (GORINI, 1971; BISWAS and GORINI, 1972). Thus the antibiotic streptomycin is thought to promote mistakes in translation, upon binding to the 30S subunit, by distorting site A and so permitting the binding of the wrong aminoacyl tRNA (DAVIES et al., 1964).

2. Linkage (Peptide Bond Formation)

After binding of the charged tRNA to the A site, peptide bond synthesis occurs between the free α-amino group of the incoming amino acid and the esterified carboxyl group of the peptidyl-tRNA located at the P site. This reaction is catalyzed by the enzyme peptidyltransferase which is an integral part of the 50S subunit (MADEN et al., 1968). No other soluble or supernatant factors are required in this reaction and GTP hydrolysis is unnecessary as the free-energy change is sufficiently negative to drive the reaction.

Puromycin, an antibiotic which can be considered a structural analogue of the aminoacyl-adenosyl moiety of charged tRNA (YARMOLINSKY and DE LA HABA, 1959), has been used extensively in studying the linkage reaction (VAZQUEZ et al., 1969). This antibiotic reacts with peptidyl-tRNA to form a peptidyl-puromycin compound only when the peptidyl-tRNA is at the P site. Peptide chain elongation is terminated as a result and the growing peptide chain is released from the ribosome (SMITH et al., 1965; GOTTESMAN, 1967).

Other antibiotics such as chloramphenicol, erythromycin, lincomycin and sparsomycin have been shown to act at the peptidyltransferase centre on the 50S subunit of the *E. coli* ribosome (MONRO and VAZQUEZ, 1967; VOGEL et al., 1971). One molecule of antibiotic is bound per subunit but the binding sites do not appear to be identical (FERNANDEZ-MUNOZ et al., 1971).

3. Translocation

After peptide bond synthesis, the newly extended peptidyl-tRNA is at the A site and uncharged tRNA is at the P site; this stage has been termed a 'pretranslocational' complex (GUPTA et al., 1971). Translocation, which involves the movement of the ribosome complex along the mRNA accompanied by the shift of the peptidyl-tRNA from the A site to the P site, gives rise to a so-called 'posttranslocational' complex. During this process the uncharged tRNA at the P site is released and the hydrolysis of GTP occurs (LUCAS-LENARD et al., 1969).

The elongation factor EF G, which has ribosome-dependent GTPase activity, is necessary for translocation and binds to the large ribosome subunit. It also appears that the EF Tu-GTP-aminoacyl-tRNA complex described earlier cannot bind to the ribosome following the addition of EF G and GDP. This finding has been interpreted to mean that the two protein elongation factors interact with the same or overlapping regions of the 50S subunit (RICHMAN and BODLEY, 1972; CABRER et al., 1972; MILLER, 1972).

Evidence has been presented that ribosome movement along the mRNA is associated with the factor EF G and GTP hydrolysis, and it is suggested that this process occurs in the same phase as the translocation of the peptidyl-tRNA (GUPTA et al., 1971). Fusidic acid, an inhibitor of translocation (THACH and THACH, 1971), was previously thought to inhibit GTP hydrolysis catalyzed by EF G (TANAKA et al., 1968), but is now thought to inhibit the dissociation of GDP and EF G from the ribosome (CUNDLIFFE, 1972).

During the translocation event the ribosome complex moves along the mRNA. Although the distance moved by the ribosome complex relative to the mRNA has long been assumed to be three nucleotides because of the triplet nature of the genetic code, positive evidence establishing this has only recently been presented. Using bacteriophage f2 RNA as artificial mRNA, pre- and post-translocational complexes, separated by a single translocation step, were isolated and treated with ribonuclease to digest those parts of the mRNA which were not protected by the attached ribosome. The sequences of the nucleotide fragments which were subsequently isolated from the two complexes differed by three nucleotides at each end (GUPTA et al., 1971; THACH and THACH, 1971).

4. Elongation in Eukaryotic Organisms

Two elongation factors EF1 and EF2 have been isolated from eukaryotes. The factor EF1 is the aminoacyl-tRNA binding protein from rabbit reticulocytes and appears to be analogous to EF T of *E. coli* (McKEEHAN and HARDESTY, 1969). It is probably identical to the aminoacyl transferase I (T_1) from rat liver (FELICETTI and LIPMANN, 1968). The other factor EF2, is the aminoacyl transferase II (T_2) from rat liver (GALASINSKI and MOLDAVE, 1969) and is apparently analogous to the bacterial EF G.

The factor EF1, like the bacterial factor EF T, is required for the binding of charged tRNA to the ribosome. This binding occurs by way of a complex between EF1, GTP and charged tRNA and is followed by GTP hydrolysis (SKOGERSON and MOLDAVE, 1968). However the situation now seems to be more complex as evidence has been obtained for two species of EF1 termed EF1A and EF1B. EF1A appears to associate with aminoacyl-tRNA and GTP to yield an EF1B-aminoacyl tRNA-GTP complex which then reacts with the small ribosomal subunit (MOON et al., 1972).

Subsequent to binding of charged tRNA, peptide bond synthesis is catalyzed by a peptidyltransferase which is an integral part of the large ribosome subunit (SKOGERSON and MOLDAVE, 1968; NETH et al., 1970). For translocation to occur, factor EF2 is required and this reaction is accompanied by an EF2-catalyzed GTP hydrolysis (SKOGERSON and MOLDAVE, 1968; CULP et al., 1969).

Supporting evidence that EF2 is important for translocation has come from studies with diphtheria toxin. This toxin inhibits factor EF2 in eukaryotic systems (COLLIER, 1967) and this inhibition apparently results from the toxin catalyzed attachment of the adenosine-diphosphate-ribose moiety of NAD to the EF2 molecule (HONJO et al., 1968; GILL et al., 1969).

C. Termination

The need for a special termination mechanism became apparent from *in vitro* studies, when it was found that polyphenylalanine remained bound to ribosomes when the end of the polyuridylic acid (poly U) molecule directing synthesis had been reached (GILBERT, 1963). Moreover the link between tRNA and the polypeptide remained intact indicating that hydrolysis of this bond did not automatically follow cessation of elongation. Specific termination signals in mRNA also seemed likely when the translation of polycistronic messenger was considered. In general it appeared improbable that a polycistronic message was translated as a unit and the resulting giant polypeptide cleaved into the appropriate proteins.

Additionally, work on the genetic code revealed that three base triplets, UAG, UAA and UGA, did not specify any amino acid and these triplets have also been recognized as nonsense codons from genetic studies (GAREN, 1968). Direct evidence that these nonsense codons are natural termination signals has been presented in the case of bacteriophage R17 RNA in which a tandem arrangement of UAA and UAG is present at the end of the coat protein cistron (NICHOLS, 1970). It is now believed that natural termination signals in *E. coli* at least, contain not one but two codons (LU and RICH, 1971); however in the case of the histidine (*his*) operon there is evidence that only a single chain termination codon exists at the end of the *his* D gene (RECHLER and MARTIN, 1970). Two termination signals may be necessary because single suppressor mutations (which change nonsense codons to sense codons) would result in the uninterrupted translation of polycistronic mRNA; the resulting proteins would almost certainly be nonfunctional. With two termination signals at the end of each cistron, the detrimental effects of suppressor mutations would be considerably lessened.

Termination, unlike initiation and elongation, does not depend on tRNA. However the process of termination is still a positive event, so that the absence of a specific aminoacyl-tRNA, which may occur in starvation, is not sufficient for termination to take place. In addition to the termination signal on the mRNA, protein release factors have been implicated (CAPECCHI, 1967).

The release factors, termed R1 and R2, have been purified to homogeneity and appear to be separate proteins differing in their specificity and acting singly in termination (SCOLNICK et al., 1968; CAPECCHI and KLEIN, 1970). These factors differ in their recognition of nonsense codons. Thus R1 recognizes UAA and UAG, while R2 recognizes UAA and UGA. Release factors will bind to 70S ribosomes but not to 50S or 30S subunits (SCOLNICK and CASKEY, 1969), and it has been suggested that they promote termination in the presence of nonsense codons by converting the peptidyltransferase enzyme into a hydrolase with resultant cleavage of the ester bond linking the tRNA to the polypeptide (Lucas-Lenard and Lipmann, 1971).

Termination in eukaryotes has received little attention to date. However Oliver and his associates, studying tyrosine aminotransferase synthesis in rat liver, have suggested that a control step in the production of this enzyme occurs at the termination step in translation (Chuah et al., 1971; Chuah and Oliver, 1971). It appears in this system that a fully synthesized enzyme is located on a polysome fraction ready to be released by the appropriate signal. Such a control mechanism it seems would be confined to cells which have already differentiated, and are awaiting a final signal before they become active in their specialized function.

D. The Ribosome Cycle

Following termination and the R factor mediated release of the completed polypeptide chain, the ribosome appears to 'run off' the mRNA as a 70S particle, although it is unclear whether the subunits have to separate to allow this to occur. Two soluble protein factors and GTP appear to be involved in this process. One factor has been identified as EF G and the other protein, termed ribosome release factor, seems to be specific for this step (Hirashima and Kaji, 1972). In any event a consideration of the process of initiation, which occurs by subunit association, demands at some point the dissociation of the 70S ribosome if this structure is to be used more than once. This dissociation into 30S and 50S subunits is thought to be mediated by a specific dissociation factor termed DF (Davis, 1971). DF is now thought to be identical to the initiation factor IF3 in the *E. coli* system (Davis, 1971) and it has been shown to bind specifically to the 30S subunit (Sabol and Ochoa, 1971). A similar dissociation factor has been detected in preparations from rabbit reticulocytes (Lubsen and Davis, 1972) and yeast (Petre, 1970). It appears then that DF dissociates the 70S ribosomes after run-off by binding to the 30S subunit and by so doing priming the dissociated small subunit to recognize an initiation site on mRNA. During initiation when the 50S subunit binds to the 30S initiation complex, the IF3 factor is released and can now act again in dissociating a 70S ribosome.

Finally it appears that several components of the translation machinery are used in a cyclic fashion and this is illustrated in Fig. 2 which serves to summarize the events discussed in the preceding sections.

E. Translation of Polycistronic Messenger

1. Occurrence of Polycistronic Messenger and Its Translation

In metabolic pathways involving the activity of several enzymes, co-ordination of synthesis of these enzymes would seem desirable for the economy of the cell. This is achieved in bacteria in some cases by having the relevant genes for a pathway clustered at a single locus on the chromosome. The transcription of these genes in bacteria may be regulated as a unit according to the operon concept of JACOB and MONOD (1961a), and it was anticipated by these authors that the genes of an operon may be transcribed as a polycistronic messenger RNA molecule. This prediction has since been shown to be justified in a number of cases, by the identification, for example, of polycistronic messengers for the nine gene histidine (*his*) operon of *Salmonella typhimurium* (MARTIN, 1963), and the three gene lactose (*lac*) operon (ATTARDI et al., 1963; GUTTMAN and NOVICK, 1963) and five gene tryptophan (*try*) operon (IMAMOTO et al., 1965) of *E. coli*.

Supporting evidence that polycistronic messengers occur has been presented in electron microscope studies of lysed *E. coli* preparations (MILLER et al., 1970). The largest actively transcribing segment of chromosome observed by these authors was about 3 μm which compares with the estimated lengths of various bacterial operons. In eukaryotes, with rare exceptions, clustering of genes does not occur, and it seems likely that higher organisms produce monocistronic messengers only.

Although evidence has accumulated that operons are transcribed as giant RNA molecules, it was still considered likely that these large molecules were cleaved to individual cistrons for translation. This idea was abandoned however when it was shown that the polycistronic *lac* messenger is translated without degradation (KIHO and RICH, 1965). The proposal that intact polycistronic messenger is translated provides an explanation for the sequential appearance of the enzymes of the *lac*, *his*, *try* and *gal* operons upon induction or derepression, as it has been proposed that translation begins only at the operator end (AMES and HARTMAN, 1963). However the sequential appearance of enzymes could possibly be explained by some form of transcriptional control (EPSTEIN and BECKWITH, 1968). For instance it could be envisaged that the translation of the first gene is necessary before transcription of subsequent genes occurs. In this regard transcription appears to be accompanied by translation and a cluster of ribosomes probably follows close behind the RNA polymerase engaged in transcription (MORSE et al., 1969).

If initiation occurs only at the operator end of a polycistronic message then equimolar proportions of gene products should be produced provided ribosomes do not detach before reaching the end of the polycistronic message. Results supporting this model have been obtained with the *gal* (WILSON and HOGNESS, 1969) and *try* operons (MORSE and YANOFSKY, 1968) of *E. coli*. However in the case of the *lac* operon of *E. coli*, the *his* operon of *S. typhimurium* and the three gene R17 RNA phage, this is not so. In the *lac* operon more β-galactosidase is produced than thiogalactoside transacetylase (BROWN et al., 1967) and in the *his* operon the first cistron is translated three times as frequently as the second, third and sixth cistrons (WHITFIELD et al., 1970). Additionally with R17 phage, the three genes coding for the coat protein, RNA synthetase and maturation protein are translated in the molar proportions of 20:6:1 (JEPPESEN et al., 1970). It is suggested that this ratio does not reflect different rates of elongation but rather different rates of initiation which in turn may be influenced by the secondary and tertiary structure of the viral RNA molecule. However as the authors point out, the structural constraints in the viral RNA molecule, necessary for packaging, may mean that these RNAs are not typical of mRNAs in general.

2. Polarity

Polarity occurs when a nonsense mutation in an operon decreases the production of proteins coded by genes distal to the operator side of the mutation. That is, when a nonsense mutation which results in premature termination of translation occurs in the first gene of an operon, the translation of subsequent genes is reduced (FRANKLIN and LURIA, 1961; JACOB and MONOD, 1961b; AMES and HARTMAN, 1963). Nonsense or frameshift mutations* can eliminate the polarity caused by the original mutation. These observations provide strong support for the concept that polycistronic messengers are translated without cleavage.

Two conflicting mechanisms have been advanced to explain polarity. The first suggests that polarity results from close coupling between translation and transcription. It is believed in this model that ribosome movement is essential for transcription of messenger from DNA and that a polar mutation, because it results in polypeptide chain termination and ribosome release, prevents transcription (IMAMOTO and KANO, 1971).

The alternative model, which has gained much recent support, suggests that because ribosomes are released at the nonsense mutations producing polarity, the untranslated portion of the messenger is rapidly degraded by nuclease enzymes (MORSE and YANOFSKY, 1969). Two groups have isolated suppressor mutants in which the survival of untranslated polycistronic mRNA is greatly increased. It is believed that some of these mutants at least have altered endonucleases (MORSE and GUERTIN, 1971; CARTER and NEWTON, 1971).

F. Antibiotics as Inhibitors of Translation

Antibiotics have been widely used as tools with which to explore molecular events associated with translation. The numbers of these compounds grow each year, and a thorough review of their structure and mode of action would require a great deal of space. The Appendix to this chapter contains a summary of the salient structural and functional properties of the more important inhibitors of translation.

G. Concluding Remarks

From the above outline it can be seen that several of the stages that combine to form the complex mechanism of mRNA translation are understood in some detail at least for the bacterium *E. coli*. Translation in higher organisms is less well understood and it is presumably these systems that will attract increasing attention, especially in relation to possible regulation mechanisms concerned

---

* Frameshift mutations occur when a single nucleotide addition or deletion is made in the transcribed mRNA. All codons on the 3′-side of the changed base sequence are then read in a different triplet "frame".

with cell differentiation. In this regard control of translation by hormones in higher organisms has received attention (TOMPKINS et al., 1969; KORNER, 1970; LEVINSON et al., 1971). Also a start has been made in examining the role of initiation factors in selective translation of different eukaryotic messengers as discussed above. In addition tRNA may have a regulatory role in translation and this possibility is discussed in Chapter 5.

References

Reviews

LENGYEL, P. L., SOLL, D.: Bacteriol. Rev. **33**, 264 (1969).
LUCAS-LENARD, J., LIPMANN, F.: Ann. Rev. Biochem. **40**, 409 (1971).
NOMURA, M.: Bacteriol. Rev. **33**, 264 (1970).
WEISBLUM, B., DAVIES, J.: Bacteriol. Rev. **32**, 493 (1968).
Cold Spring Harbor Symp. Quant. Biol. (1969).
Translation: its mechanism and control. Fogharty International Center, National Institute of Health (in press).

Other References

AMES, B. N., HARTMAN, P. E.: Cold Spring Harbor Symp. Quant. Biol. **28**, 349 (1963).
ATTARDI, G., NAONO, S., ROUVIERE, J., JACOB, F., GROS, F.: Cold Spring Harbor Symp. Quant. Biol. **28**, 363 (1963).
BERISSI, H., GRONER, Y., REVEL, M.: Nature New Biol. **234**, 44 (1971).
BISWAS, D. K., GORINI, L.: J. Mol. Biol. **64**, 119 (1972).
BLUMENTHAL, T., LANDERS, T. A., WEBER, K.: Proc. Nat. Acad. Sci. U.S. **69**, 1313 (1972).
BROWN, J. L., BROWN, D. M., ZABIN, I.: J. Biol. Chem. **242**, 4254 (1967).
BURGESS, A. B., MACH, B.: Nature New Biol. **233**, 209 (1971).
CABRER, B., VAZQUEZ, D., MODOLELL, J.: Proc. Nat. Acad. Sci. U.S. **69**, 733 (1972).
CAPECCHI, M. R.: Proc. Nat. Acad. Sci. U.S. **58**, 1144 (1967).
CAPECCHI, M. R., KLEIN, H. A.: Nature **226**, 1029 (1970).
CARTER, T., NEWTON, A.: Proc. Nat. Acad. Sci. U.S. **68**, 2962 (1971).
CASKEY, C.T., REDFIELD, B., WEISSBACH, H.: Arch. Biochem. Biophys. **120**, 119 (1967).
CASKEY, T., LEDER, P., MOLDAVE, K., SCHLESSINGER, D.: Science **176**, 195 (1972).
CERNA, J., RYCHLIK, I., PULKRABEK, P.: Eur. J. Biochem. **9**, 27 (1969).
CHAE, Y.-B., MAZUMDER, R., OCHOA, S.: Proc. Nat. Acad. Sci. U.S. **62**, 1181 (1969).
CHUAH, C. C., HOLT, P. G., OLIVER, I.T.: Intern. J. Biochem. **2**, 193 (1971).
CHUAH, C. C., OLIVER, I. T.: Biochemistry **10**, 2990 (1971).
CIFERRI, O., PARISI, B.: Progr. Nucl. Acid Res. Mol. Biol. **10**, 121 (1970).
CLARK, B. F. C., MARCKER, K. A.: J. Mol. Biol. **17**, 394 (1966).
CLARK-WALKER, G. D., LINNANE, A. W.: Biochem. Biophys. Res. Commun. **25**, 8 (1966).
COLLIER, R. J.: J. Mol. Biol. **25**, 83 (1967).
CRYSTAL, R. G., ANDERSON, W. F.: Proc. Nat. Acad. Sci. U.S. **69**, 706 (1972).
CRYSTAL, R. G., SHAFRITZ, D. A., PRICHARD, P. M., ANDERSON, W. F.: Proc. Nat. Acad. Sci. U.S. **68**, 1810 (1971).
CULP, W. J., MCKEEHAN, W. L., HARDESTY, B.: Proc. Nat. Acad. Sci. U.S. **64**, 388 (1969).
CUNDLIFFE, E.: Biochem. Biophys. Res. Commun. **46**, 1794 (1972).
DAVIES, J., GILBERT, W., GORINI, L.: Proc. Nat. Acad. Sci. U.S. **51**, 883 (1964).
DAVIS, B. D.: Nature **231**, 153 (1971).
DE GROOT, N., PANET, A., LAPIDOT, Y.: Eur. J. Biochem. **23**, 523 (1971).
EPSTEIN, W., BECKWITH, J. R.: Ann. Rev. Biochem. **37**, 411 (1968).
FELICETTI, L., LIPMANN, F.: Arch. Biochem. Biophys. **125**, 185 (1968).
FERNANDEZ-MUNOZ, R., MONRO, R. E., TORRES-PINEDO, R., VAZQUEZ, D.: Eur. J. Biochem. **23**, 185 (1971).

Franklin, N. C., Luria, S. E.: Virology **15**, 299 (1961).
Galasinski, W., Moldave, K.: J. Biol. Chem. **244**, 6527 (1969).
Garen, A.: Science **160**, 149 (1968).
Gilbert, W.: J. Mol. Biol. **6**, 389 (1963).
Gill, D. M., Pappenheimer, A. M., Brown, R., Kurnick, J.T.: J. Exp. Med. **129**, 1 (1969).
Gorini, L.: Nature New Biol. **234**, 261 (1971).
Gottesman, M. E.: J. Biol. Chem. **242**, 5564 (1967).
Groner, Y., Revel, M.: Eur. J. Biochem. **22**, 144 (1971).
Gupta, S. L., Waterson, J., Sopori, M. L., Weissman, S. M., Lengyel, P.: Biochemistry **10**, 4410 (1971).
Gurdon, J. B., Lane, C. D., Woodland, H. R., Marbaix, G.: Nature **233**, 177 (1971).
Guttman, B. S., Novick, A.: Cold Spring Harbor Symp. Quant. Biol. **28**, 373 (1963).
Hershey, J.W. B., Dewey, K. F., Thach, R. E.: Nature **222**, 944 (1969).
Heywood, S. M.: Cold Spring Harbor Symp. Quant. Biol. **34**, 799 (1969).
Heywood, S. M.: Nature **225**, 696 (1970).
Hierowski, M.: Proc. Nat. Acad. Sci. U. S. **53**, 594 (1965).
Hirashima, A., Kaji, A.: J. Mol. Biol. **65**, 43 (1972).
Honjo, T., Nishizuka, Y., Hayaishi, O., Kato, I.: J. Biol. Chem. **243**, 3553 (1968).
Housman, D., Gillespie, D., Lodish, H. F.: J. Mol. Biol. **65**, 163 (1972).
Hsu, W.T., Weiss, S. B.: Proc. Nat. Acad. Sci. U. S. **64**, 345 (1969).
Hunter, A. R., Jackson, R. J.: Eur. J. Biochem. **19**, 316 (1971).
Imamoto, F., Kano, Y.: Nature New Biol. **232**, 169 (1971).
Imamoto, F., Morikawa, N., Sato, K.: J. Mol. Biol. **13**, 169 (1965).
Ito, J., Crawford, I. P.: Genetics **52**, 1303 (1965).
Jacob, F., Monod, J.: J. Mol. Biol. **3**, 318 (1961a).
Jacob, F., Monod, J.: Cold Spring Harbor Symp. Quant. Biol. **26**, 193 (1961b).
Jeppesen, P. G. N., Steitz, J. A., Gesteland, R. F., Spahr, P. F.: Nature **226**, 230 (1970).
Kiho, Y., Rich, A.: Proc. Nat. Acad. Sci. U. S. **54**, 1751 (1965).
Kolakofsky, D., Dewey, K., Thach, R. E.: Nature **223**, 694 (1969).
Kolakofsky, D., Ohta, T., Thach, R. E.: Nature **220**, 244 (1968).
Korner, A.: Proc. Roy. Soc. (London), Ser. B **176**, 287 (1970).
Lamb, A. J., Clark-Walker, G. D., Linnane, A.W.: Biochim. Biophys. Acta **161**, 415 (1968).
Lee-Huang, S., Ochoa, S.: Nature New Biol. **234**, 236 (1971).
Levinson, B. B., Tomkins, G. M., Stellwagen, R. H.: J. Biol. Chem. **246**, 6297 (1971).
Lipmann, F.: Science **164**, 1024 (1969).
Lockwood, A. H., Chakraborty, P. R., Maitra, U.: Proc. Nat. Acad. Sci. U. S. **68**, 3122 (1971).
Lodish, H. F.: Nature **226**, 705 (1970).
Lu, P., Rich, A.: J. Mol. Biol. **58**, 513 (1971).
Lubsen, N. H., Davis, B. D.: Proc. Nat. Acad. Sci. U. S. **69**, 353 (1972).
Lucas-Lenard, J., Tao, P., Haenni, A.L.: Cold Spring Harbor Symp. Quant. Biol. **34**, 455 (1969).
Maden, B. E. H., Traut, R. R., Monro, R. E.: J. Mol. Biol. **35**, 333 (1968).
Marcker, K.: J. Mol. Biol. **14**, 63 (1965).
Marcker, K., Sanger, F.: J. Mol. Biol. **8**, 835 (1964).
Martin, R. G.: Cold Spring Harbor Symp. Quant. Biol. **28**, 357 (1963).
Mathews, M. B., Osborn, M., Lingrel, J. B.: Nature New Biol. **233**, 206 (1971).
McKeehan, W. L., Hardesty, B.: J. Biol. Chem. **244**, 4330 (1969).
Miller, D. L.: Proc. Nat. Acad. Sci. U. S. **69**, 752 (1972).
Miller, D. L., Weissbach, H.: Biochem. Biophys. Res. Commun. **38**, 1016 (1970).
Miller, O. L., Hamkalo, B. A., Thomas, C. A.: Science **169**, 392 (1970).
Moldave, K., Gasior, E.: Fed. Proc. **30**, 1290 (1971).
Monro, R. E., Vazquez, D.: J. Mol. Biol. **28**, 161 (1967).
Moon, H. M., Redfield, B., Weissbach, H.: Proc. Nat. Acad. Sci. U. S. **69**, 1249 (1972).
Morse, D. E., Guertin, M.: Nature New Biol. **232**, 165 (1971).
Morse, D. E., Mosteller, R. D., Yanofsky, C.: Cold Spring Harbor Symp. Quant. Biol. **34**, 725 (1969).
Morse, D. E., Yanofsky, C.: J. Mol. Biol. **38**, 447 (1968).
Morse, D. E., Yanofsky, C.: Nature **224**, 329 (1969).
Naora, H., Kodaira, K.: Biochim. Biophys. Acta **209**, 196 (1970).
Neth, R., Monro, R. E., Heller, G., Battaner, E., Vazquez, D.: FEBS Lett. **6**, 198 (1970).

NICHOLS, J.L.: Nature **225**, 147 (1970).

PARENTI-ROSINA, R., EISENSTADT, A., EISENSTADT, J.M.: Nature **221**, 363 (1969).

PETERMAN, M. L.: The physical and chemical properties of ribosomes. Amsterdam: Elsevier 1964.

PETRE, J.: Eur. J. Biochem. **14**, 399 (1970).

RECHLER, M.M., MARTIN, R.G.: Nature **226**, 908 (1970).

REVEL, M., GREENSHPAN, H., HERZBERG, M.: Eur. J. Biochem **16**, 117 (1970).

RICHMAN, N., BODLEY, J.W.: Proc. Nat. Acad. Sci. U.S. **69**, 686 (1972).

RUDLAND, P.S., WHYBROW, W.A., CLARK, B.F.C.: Nature New Biol. **231**, 76 (1971).

SABOL, S., OCHOA, S.: Nature New Biol. **234**, 233 (1971).

SCOLNICK, E., TOMPKINS, R., CASKEY, C., NIRENBERG, M.: Proc. Nat. Acad. Sci. U.S. **61**, 768 (1968).

SCOLNICK, E.M., CASKEY, C.T.: Proc. Nat. Acad. Sci. U.S. **64**, 1235 (1969).

SHAERITZ, D.A., LAYCOCK, D.G., CRYSTAL, R.G., ANDERSON, W.F.: Proc. Nat. Acad. Sci. U.S. **68**, 2246 (1971).

SKOGERSON, L., MOLDAVE, K.: Arch. Biochem. Biophys. **125**, 497 (1968).

SMITH, A.E., MARCKER, K.A.: Nature **226**, 607 (1970).

SMITH, J.D., TRAUT, R.R., BLACKBURN, G.M., MONRO, R.E.: J. Mol. Biol. **13**, 617 (1965).

STANLEY, W.M., SALAS, M., WAHBA, A.J., OCHOA, S.: Proc. Nat. Acad. Sci. U.S. **56**, 290 (1966).

TAKEDA, M., WEBSTER, R.E.: Proc. Nat. Acad. Sci. U.S. **60**, 1487 (1968).

TAKEISHI, K., UKITA, T., NISHIMURA, S.: J. Biol. Chem. **243**, 5761 (1968).

TANAKA, N., KINOSHITA, T., MASUKAWA, H.: Biochem. Biophys. Res. Commun. **30**, 278 (1968).

THACH, S.S., THACH, R.D.: Proc. Nat. Acad. Sci. U.S. **68**, 1791 (1971).

TOMKINS, G.M., GELEHRTER, T.D., GRANNER, D., MARTIN, D., SAMUELS, H.H., THOMPSON, E.B.: Science **166**, 1474 (1969).

VAZQUEZ, D., STAEHELIN, T., CELMA, M.L., BATTANER, E., FERNANDEZ-MUNOZ, R., MONRO, R.E.: In Inhibitors: tools in cell research (BUCHER, T., SIES, H., eds.), p. 100. Berlin-Heidelberg-New York: Springer 1969.

VOGEL, Z., VOGEL, T., ZAMIR, A., ELSON, D.: J. Mol. Biol. **60**, 339 (1971).

WAHBA, A.J., IWASAKI, K., MILLER, J.J., SABOL, S., SILLERO, M.A.G., VASQUEZ, C.: Cold Spring Harbor Symp. Quant. Biol. **34**, 291 (1969).

WATERSON, J., BEAUD, G., LENGYEL, P.: Nature **227**, 34 (1970).

WEISSBACH, H., MILLER, D.L., HACHMANN, J.: Arch. Biochem. Biophys. **137**, 262 (1970).

WHITFIELD, H.J., GUTNICK, D.L., MARGOLIES, M.N., MARTIN, R.G., RECHLER, M.M., VOLL, M.J.: J. Mol. Biol. **49**, 245 (1970).

WILSON, D.B., DINTZIS, H.M.: Proc. Nat. Acad. Sci. U.S. **66**, 1282 (1970).

WILSON, D.B., HOGNESS, D.S.: J. Biol. Chem. **244**, 2143 (1969).

YARMOLINSKY, M., DE LA HABA, G.: Proc. Nat. Acad. Sci. U.S. **45**, 1721 (1959).

ZASLOFF, M., OCHOA, S.: Proc. Nat. Acad. Sci. U.S. **68**, 3059 (1971).

Inhibitors of Translation

P. R. STEWART

Introduction

Inhibitors of protein synthesis have been widely used as tools to explore the molecular events involved in the translation of mRNA. To localize antibiotic action to a particular event in translation, or to a site on the ribosome, a number of criteria may be used. These include: binding of labelled antibiotic to a particular subunit, or competition with the binding of an antibiotic with known properties; interference with the binding of a particular initiation or elongation factor to the ribosome; inhibition of hydrolytic and transacylation reactions that accompany peptide elongation; and use of antibiotic-resistant mutants.

Inhibitors of translation can be classified on the basis of their site and mode of action on the ribosome, and on the type of ribosomal systems (prokaryotic, eukaryotic) that they inhibit. Inhibition of protein synthesis may reflect an effect on any of the numerous steps comprising initiation, elongation or termination. Examination of polysome patterns in cells treated with antibiotic may provide information as to where in the ribosome cycle an antibiotic acts. Thus, inhibition of initiation would be expected to result in polysome breakdown. Inhibition of translocation should result in preservation of the polysomes of the cell. Inhibition of termination may cause ribosomes to accumulate along the mRNA, and result in the accumulation of polysomes of abnormally large size. However, there are exceptions, and it is unwise to rely too much on indirect data of this sort.

The properties and probable mode of action of the most important antibiotic inhibitors of translation at the ribosome level have been reviewed by PESTKA (1971) and by WEISBLUM and DAVIES (1968). Useful general information on the synthesis, isolation and chemical properties of these antibiotics is provided by KORZYBSKI et al. (1966) and by GOTTLIEB and SHAW (1967); the Merck Index is also useful in this respect.

This appendix is an attempt to condense the information on inhibitors of translation contained in those sources, plus more recent information, where available, to cover the inevitable gaps in this material. It should be read in conjunction with the description of protein synthesis given in Chapter 7.

A. General Properties of Translation Inhibitors

The general chemical and biochemical features of translation inhibitors are given in Table 1 and Fig. 1. These antibiotics fall into several groups: inhibitors of

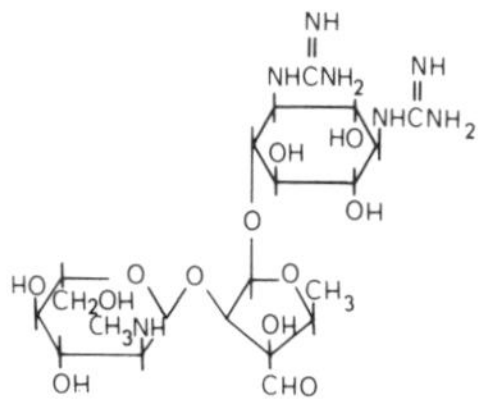

I   STREPTOMYCIN   (582)

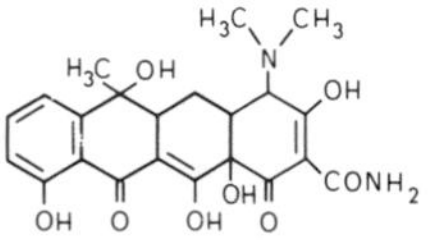

II   TETRACYCLINE   (444)

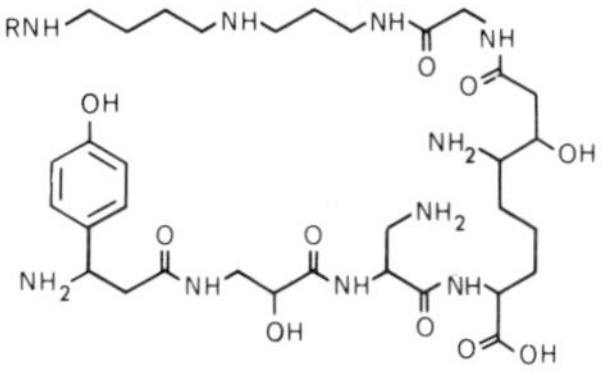

III   EDEINE   (747)

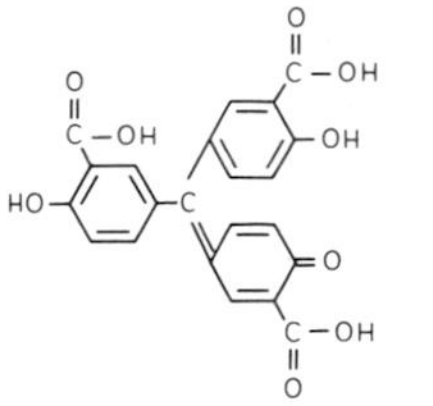

IV   AURINTRICARBOXYLIC ACID   (425)

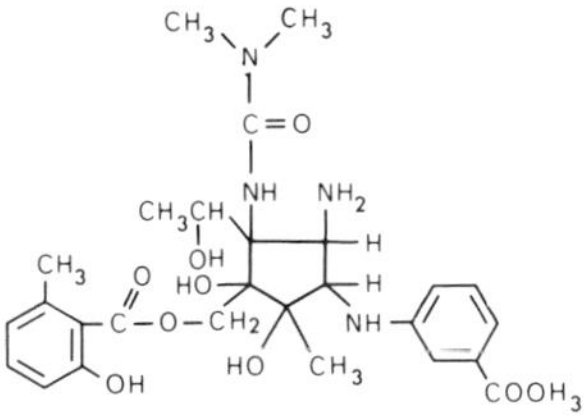

V   PACTAMYCIN   (558)

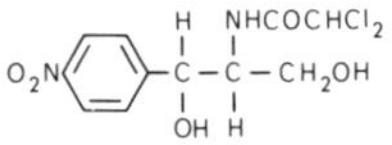

VI   CHLORAMPHENICOL   (323)

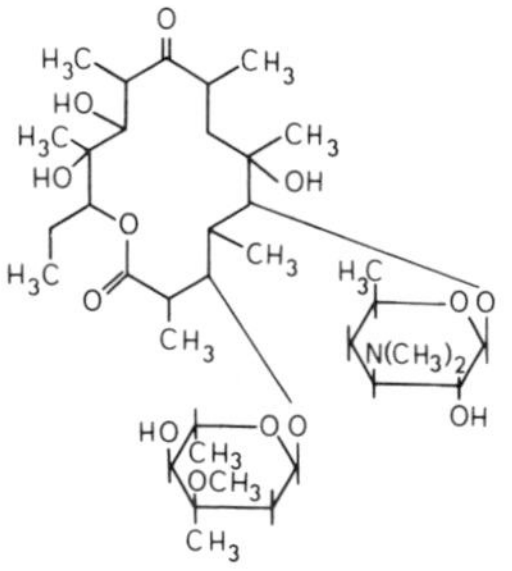

VII   ERYTHROMYCIN   (734)

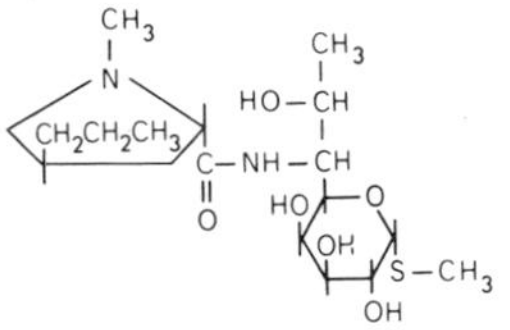

VIII   LINCOMYCIN   (407)

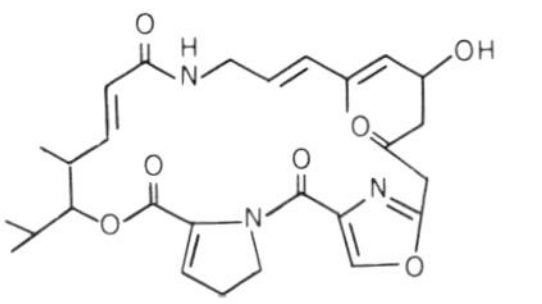

IX   OSTREOGRYCIN A   (523)

Fig. 1. Chemical structures of antibiotics representative of the groups shown in Table 1. Molecular weights are given in parentheses

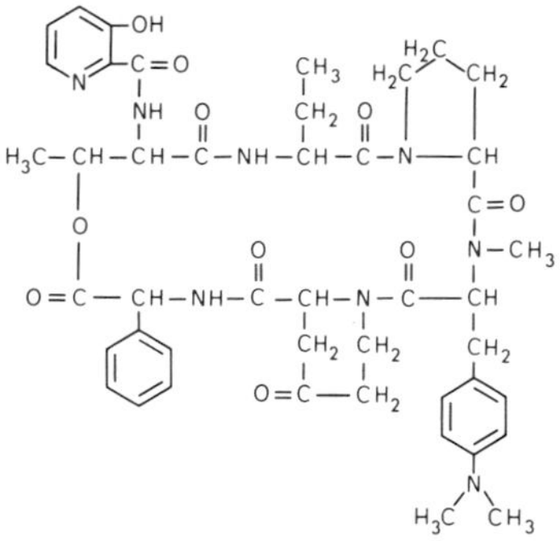

X OSTREOGRYCIN B (866)

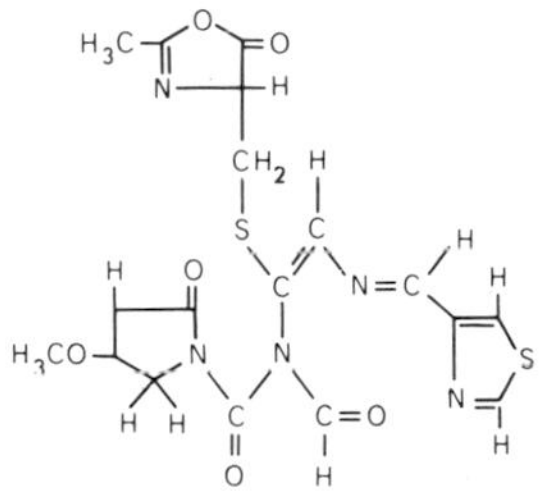

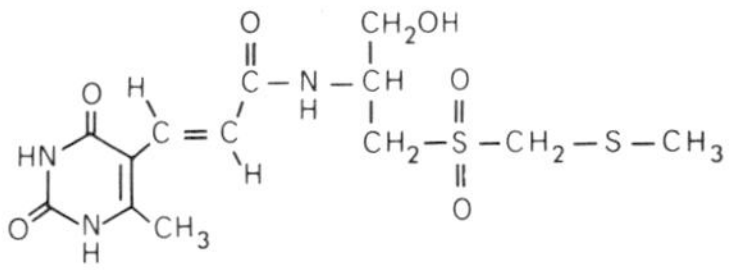

XIV SPARSOMYCIN (379)

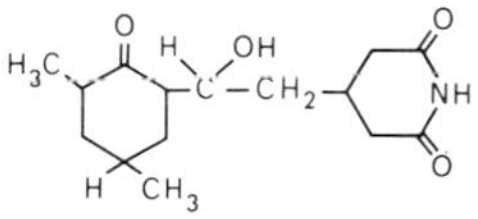

XI ACETYLALTHIOMYCIN (463)

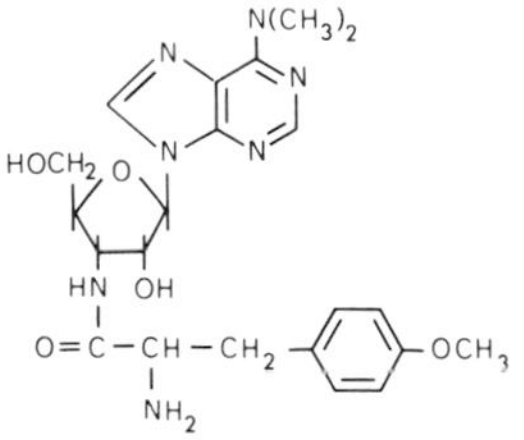

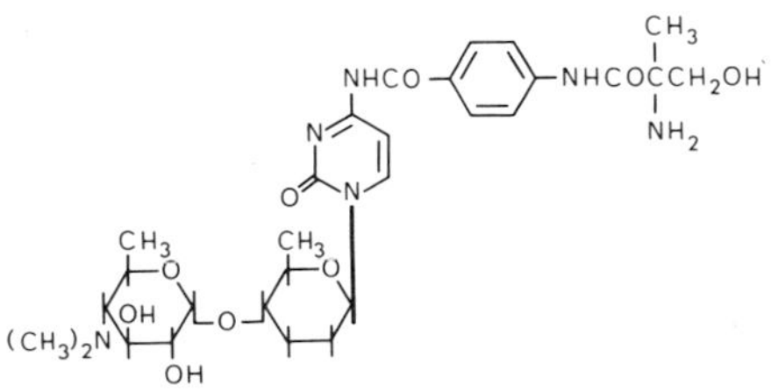

XV CYCLOHEXIMIDE (281)

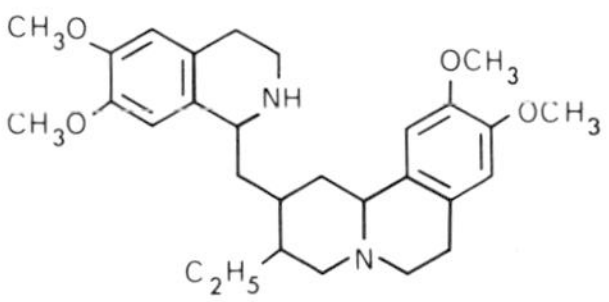

XII PUROMYCIN (472)

XVI EMETINE (481)

XIII AMICETIN (619)

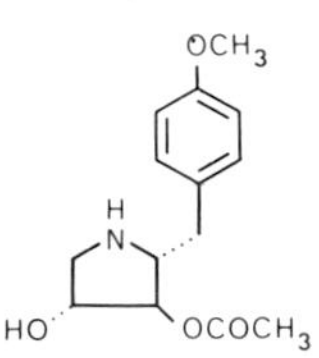

XVII ANISOMYCIN (265)

the functions of small ribosomal subunits, or of the functions of the large subunits. These can then be subdivided depending on whether eukaryote or prokaryote systems, or both, are inhibited. It is notable that except perhaps for pactamycin,

Table 1. Mode of action of translation inhibitors

Antibiotic	Chemical structure	Ribosome subunit affected	Effective concentration in vitro (μM)	Ribosome function inhibited				
				Initiation	Codon recognition	Transpeptidation	Translocation	Termination
Streptomycin	I	30S	30	+ +	+ +	+	+ +	+
Neomycin								
Kanamycin								
Spectinomycin			1	−	−	−	+ +	
Kasugamycin				−	+ +			
Tetracycline	II	30S	50	+	+ +	+	+ +	+ +
Edeine	III	30S	0.1	+ +	+	−	+ +	
Aurin tricarboxylic acid	IV	30S, 40S	10	+ +	−	−	−	−
Pactamycin	V	30S, 40S	1	+ +	+	−		
Chloramphenicol	VI	50S	20	−	−	+ +	+	+
Erythromycin	VII	50S	1	−	−	+ +	+ +	
Spiramycin								
Tylosin								
Niddamycin								
Carbomycin								
Lincomycin	VIII	50S	50		−	+ +		+
Streptogramins A	IX	50S	10[a]	+ +		+ +	+	
Streptogramins B	X	50S	10[a]				+ +	
Thiostrepton	XI	50S	0.5	−	−	+ +		
Puromycin	XII	50S, 60S	20	−	−	+ +	+ +	+
Amicetin	XIII	50S, 60S	2	−	−	+ +	+	+
Gougerotin								
Sparsomycin	XIV	50S, 60S	0.5	−	−	+ +	+ +	+
Cycloheximide	XV	60S	10	+ +		−	+ +	
Emetine	XVI	60S	50			−	+ +	
Anisomycin	XVII	60S	1			+ +		

Notes: Chemical structures: roman numerals refer to Fig. 1. Ribosome subunits: 50S, 30S large and small subunits respectively of prokaryote ribosomes; 60S, 40S large and small subunits of eukaryote ribosomes. Effective concentration *in vitro*: approximate concentration required to inhibit protein synthesis by 50% or more. Ribosome function inhibited: + + =inhibits function substantially at concentrations similar to those shown in previous column; + =small effect at low concentration, or effective only at high concentrations (>100 μM); − =no effect measurable; blank space indicates insufficient information available to assess inhibitory effects.

[a] Antibiotics of the streptogramin A and streptogramin B groups exhibit synergism (ENNIS, 1965; PESTKA, 1971). When present together, the concentration required for effective inhibition decreases about ten-fold (ENNIS, 1965) from that shown in this Table.

none of the antibiotics described specifically affect functions of the small (40S) eukaryote subunit.

The ribosome functions that are affected have been outlined earlier (Chapter 7). Three points need to be made in this respect. Firstly, antibiotics may produce effects remote from the primary event inhibited, as is described below. Secondly, the binding and efficacy of an antibiotic may be very dependent on the "ribosomal state", that is, the conformation of the ribosome-nucleic acid-polypeptide complex at any particular point in the translation cycle. The conformation of the complex determines the availability of antibiotic binding sites on the complex. Thirdly, and particularly relevant to experiments *in vitro*, ionic conditions are important determinants of ribosome structure; this applies particularly to mono- and divalent cation type and concentration. PESTKA (1971) has summarized these effects very adequately.

It should also be mentioned that effects established *in vitro* may not necessarily be relevant quantitatively and qualitatively to the situation *in vivo*. Emetine, for example, is concentrated by the cell (PESTKA, 1971); the absence of an effect of an antibiotic *in vivo*, on the other hand, may be due to permeability barriers at the cell surface.

B. Summary of Primary Effects *in vitro*

Table 1 summarizes the effects of the most common inhibitors of translation. The notes below should be read in conjunction with that Table, since they are intended as a summary of the probable *primary* effects of these antibiotics.

1. Inhibitors of Small Subunit Function

a) Aminoglycosides. Streptomycin, neomycin and kanamycin inhibit the binding of aminoacyl-tRNA to the A-site of the 30S subunit; they also induce codon misreading. Spectinomycin and kasugamycin inhibit mRNA interaction with the 30S subunit and aminoacyl-tRNA binding, but do not cause miscoding, presumably because they do not contain the streptamine residue in their structure. Other aminoglycosides which are not so well characterized are paromomycin, gentamycin and bluensomycin.

b) Tetracycline (Chlortetracycline). Inhibits the binding of aminoacyl-tRNA to the A-site of the 30S subunit.

c) Edeine. Inhibits aminoacyl-tRNA and fmet-tRNA binding to A- and P-sites of 30S subunit.

d) Aurin Tricarboxylic Acid. Prevents attachment of mRNA to ribosomes of prokaryotes and eukaryotes.

e) Pactamycin. Inhibits binding of initiator tRNA to 30S and 40S initiation complexes.

2. Inhibitors of Large Subunit Function

a) Chloramphenicol. Inhibits transpeptidation by preventing attachment of aminoacyl-tRNA to the P-site of 50S subunits.

b) Macrolides. In general these antibiotics inhibit transpeptidation by inhibiting attachment of peptidyl-tRNA and aminoacyl-tRNA to P- and A-sites of 50S subunits respectively.

c) Lincomycin. Interferes with correct binding of peptidyl-tRNA to the 50S subunit, and may also affect aminoacyl-tRNA binding; hence inhibits transpeptidation. This antibiotic is unusual, however, in that inhibition of protein synthesis *in vivo* may be due to a different effect from that exhibited *in vitro*, perhaps by interference with initiation (Pestka, 1971).

d) Streptogramins. Two groups of antibiotics which act synergistically. A group (streptogramin A, ostreogrycin A, synergistin A (PA 114A), vernamycin A, mikamycin A) inhibit aminoacyl-tRNA binding that is associated with initiation; they are thus less effective after protein synthesis has begun. The B group antibiotics (ostreogrycin B, synergistin B (PA 114B), etamycin, mikamycin B) enhance the binding of the A group to 50S subunits, and also act separately by preventing the release of peptidyl- or aminoacyl-tRNA from ribosomes.

e) Thiostrepton. Inhibitory action not well characterized; appears to inhibit ribosomal functions associated with 50S subunits that are integral to translocation. Others in this group include siomycin A, althiomycin, multhiomycin.

f) Puromycin. Resembles the aminoacyl-adenylyl end of aminoacyl-tRNA, and competing with charged tRNA accepts the nascent peptide transferred from the P-site on 50S or 60S subunits. This results in the premature release of incomplete polypeptide.

g) 4-Aminohexose Pyrimidine Nucleosides (gougerotin, amicetin, blasticidin S, plicacetin, bamicetin). Inhibit transpeptidation, by undefined mechanism, in eukaryote and prokaryote systems.

h) Sparsomycin. Inhibits binding of aminoacyl moiety of aminoacyl-tRNA to 50S or 60S subunit.

i) Glutarimide Antibiotics (cycloheximide, streptovitacin A). Cycloheximide inhibits the binding of initiator-tRNA, and also the release of deacylated tRNA from the A-site of 60S subunits. It thus inhibits initiation and elongation, though the latter process is less sensitive.

j) Emetine. The mode of action of this alkaloid is not definitely established, but it probably inhibits binding of aminoacyl-tRNA to the ribosome-mRNA complex.

Two other important inhibitors, whose structure or mode of action have not been completely characterized, should also be mentioned. Fusidic acid, a steroid antibiotic (Gottlieb and Shaw, 1967), appears to inhibit translocation on 70S ribosomes by preventing the dissociation of a ribosome—translocation factor (EF G)—GDP complex (Bodley et al., 1970). Diphtheria toxin, a polypeptide with molecular weight 62,000 daltons (Uchida et al., 1972), inactivates aminoacyl transferase II (EF2) in eukaryote cells by catalyzing the covalent attachment of NAD to the transferase (Honjo et al., 1969); translocation is thus inhibited.

C. Inhibitors of Organelle Protein Synthesis

The sensitivity of mitochondrial and chloroplast protein synthesis to chloramphenicol on the one hand, and insensitivity to cycloheximide on the other, now appears established as a general phenomenon. At the outset, it was thought that this indicated a close similarity of organelle and prokaryote protein synthesis. As further evidence has accumulated however, it appears that mitochondrial protein synthesis differs from prokaryotic protein synthesis in a number of ways, the most evident of these being the response to translation inhibitors. In primitive eukaryotes such as yeast, for example, the mitochondrial system is insensitive to kanamycin and streptomycin (DAVEY et al., 1970). The phenanthrene alkaloids cryptopleurine and tylocrebrine, on the other hand, inhibit mitochondrial but not bacterial protein synthesis (HASLAM et al., 1968).

However, it is also becoming evident that mitochondrial protein synthesizing systems differ from one another, depending on the organism from which they are isolated. Table 2 summarizes some of this information, and it can be seen that mitochondria from higher eukaryotes generally are more resistant to inhibitors of prokaryote protein synthesis, than are mitochondria from lower organisms. Thus, of the inhibitors shown in Table 2, yeast mitochondria are sensitive to them all, whereas mammalian mitochondria are insensitive to erythomycin, lincomycin, neomycin and paromomycin (TOWERS et al., 1972; cf. KROON and

Table 2. Sensitivity of organelle protein synthesis to ribosomal inhibitors

Protein synthesizing system	Cyclo-heximide	Eme-tine	Chlor-amphenicol	Macrolides			Aminoglycosides		Linco-mycin
				Erythro-mycin	Spira-mycin	Carbo-mycin	Par-omo-mycin	Neo-mycin	
Eukaryote 80S[a]	+	+	−	−	−	−	−	−	−
Prokaryote 70S	−	−	+	+	+	+	+	+	+
Yeast mito-chondria 70–80S	−		+	+	+	+	+	+	+
Insect mito-chondria 60S	−	±	+	−			+	+	−
Mammalian mitochondria 50–60S	−	+	+	−	+	+	−	−	−
Chloroplast 70S	−		+	+			+ (Spectinomycin)		+

Notes: Protein synthesizing system: S values indicate sedimentation coefficients of ribosomes in the sytem. Inhibition: + =antibiotic inhibits protein synthesis by more than 50% in isolated cytoplasmic ribosomal system (eukaryote, prokaryote) or by intact organelle system, at concentrations less than 100 µM; − =antibiotic does not inhibit significantly at this concentration; blank space indicates insufficient information available.

De Vries, 1971). Protein synthesis in mitochondria from the blowfly *Lucilia cuprina* is insensitive to erythromycin and lincomycin, but sensitive to the amino-glycosides listed; these mitochondria thus have properties apparently inter-mediate between the yeast and mammalian systems (Williams and Birt, 1972; Williams, 1972).

The basis for these differences between mitochondrial systems has not been determined. As noted in Chapter 8, the physical size of mitochondrial ribosomes is 70–80S in yeast and 50–60S in animals. The decreased size of ribosomes in animal mitochondria is in part due to a decrease in size of the ribosomal RNA, but it is likely that the ribosomal proteins have also been modified during their evolution, perhaps from a prokaryotic progenitor. This could account for the progressive loss of sensitivity to antibiotics apparent in the series yeast-insect-mammal (Table 2), since in bacterial systems it is known that binding sites for some antibiotics are associated with ribosomal proteins.

In the case of chloroplasts, protein synthesis is inhibited by the same anti-biotics as those that inhibit prokaryote and yeast mitochondrial protein synthesis (Table 2), though the information available in this respect is incomplete (Boulter et al., 1972).

References

Bodley, J.W., Lin, L., Sal, M.L., Tao, M.: FEBS Lett. **11**, 153 (1970).

> Boulter, D., Ellis, R.J., Yarwood, A.: Biol. Rev. (Cambridge) **47**, 113 (1972).

Davey, P.J., Haslam, J.M., Linnane, A.W.: Arch. Biochem. Biophys. **136**, 54 (1970).

Ennis, H.L.: J. Bacteriol. **90**, 1109 (1965).

Gottlieb, D., Shaw, P.D.: Antibiotics. Berlin-Heidelberg-New York: Springer 1967.

Haslam, J.M., Davey, P.J., Linnane, A.W., Atkinson, M.R.: Biochem. Biophys. Res. Commun. **33**, 368 (1968).

Honjo, T., Nishizuka, Y., Hayaishi, O.: Cold Spring Harbor Symp. Quant. Biol. **34**, 603 (1969).

Korzybski, T., Kowszyk-Gindifer, Z., Kurylowicz, W.: Antibiotics: Origin, nature and properties. Oxford: Pergamon Press 1967.

Kroon, A.M., De Vries, H.: In Autonomy and biogenesis of mitochondra and chloroplasts (Board-man, N.K., Linnane, A.W., and Smillie, R.M., eds.), p. 318. Amsterdam: North Holland 1971.

Pestka, S.: Ann. Rev. Microbiol. **25**, 487 (1971).

Towers, N.R., Dixon, H., Kellerman, G.M., Linnane, A.W.: Arch. Biochem. Biophys. **151**, 361 (1972).

Uchida, T., Pappenheimer, A.M., Harper, A.A.: Science **175**, 901 (1972).

Weisblum, B., Davies, J.: Bacteriol. Rev. **32**, 493 (1968).

Willams, K.L.: Ph. D. Thesis, Australian National University, Canberra, Australia (1972).

Willams, K.L., Birt, L.M.: FEBS Lett. **22**, 327 (1972).

CHAPTER 8

Mitochondrial RNA

P. R. STEWART

Introduction

Early studies of the origin and autonomy of cellular organelles were strongly though indirectly suggestive of the presence of nucleic acids in mitochondria (for reviews see NASS, 1970; RABINOWITZ and SWIFT, 1970). A serious consideration of this possibility, however, did not come until the early 1960s, partly because of the preoccupation with mitochondria as rigidly compartmentalized sites of energy transduction which was so characteristic of research on this organelle in the 1950s, and partly because experimentally the possibility was difficult to test. Thus, the first unequivocal demonstration that mitochondria contain DNA required histochemical analysis of organelles *in situ* (NASS et al., 1965). From this followed the proposal that information encoded in this DNA is expressed as the 'phenotype' of the organelle through autonomous systems of transcription and translation.

Substantial grounds for the belief that mitochondria contain functional RNA existed before any conclusive identification of unique species of RNA was made. Thus, MAGER's (1960) finding that mitochondria isolated from *Tetrahymena* were capable of protein synthesis was very suggestive. Electron microscopists had also noted particles in mitochondria that resembled the ribosomes of the surrounding cytoplasm (for review, see RABINOWITZ and SWIFT, 1970).

From the vantage of hindsight it is clear now that many of the early reports on the isolation of RNA or ribosomes from mitochondria were incorrect, at least to the extent that the material extracted probably came from contaminating cytoplasmic ribosomes and microsomes. It was not until 1967 that the extraction and characterization of RNA and ribosomes from mitochondria was detailed in reports which were subsequently confirmed by other workers and alternative techniques (ROGERS et al., 1967; O'BRIEN and KALF, 1967). It is from that point in time that this survey starts.

The question of whether mitochondria originated from prokaryotic symbionts will not be considered here, since it is effectively dealt with by NASS (1970) and by BORST and GRIVELL (1971). It is an interesting question, and as a premise has generated a most useful stimulus for novel and fruitful experimental approaches.

A. Experimental Problems Associated with the Isolation and Characterization of RNA from Mitochondria

The RNA found in mitochondria, so far as can be determined, is involved in the same events as those that have been characterized in the nucleus and cytoplasm: transcription of information from DNA, and its subsequent translation into particular sequences of amino acids as polypeptides. Protein synthesis involves all species of RNA that have so far been identified in mitochondria, and demonstration of this activity has sometimes been the basis for concluding that RNA species found associated with these organelles are intrinsic to them. It is therefore relevant to consider some of the problems and criteria involved in identifying protein synthesis in isolated mitochondria.

In the first place, it is necessary to exclude the possibility that the protein synthesis observed is due to contamination by cytoplasmic ribosomes. The cytoplasm usually is densely packed with free and membrane-bound ribosomes; ribosomal particles are less frequent in mitochondria by comparison (RABINOWITZ and SWIFT, 1970). Contamination by a small fraction of the total cytoplasmic ribosomes or microsomes is potentially capable of swamping the activity of the mitochondrial system and seriously interfering with the isolation of specific components. Fortunately for the experimenter, protein synthesis by intact mitochondria is "self-contained" in that it requires no added factors such as the soluble proteins and transfer RNA required for reconstituting the activity of washed cytoplasmic ribosomes. Furthermore, the response of mitochondrial protein synthesis towards inhibitors such as cycloheximide and chloramphenicol (discussed in more detail in Chapter 7: Appendix) is opposite to that of the cytoplasmic system. Response to inhibitors is probably the most generally used criterion of contamination by active cytoplasmic ribosomes. This criterion does not, however, account for contaminating but inactive cytoplasmic ribosomes, which would nevertheless continue to contribute contaminating RNA to mitochondria. Functional criteria must therefore be used with caution in assessing the physical purity of organelle preparations.

Another technical problem met with when measuring mitochondrial protein synthesis is the possible contamination of preparations by bacteria. Many bacteria are similar in size and density to mitochondria and thus if present become concentrated in the mitochondrial fraction during isolation and purification. Moreover, bacterial and mitochondrial protein synthesis are similar in a number of ways, particularly with respect to autonomy and response towards inhibitors, and thus discrimination between them is difficult. Contamination of mitochondrial fractions by bacteria during preparation and assay can usually be limited to 10^3 organisms per ml or less if appropriate precautions are taken. Since there are 10^9–10^{10} mitochondria per mg of protein (NASS, 1970), an amount that might be used in a routine assay of protein synthesis, bacteria at the level indicated would need to be at least one million times more active than mitochondria to make a significant contribution. It is unlikely that such a large difference exists (SWANSON and DAWID, 1970; GRIVELL et al., 1971 b), and there are other reasons for believing that at this level of contamination, bacteria do not account for the bulk of the activity observed in mitochondrial preparations (NASS, 1970; ASHWELL and WORK, 1970). More important to the matter of this chapter is that on the assumption that a bacterium contains approximately ten times as much RNA as a mitochondrion (NASS, 1970), contamination of up to 10^5–10^6 organisms per mg of mitochondrial protein could be tolerated in mitochondrial preparations without bacterial RNA species becoming quantitatively detectable.

The general conclusions that have so far been made concerning the physical and chemical properties of mitochondrial RNA and of mitochondrial ribosomes will now be considered. These should serve as a guide to investigators about to start an examination of mitochondrial RNA in new organisms. Further information of this type is provided by BORST and GRIVELL (1971).

B. Mitochondrial Ribosomes

1. Location within the Organelle

There has been no systematic examination of the location within mitochondria of individual RNA species. Since it is probable that the greater part of the RNA in mitochondria is a constituent of, or functionally associated with ribosomes, it is pertinent to consider where ribosomes may be located within the organelle. Electron-microscopic examination of isolated mitochondria and of mito- chondria *in situ* indicates that ribosomes may be free in the matrix, or associated with the cristae (VIGNAIS et al., 1969; RABINOWITZ and SWIFT, 1970). Early studies on the fractionation of mitochondria indicate that RNA sediments with membrane fragments (NASS, 1970); mitochondrial DNA and DNA-dependent RNA poly- merase are probably located in this fraction as well (see, for example, NASS, 1970; TSAI et al., 1971; VAN TUYLE and KALF, 1972). Extraction of ribosomes from yeast mitochondria has been reported to require both high ionic strength and detergent (SCHMITT, 1970). Under these conditions some residual RNA is left with the membranes, however, and there also may be loss of rRNA from the ribosomal fraction by solubilization (YU et al., 1972a).

Mouse-liver mitochondria, *in vivo* and *in vitro*, incorporate uridine into RNA that is localized on the periphery of the organelle, presumably on inter-cristal regions of the inner-membrane (CURGY, 1970). WERNER and NEUPERT (1972) report that in rat-liver, protein synthesis occurs on the inner peripheral membrane rather than on the cristae. These findings suggest that "growth sites" may exist on the inner membrane, where organellar protein and RNA synthesis are located.

2. Physical and Chemical Properties

Until perhaps two years ago, no consistent pattern of size of mitochondrial ribosomes was evident (ASHWELL and WORK, 1970); mitochondrial ribosomes with sedimentation coefficients ranging from 40S through 80S or more had been described. It is now clear that phylogenetic factors account for many of these differences. Thus, higher organisms such as mammals, amphibia, birds and insects contain mitochondrial ribosomes with sedimentation values of 50–60S. Primitive eukaryotes, typified by yeast and fungi, yield mitochondrial ribosomes with sedimentation values generally in the region 70–80S, that is, similar to values for cytoplasmic ribosomes of eukaryotes and bacteria. The dissociation of mitochondrial ribosomes into subunits generally requires lower concentrations of magnesium ion than is the case with bacterial ribosomes; in this respect they resemble the cytoplasmic ribosomes of eukaryotes (ASHWELL and WORK, 1970). A summary of the sizes of monosomes and subunits of mitochondrial and cytoplasmic ribosomes is given in Table 1. In organisms with the smaller mitochondrial ribosomes, the molecular size of mitochondrial DNA is also much smaller than in those with the larger mitochondrial ribosomes (ASHWELL and WORK, 1970). This is true also of the RNA isolated from the ribosomes, as will be discussed below.

C. Ribosomal RNA of Mitochondria from Yeast, Fungi and Protozoa

Among the yeasts, species of *Saccharomyces* have been studied most. Mitochondrial ribosomes from these organisms have been reported to have sedimentation coefficients of 72–73S (BORST and GRIVELL, 1971) or 80S (MORIMOTO and HALVORSON, 1971; YU et al., 1972b). Two major species of RNA are found in these ribosomes, with sedimentation coefficients approximating 22S and 15S. This compares with 26S and 17S RNA of the 80S cytoplasmic ribosomes from the same cells. The electrophoretic mobility on polyacrylamide gels of mitochondrial RNA is lower than would be predicted from the sedimentation coefficients (FORRESTER et al., 1970). S_E values* of 29 and 21 are found at room temperature, whereas at 2–5° these values decrease to 25 and 17; the mobility of cytoplasmic ribosomal RNA is less influenced by temperature (GRIVELL et al., 1971a; YU et al., 1972b). The proportion of guanine plus cytosine (G+C) in these RNA species has been variously reported as 24–35 mole %, compared with 46–48 mole % in cytoplasmic ribosomal RNA (ASHWELL and WORK, 1970; BORST and GRIVELL, 1971). This low content of guanine plus cytosine in mitochondrial rRNA probably accounts for the lower thermal denaturation temperature of mitochondrial ribosomes from *Saccharomyces* (MORIMOTO and HALVORSON, 1971).

Mitochondrial ribosomes have also been described from species of *Candida*, an obligately aerobic yeast. In *C. utilis* and *C. parapsilosis* mitochondrial ribosomes of 77–80S and 70S respectively have been reported (VIGNAIS et al., 1969; YU et al., 1972a). High molecular weight RNA species from mitochondrial ribosomes of *C. parapsilosis* have a lower G+C content, and a higher electrophoretic mobility (at 22°) than rRNA from cytoplasmic ribosomes (YU et al., 1972a).

In *Neurospora*, mitochondrial ribosomes have sedimentation values of 73S (KUNTZEL and NOLL, 1967) or 80S (RIFKIN et al., 1967), compared with 77–80S for cytoplasmic ribosomes. Again, the S_E values of mitochondrial ribosomal RNAs are unusually high (26 and 18) compared with sedimentation values determined by sedimentation velocity (23S and 16S contrasting with 26S and 17S for cytoplasmic ribosomal RNA). The G+C content (35–38 mole %) is

---

* S_E denotes the electrophoretic mobility of species relative to other species (usually cytoplasmic rRNA) with known sedimentation coefficients, S (LOENING, 1969; BORST and GRIVELL, 1971).

lower than that for cytoplasmic RNA (49–50 mole %). 5S RNA, which is found associated with the large subunit of eukaryotic and bacterial ribosomes and is thought to be essential for protein synthesis in those systems, is not detected in mitochondrial ribosomes from *Neurospora* (LIZARDI and LUCK, 1971).

In *Tetrahymena*, 80S-type mitochondrial ribosomes have been described. These could be distinguished from cytoplasmic particles by their resistance to dissociation at low magnesium ion concentration, a lower density, and the presence of 21S and 14S RNA containing 29 mole % G+C, compared with 26S and 17S RNA (46 mole % G+C) from cytoplasmic ribosomes (CHI and SUYAMA, 1970). An unusual feature is that both species of RNA can be extracted from a 55S subunit obtained by dissociation of the 80S particle with EDTA. This might be taken to mean that the 80S mitochondrial particle isolated is a dimer of the 55S particle, or a ribosome-membrane complex. Attached membrane fragments would also explain the anomalously low density of the 80S particle (BORST and GRIVELL, 1971). In this respect it is of interest that 55S particles appear to be significantly active in protein synthesis, forming 80S particles in the presence of polyuridylic acid (ALLEN and SUYAMA, 1972). This has been interpreted to mean that the 55S particles represent subunits that become active 80S monosomes in the presence of messenger RNA. However, it is also possible that this indicates disome formation from 55S monosomes.

Physico-chemical properties of mitochondrial RNA from three species of filamentous fungi (*Aspergillus nidulans*, *Neurospora crassa*, *Trichoderma viride*) have been compared by EDELMAN et al. (1971). The general features are similar, in that they have lower G+C content than their cytoplasmic counterparts, show lower thermal denaturation temperatures, and exhibit anomalous behaviour during gel electrophoresis at low salt and low magnesium ion concentrations. However there are considerable differences in the detailed behaviour of mitochondrial RNA from these species. For example, RNA from *Neurospora* mitochondria has the highest G+C content (38 mole %). The sedimentation and thermal denaturation properties of this RNA are more similar to the corresponding cytoplasmic species than is the case in *Aspergillus* or *Trichoderma*. The high molecular weight RNA from *Aspergillus* mitochondria derives from 70S-type ribosomes (EDELMAN et al., 1970).

Two species of RNA (14S and 11S), with low G+C content, have been isolated from *Euglena* mitochondria (KRAWIEC and EISENSTADT, 1970). General features of mitochondrial ribosomes from primitive eukaryotes, and their constituent rRNA species, are summarized in Table 1.

D. Ribosomal RNA of Mitochondria from Higher Organisms

It is now generally accepted that ribosomes from animal mitochondria (human and other mammalian tissue culture, rat, cow, pig, rabbit, toad, chicken, locust) are 50–60S type, with 25–35S and 33–45S small and large subunits (Table 1; O'BRIEN, 1971). Major RNA species from these ribosomes have sedimentation

Table 1. Physical and biochemical properties of mitochondrial ribosomes and ribosomal RNA

Organism	Mono-some	Sub-units	RNA	G+C content	References
	sedimentation coefficient			mole %	
Man (HeLa cells)	60	45, 35	16, 12	45 (65)[a]	ATTARDI and ATTARDI (1971)
Rat (liver)	55	40, 30	16, 13	47 (64)	BORST and GRIVELL (1971)
Mouse (L-cells)	—	—	16, 13	37 (59)	MONTENECOURT et al. (1970)
Hamster (BHK-21 cells)	50	33, 25	17, 13	38 (57)	DUBIN and MONTENECOURT (1970)
Chicken (liver)	55	—	18, 12[b]	—	RABBITS and WORK (1971)
Xenopus (eggs)	60	43, 32	17, 13	38	DAWID (1972)
Locust (flight muscle)	60	40, 25	16, 13[b]	—	KLEINOW et al. (1971)
Tetrahymena	80	55	21, 14	29 (46)	CHI and SUYAMA (1971)
Neurospora	73–80	50, 37	23, 16	37 (50)	RIFKIN and LUCK (1967) KUNTZEL and NOLL (1967)
Aspergillus	70	50, 32	24, 16	32 (51)	EDELMAN et al. (1970)
Saccharomyces	72–80	55, 38	21, 15	28 (46)	SCHMITT (1970) GRIVELL et al. (1971 b) MORIMOTO and HALVORSON (1971) YU et al. (1972 b)
Candida	70–77	50, 38	24, 16[b]	35 (46)	VIGNAIS et al. (1969) YU et al. (1972 a)

Sedimentation coefficients are calculated relative to monosome (70S) and subunits (30S, 50S) of E. coli, and of RNA derived from these (23S, 16S), or by comparison with cytoplasmic ribosomes (in most cases assumed to be 80S), subunits (60S, 40S) and RNA (25–29S, 17–19S) from the cells concerned

[a] Figures in parentheses indicate G+C content of cytoplasmic ribosomal RNA where these are known.
[b] Calculated from polyacrylamide gel electrophoresis.

values correspondingly lower than their cytoplasmic counterparts. Thus, the larger rRNA species from mitochondria has a sedimentation coefficient of 16–19S, which is similar to, or slightly lower than the smaller of the cytoplasmic rRNA species. The smaller rRNA species from animal mitochondria is generally 12–13S (BORST and GRIVELL, 1971). Sedimentation coefficients and electrophoretic mobilities of RNA from ribosomes of rat-liver mitochondria show temperature dependence similar to that of yeast mitochondrial rRNA (GROOT et al., 1970). The G+C content of mitochondrial rRNA from higher organisms is also lower than that of cytoplasmic species, though generally not as low as that of mitochondrial RNA from primitive eukaryotes. These features of animal mitochondrial rRNA are summarized in Table 1.

E. Unusual Properties of Mitochondrial rRNA

From the observations summarized above, it can be inferred that regardless of the source, mitochondrial ribosomal RNAs have two unusual physical features when compared with the cytoplasmic counterparts. In the first place, the mobility of mitochondrial RNA in acrylamide gels is lower than that predicted from sedimentation data. Secondly, sedimentation and electrophoretic characteristics, particularly the latter, are very sensitive to changes in ionic strength and temperature. An illustration of the practical consequences of this is given in Fig. 1. When mixtures of mitochondrial and cytoplasmic rRNA from *Candida* or *Saccharomyces* are subjected to electrophoresis on acrylamide gels at room temperature (22°), the mitochondrial RNA species from *Candida* have higher mobilities than the corresponding cytoplasmic species, as would be expected from their lower sedimentation coefficients (Table 1). In the case of *Saccharomyces*, however, the mitochondrial species have *lower* mobilities, as well as lower sedimentation coefficients. This state of affairs changes as the temperature is raised or lowered. Thus, the

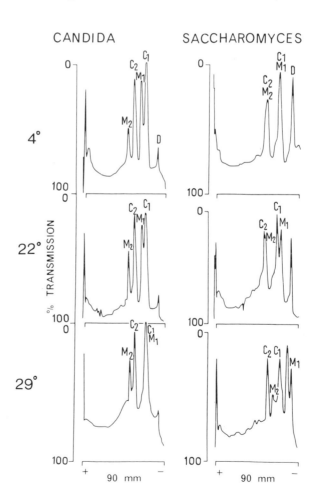

Fig. 1. Effect of temperature on the electrophoretic mobilities of mitochondrial and cytoplasmic RNA species from yeasts. RNA was extracted from mitochondria isolated from *Candida* and *Saccharomyces*, then subjected to electrophoresis at the temperatures indicated. D: DNA; M_1, M_2: large and small rRNA from mitochondria; C_1, C_2: large and small rRNA from cytoplasmic ribosomes contaminating the mitochondria (from YU et al., 1972b)

resolution of mitochondrial RNAs from cytoplasmic species increases at lower temperature in the case of *Candida*, but the converse is true in the case of *Saccharomyces*. This apparently opposing effect of temperature on the resolution of mixtures of cytoplasmic and mitochondrial RNA in polyacrylamide gels, however, does not reflect opposing effects of temperature on the intrinsic mobilities of mitochondrial RNA from the two organisms. At lower temperature the mobility of mitochondrial species increases, presumably due to more compact folding of the molecules. The mobility of cytoplasmic rRNA species on the other hand is negligibly affected by the same temperature variation (Fig. 1). This, together with the anomalous mobility of *Saccharomyces* mitochondrial rRNA relative to cytoplasmic rRNA, accounts for the differential effect of temperature on resolution of RNA species from these two organisms (YU et al., 1972b).

An explanation for these unusual characteristics is possibly provided by the low G+C content of mitochondrial RNA, which would allow greater unfolding of these molecules in low concentrations of salt and divalent cation, and at high temperatures. But, as GRIVELL et al. (1971a) point out, this may be an oversimplification, and it may be that in mitochondrial rRNA the melting of a few critical base-paired regions leads to very large changes in conformation of the molecules. Whatever the explanation, this unusual property of mitochondrial RNA has so far made it difficult to obtain accurate estimates of molecular weight using hydrodynamic techniques.

Two laboratories have reported molecular weight determinations on mitochondrial rRNA by contour length measurements. RNA is spread under denaturing conditions, dried and shadowed, then directly visualized in an electron microscope. In *Aspergillus* mean chain lengths of 0.47 and 0.91 μm have been determined in high concentrations of urea, compared with 0.52 and 1.10 μm for cytoplasmic species (VERMA et al., 1970). Assuming an internucleotide distance of 2.45 Å, molecular weights of 0.66 and 1.27×10^6 daltons for mitochondrial RNA, and 0.73 and 1.54×10^6 daltons for cytoplasmic species were calculated; these agree closely with values obtained from gel electrophoresis. Using 4M urea and high concentrations of formamide, the molecular weights of 12S and 16S RNA from mitochondria of HeLa cells were calculated to be 0.35 and 0.54×10^6 daltons, by direct comparison with 18S cytoplasmic RNA (assumed to be 0.71×10^6 daltons) (ROBBERSON et al., 1971). These were significantly lower than values deduced from gel electrophoresis, and suggested an internucleotide spacing of 2.6–2.7 Å. The latter authors commented that strongly denaturing conditions are needed to achieve complete rupture of the secondary structure of the RNA before an effective comparison of molecules with varying G+C content can be made.

The bases of cytoplasmic rRNA from HeLa cells are methylated to the extent of 1.4 methyl groups per 100 nucleotides; mitochondrial rRNA from the same cells contains about one percent methylated ribose (ATTARDI and ATTARDI, 1971). VESCO and PENMAN (1969) and DUBIN and CZAPLICKI (1971) however have reported that methylation is very low or absent from mammalian mitochondrial rRNA. NOLL (1970) reports more than 5 percent ribose methylation of mitochondrial rRNA from *Neurospora*. More experiments are clearly required to decide whether there is significant methylation of mitochondrial rRNA.

F. Transfer RNA

Mitochondria from *Saccharomyces cerevisiae* contain tRNA species for formyl-methionine, methionine, isoleucine, glycine, alanine, phenylalanine, tyrosine, leucine and valine which hybridize specifically with mitochondrial DNA (CASEY et al., 1972). Calculations from the hybridization data suggest that there is no more than one copy of each tRNA gene per copy of the genome. A full complement of tRNA species has been found in mitochondria from *Neurospora*. Most of these appear to be distinct from tRNA species occurring in the surrounding cytoplasm (REGER et al., 1970; FAIRFIELD and BARNETT, 1971). Mitochondrial tRNA appears to have similar mobility and sedimentation properties as cytoplasmic and bacterial tRNA; it resembles the latter in being devoid of a certain type of fluorescent base (FAIRFIELD and BARNETT, 1971).

Mitochondrial DNA and tRNA from rat liver, *Xenopus* eggs, and mammalian tissue culture cells are also homologous and as discussed below, mitochondrial tRNA from HeLa and *Xenopus* cells hybridizes with mitochondrial DNA to an extent that would suggest that there is one copy of each tRNA species per copy of the genome.

The relationship between cytoplasmic and mitochondrial amino acid-tRNA ligases is an interesting one. The cytoplasmic enzyme from rat liver appears to have a high specificity towards tRNA acceptor species of the cytoplasm in that the enzyme is 17–75 times more active with these homologous substrates than with mitochondrial tRNA. The mitochondrial enzyme, on the other hand, shows ratios of activity towards cytoplasmic and mitochondrial tRNA species of 0.7–2, that is, it is about equally active with tRNA species from either cell compartment (BUCK and NASS, 1969).

Formylmethionyl-tRNA has been identified in yeast and rat-liver mito-chondria, but not in the cytoplasm from these cells (SMITH and MARCKER, 1969). Its presence in mitochondria has since been confirmed in human culture and *Neurospora* cells (GALPER and DARNELL, 1969; EPLER et al., 1970).

tRNA from all sources examined is methylated (Chapter 5). The extent of ribose methylation of mitochondrial tRNA from HeLa cells is less than one half that of the cytoplasmic tRNA (10 methyls per 100 nucleotides) from the same cells (ATTARDI and ATTARDI, 1971).

G. The Synthesis of Mitochondrial RNA

1. Genes for Mitochondrial RNA

The presence of DNA in mitochondria has been established by electron micro-scopy, autoradiography and isolation techniques (ASHWELL and WORK, 1970). In all cases the DNA isolated is double stranded, and is almost certainly circular. In many cases mitochondrial DNA differs significantly from nuclear DNA in buoyant density, implying a difference in base composition; this has been con-firmed by direct analysis. The $G+C$ content of mitochondrial DNA may be higher, lower, or the same as nuclear DNA from the same organism (Ashwell and Work, 1970). In animals, mitochondrial DNA with a molecular weight of approximately 10×10^6 daltons, is present as supercoiled circles of 4–6 μm. In lower eukaryotes such as yeast and *Neurospora* the molecular weight is 50–55×10^6, in circles of 25–26 μ (Ashwell and Work, 1970). In *Tetrahymena*, linear molecules about 17 μ long have been described (Suyama and Miura, 1968). Separation of mitochondrial DNA into heavy and light strands, that is, complementary strands with different base composition, can be achieved in alkaline cesium chloride gradients (Clayton and Vinograd, 1967; Borst and Aaij, 1969).

Mitochondrial DNA is generally considered to be the site of information that encodes most mitochondrial RNA. In all cases so far tested, high molecular weight RNA, and certain species of tRNA, hybridize with mitochondrial DNA. Because of the small size of mitochondrial DNA, the number of copies of genes for these species of RNA in each molecule of organellar DNA is limited. There may be duplication of the genes for certain mitochondrial RNA species in the nuclear genome, though proof of this one way or the other is difficult.

The evidence for these conclusions has been derived predominantly from studies of animal mitochondria, though the work of Fukuhara with yeast is also important. Fukuhara (1970) examined carefully whether sequence homologies for mitochondrial RNA existed in nuclear and mitochondrial DNA, as had been reported by earlier workers (for review see Ashwell and Work, 1970). Fukuhara considered that apparent homology could be due to the presence in mitochondrial preparations of a mixture of transcripts of mitochondrial and nuclear DNA. This is particularly likely when the purity of the mitochondria has not been rigorously established, so that cytoplasmic nuclear transcripts could contaminate the isolated mitochondria. Fukuhara showed that treatment with RNAase of hybrids of "mitochondrial" RNA/mitochondrial DNA eliminated most of the RNA that was crosshybridizable with nuclear DNA. The fraction of "mitochondrial" RNA that formed true hybrid with mitochondrial DNA was resistant to the RNAase treatment, and could subsequently be dehybridized from the mitochondrial DNA. This RNA was found to be rehybridizable with mitochondrial DNA, but did not hybridize significantly with nuclear DNA. The corresponding experiment with nuclear DNA/"mitochondrial" RNA, showed that after RNAase treatment of the hybrid, the dehybridized, intact RNA had no affinity for mitochondrial DNA. It was concluded therefore that there could be at most only a very small degree of sequence homology between nuclear and mitochondrial DNA in terms of transcripts of mitochondrial RNA. Nevertheless, it is highly probable that mitochondria *in situ* may contain RNA transcribed from both nuclear DNA and mitochondrial DNA, and this possibility of mixtures of transcripts must be borne in mind in considering evidence as to the genetic origin of mitochondrial RNA.

The comprehensive study of the properties, origin and synthesis of mito-chondrial RNA in HeLa cells by Attardi and co-workers must be considered an exemplar for investigators in this field. By combining methods of chemical characterization and kinetic labelling with techniques for hybridization and ultrastructural analysis by electron microscopy, the biochemistry of the major stable species of mitochondrial RNA have been advanced further in this organism than in any other. Application of this general approach to organellar RNA in other organisms should soon be reported, and the results will be of the utmost interest in giving meaning to the phylogenetic differences discussed earlier. It is appropriate to consider the work of ATTARDI's group in some detail.

ALONI and ATTARDI (1971b) found that total mitochondrial RNA, uniformly labelled by growing HeLa cells for long periods in the presence of ^{3}H-uridine, hybridized almost exclusively with the heavy strand of mitochondrial DNA. Since the base composition of the heavy strand is more nearly complementary to mitochondrial RNA than is the light strand (Table 2), this result was not unexpected. Between 85 and 100% of the heavy strand was hybridized by the total

Table 2. Base composition (mole %) of heavy and light strands of mitochondrial DNA and of RNA transcribed from these strands in HeLa cells

		A	T or U	G	C
DNA	H-strand	22.2	31.8	46	
	L-strand	31.8	22.2	46	
Stable RNA	9–25S	33.9	22.6	24.5	18.9
				43.4	
	26–48S	31.4	25.3	23.9	19.4
				43.3	

From ALONI and ATTARDI (1971b).

complement of mitochondrial RNA, indicating that all or almost all of the sequences of the heavy strand of mitochondrial DNA are actively transcribed into mitochondrial species of RNA. Titration of the heavy strand of DNA with purified 16S and 12S mitochondrial rRNA gave saturation values of 13% and 9% respectively, or 19% in combination (ALONI and ATTARDI, 1971c). Assuming that the molecular weight of the heavy strand of HeLa mitochondrial DNA is 5×10^6 daltons, and that the preparations were homogeneous, this level of hybridization amounts to about 10^6 daltons of RNA. The combined saturation level, though not quite corresponding to that expected if the hybridization was strictly additive, does indicate that the two species of RNA contain predominantly different sequences. This, together with kinetic labelling evidence showing no precursor-product relationship between the two species of ribosomal RNA (ATTARDI and ATTARDI, 1971), was taken to mean that the 16S and 12S cistrons are distinct. A more direct proof of this is given by electron microscopy of hybrids of DNA heavy strand and 12S and 16S RNA (ROBBERSON et al., 1972). From the micrographs presented it appears that the length of hybrid formed between 12S plus 16S RNA and mitochondrial DNA equals, within error, the sum of the lengths of DNA hybrids formed with 12S and 16S RNA separately. That the genes for 12S and 16S RNA are adjacent or very close to each other is shown by the absence of unhybridized DNA "bushes" internally in the duplex rRNA-DNA structures formed.

Furthermore, since the molecular weights of 16S and 12S RNA are 0.50×10^6 and 0.35×10^6 daltons (ROBBERSON et al., 1971), the extent of hybridization described earlier indicates that a single gene exists for each of these species of RNA per molecule of mitochondrial DNA. It should be added that this does not necessarily mean one gene copy per mitochondrion, since there are almost certainly multiple copies of DNA in each organelle (ASHWELL and WORK, 1970).

Low molecular weight RNA (4S) from mitochondria gave a saturation level of 4% with heavy strand of mitochondrial DNA, indicating eight genes for molecules of molecular weight 25,000. Light strand mitochondrial DNA gave small, probably insignificant, hybridization values at saturation with the stable high molecular weight RNA species (ALONI and ATTARDI, 1971c). With 4S RNA, however, hybridization was more significant at 1.5%, which would correspond to three genes for molecules of the size of tRNA. ALONI and ATTARDI (1971c) suggested that these few genes on the light strand of DNA may represent an initiation or termination region for the transcription of the circular mitochondrial DNA duplex.

Summation of the hybridization figures for stable species (that is, ribosomal plus 4S RNA) hybridizable with the heavy strand of DNA gives only about 25% of the total base pairs available on this DNA strand. The remainder presumably is concerned with unstable species, for example mRNA, or spacer nucleotides, since most if not all of this DNA is transcribed, as mentioned earlier. Furthermore, transcription appears to result in the formation of very large single molecules, which presumably are subsequently processed into the smaller functional RNA units.

These results showing that transcription of rRNA and tRNA in mitochondria is from the heavy strand of the DNA of the organelle, confirmed and extended the findings of BORST and AAIJ (1969). ATTARDI's group proceeded further to show that the light strand of mitochondrial DNA is not inert so far as transcription is concerned. Apart from the small amount of information transcribed in the three tRNA species mentioned above, it was found that when labelling of transcribed RNA was examined for very short (pulse) periods, RNA was synthesized in mitochondria which hybridized extensively with the light strand of mitochondrial DNA, as well as with the heavy strand (ALONI and ATTARDI, 1971a). As the labelling period was increased, the relative labelling of RNA from heavy strand compared with light strand also increased.

The RNA formed on light strand was found to have a very high molecular weight. Failure to recover it from mitochondria except immediately subsequent to its synthesis means either that it is exported from the organelle, or is rapidly degraded. It is notable that earlier work from ATTARDI's laboratory describes RNA of mitochondrial origin associated with a membranous protein synthesizing fraction located elsewhere in the cytoplasm (ATTARDI and ATTARDI, 1968).

The general pattern of RNA synthesis observed in mitochondria from HeLa cells has also been demonstrated in the toad *Xenopus laevis* (DAWID, 1972; DAWID and CHASE, 1972). Mitochondrial rRNA and 4S RNA were found to be complementary to mitochondrial DNA, and to have no detectable sequence homology with the corresponding cytoplasmic species of RNA. Each circle of mitochondrial DNA duplex (12×10^6 daltons) contains one gene each for the large and small rRNA, and about 15 genes for tRNA. Together these account for 20% of the information content of the heavy strand of mitochondrial DNA. Uncharacterized mitochondrial RNA represents an additional 16% of the mitochondrial information. The remaining two thirds of the sequence of mitochondrial DNA is represented very rarely or not at all in the stable RNA recovered from mitochondria isolated from *Xenopus*.

Single genes for rRNA on mitochondrial DNA also appears to be the case for *Tetrahymena* (CHI and SUYAMA, 1970), yeast (MORIMOTO et al., 1971) and *Neurospora* (SCHAFER and KUNTZEL, 1972).

A scheme indicating the essential features of the relationship between mitochondrial DNA and mitochondrial RNA in animal cells is shown in Fig. 2.

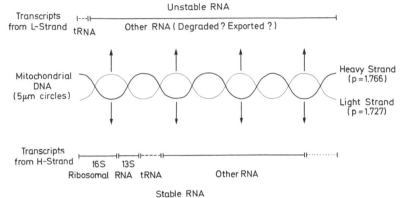

Fig. 2. Synthesis of mitochondrial RNA from mitochondrial DNA in HeLa cells. One copy each of the two rRNA species, and 8 molecules of tRNA are transcribed from the heavy strand of DNA. Other, uncharacterized, RNA species of high molecular weight are also transcribed, while up to 15% of the information in this strand is not recovered in transcription products. From the light strand, about 3 copies of tRNA are transcribed together with unstable RNA of very high molecular weight. The mitochondrial DNA duplex is circular; the figures in parentheses indicate the densities of the two strands (from ALONI and ATTARDI, 1971a, b, c)

2. Transcription in Mitochondria

a) RNA Polymerase and Inhibitors of Transcription. A substantial amount of evidence indicates that mitochondrial DNA is transcribed inside the organelle by a mitochondrial DNA-dependent RNA polymerase (SACCONE et al., 1967; TSAI et al., 1971; REID and PARSONS, 1971). The enzyme from mitochondria of yeast (TSAI et al., 1971), *Neurospora* (KUNTZEL and SCHAFER, 1971) and rat liver (REID and PARSONS, 1971) has been solubilized and purified. In each case, the enzyme is insensitive to α-amanitin, which inhibits nuclear RNA polymerase. The yeast and *Neurospora* enzymes are more active with mitochondrial DNA as template, compared with calf-thymus DNA. The enzyme from yeast mitochondria is not inhibited by rifampicin, whereas the enzymes from *Neurospora* and rat liver are sensitive to rifamycin, a closely allied antibiotic.

RNA synthesis by isolated mitochondria, variously damaged or fragmented, has been found in some but not all cases to be sensitive to rifampicin; membrane barriers or crypticity may be an important factor in these experiments.

Preferential inhibition *in vivo* of the synthesis of mitochondrial RNA, compared with cytoplasmic RNA, by ethidium bromide has been widely observed. Actinomycin D, on the other hand, appears to inhibit preferentially the synthesis of cytoplasmic rRNA precursor in the nucleolus (see Chapter 3), and this antibiotic has been used to amplify the labelling of RNA by mitochondria *in situ* (BORST and GRIVELL, 1971). Cordycepin (3'-deoxyadenosine) has also been used as a preferential inhibitor of mitochondrial RNA synthesis (PENMAN et al., 1970).

b) Turnover of Mitochondrial RNA. The turnover, or metabolic stability, of mitochondrial RNA has been measured using a number of indirect techniques. In the presence of ethidium bromide, 12S and 16S RNA disappear rapidly from HeLa mitochondria (PENMAN et al., 1970; ATTARDI and ATTARDI, 1971). However, the decay kinetics are unusual, and in the absence of inhibitor the kinetics of labelling of these RNA species is not a great deal faster than that for cytoplasmic rRNA (ATTARDI and ATTARDI, 1971). It is thus possible that the decay of the mitochondrial species of RNA in these conditions is to some extent an artefact induced by the drug treatment.

The rate at which mitochondrial protein synthesis decays either *in vivo* or *in vitro* in the presence of inhibitors of mitochondrial RNA synthesis has also been used in an attempt to measure RNA turnover. In intact yeast cells, the half-life of mitochondrial protein synthesis in the presence of ethidium bromide or euflavin is 10 minutes or less (WEISLOGEL and BUTOW, 1971; GROOT et al., 1972). In HeLa cells it is 1–3 hours (LEDERMAN and ATTARDI, 1970; ZYLBER et al., 1971). With isolated animal mitochondria, values of less than 10 minutes have been found (GAMBLE and MCCLUER, 1970). These values do not necessarily reflect turnover of mitochondrial RNA, since the inhibitors of RNA synthesis used may inhibit protein synthesis directly, or inhibit other processes on which protein synthesis is dependent.

c) Relationship of Transcription to the Cell Cycle. Examination of mitochondrial RNA synthesis *in vivo* in synchronized human and mouse tissue cultures indicates that there are peaks of activity in the S and G2 phases of the cell cycle (BOSMANN, 1971; PICA-MATTOCCIA and ATTARDI, 1971). However, mitochondria isolated from mouse cells show enhanced activity only in G2 phase. Mitochondrial protein synthesis is maximal in G1 and G2 phases *in vivo*, and in G1 phase only *in vitro* (BOSMANN, 1971).

3. Synthesis of Mitochondrial Ribosomal Proteins

Earlier studies suggested that the proteins of mitochondrial ribosomes are made by the cytoplasmic ribosomes (NEUPERT et al., 1969; KUNTZEL, 1969; DAVEY et al., 1969). MILLIS and SUYAMA (1972) have re-examined this matter more carefully and conclude that some mitochondrial ribosomal proteins may be made by the mitochondrial protein synthesizing system, though apparently under very stringent regulatory control by the cytoplasmic system. This control is such that inhibition of cytoplasmic protein synthesis by cycloheximide results in complete cessation of synthesis of mitochondrial ribosomal proteins and with this a very substantial inhibition of synthesis of mitochondrial rRNA.

4. Mitochondrial RNA and Development

A number of studies have been carried out to relate changes in the rate of mitochondrial RNA synthesis, or in the types of RNA synthesized, to changes in developmental patterns in the cell or organism concerned. These studies have naturally focused on organisms in which there is good reason to believe that altered mitochondrial function is closely related to the developmental response observed.

a) Effects of Hormones in Animal Systems. Several studies have been made of effects of corticoid and thyroid hormones on the synthesis of mitochondrial RNA. It has been proposed that certain short-term effects of hormones, such as the thyroxine stimulation of liver microsomal protein synthesis, are mediated by a primary effect on mitochondria in the cells (SOKOLOFF et al., 1968). SCHIMMELPFENNIG et al. (1970) examined RNA synthesis in rat liver mitochondria and found that uridine incorporation into RNA was inhibited by thyroidectomy, and stimulated by subsequent treatment with thyroid hormone. However, a similar response was noted in liver nuclei, isolated from treated rats, indicating that the mitochondrial effect was not specific. NUSSDORFER and MAZZOCHI (1971) noted similar effects of adrenocorticotrophic hormone on nuclear and mitochondrial RNA synthesis in rat adrenocortical cells. They suggested that this might indicate a co-ordinate synthesis of RNA in the cells to program protein synthesis, both mitochondrial and cytoplasmic, which result in altered development of mitochondria in these cells.

MANSOUR and NASS (1970) found that physiological concentrations of cortisol in rats specifically stimulated RNA polymerase in liver mitochondria. YU and FEIGELSON (1970) observed a stimulation of incorporation of uridine, but not guanine, into mitochondrial RNA in rat liver up to 4 hours after treatment with cortisone; similar effects were seen with cytoplasmic and nuclear RNA, suggesting again that the effect of the hormone either is not specific for mitochondria, or induces both mitochondrial and extramitochondrial synthesis that leads to altered mitochondrial development.

b) Induced Development of Mitochondria in Yeast. In *Saccharomyces cerevisiae*, release of catabolite repression of respiratory function is accompanied by an increased synthesis of many mitochondrial enzymes (for review see ROODYN and WILKIE, 1968). Preceding this synthesis of protein is an increase in the RNA polymerase activity of mitochondria (SOUTH and MAHLER, 1968). The polymerase activity of isolated mitochondria reaches a maximum 3–4 hours after derepression begins, then rapidly declines. The activity of cytochrome oxidase and malate dehydrogenase in these preparations, on the other hand, is maximal after 5–7 hours. The incorporation of labelled uracil during short periods of incubation (10 minutes) of the mitochondria was found to be predominantly into 11S RNA. When the incorporation period was increased to 15 minutes, most of the counts were in the 16S region on sucrose gradients, which corresponds to the smaller mitochondrial ribosomal RNA species from this organism. The activity of the polymerase was found to be sensitive to low concentrations of proflavine and ethidium bromide, which are potent inhibitors of respiratory adaptation and bind very effectively to mitochondrial DNA from this yeast

(ROODYN and WILKIE, 1968). Glucose repression may thus regulate the synthesis of messenger RNA species that code either for the synthesis of mitochondrial proteins on cytoplasmic ribosomes, or for membrane proteins synthesized on mitochondrial ribosomes. The latter could then regulate the synthesis of other mitochondrial proteins on cytoplasmic ribosomes.

In *Saccharomyces*, the transition from anaerobic to aerobic growth conditions is accompanied by very substantial changes in the constitution and function of mitochondrial membrane in the cells. This system has been investigated to establish whether new patterns of mitochondrial RNA synthesis are also a consequence of exposure of these cells to oxygen (FUKUHARA, 1967). It was found that aerobically and anerobically grown yeast both contain RNA hybridizable with mitochondrial DNA. However, although the total content of mitochondrial DNA is little different between the two cell types (FUKUHARA, 1969), a substantially larger amount of this mitochondrial DNA could be hybridized with RNA from aerobically grown cells than with RNA from the anaerobes. A proportion of this RNA consisted of metabolically stable RNA with a rather broad sedimentation coefficient centred around 13S. On the basis of more recent findings with this organism, this probably represents partially degraded mitochondrial rRNA. Metabolically unstable RNA (RNA whose radioactivity declines rapidly in a chase experiment) may also have been present but was not characterized. The synthesis of both high molecular weight and transfer RNA from yeast mitochondria during aeration of anaerobically grown cells is specifically inhibited by ethidium bromide and euflavine (FUKUHARA and KUJAWA, 1970).

c) Cytoplasmic Respiratory Mutants of Yeast and Neurospora. Mutations that affect respiratory function and which are inherited in a non-Mendelian fashion have been described in *Saccharomyces (petite)* and *Neurospora (poky)*. The *petite* mutation is frequently accompanied by changes in the type and amount of mitochondrial DNA present in the cell. Since rRNA, tRNA and possibly mRNA are transcribed from this DNA, these mutants provide a most useful means to examine questions concerning the nature of transcripts formed in the organelle, and the control mechanisms exerted on transcription both within and outside the mitochondrion.

WINTERSBERGER (1967) showed that a *petite* mutant of *Saccharomyces cerevisiae* contained no high molecular weight RNA. Since mutant mitochondria contain RNA polymerase, though possibly in smaller amounts (WINTERSBERGER, 1970; TSAI et al., 1971), absence of RNA could be due to the absence of DNA from the organelle, to faulty transcription, or to loss of the products of transcription during isolation and purification of the mitochondria and RNA. The absence of intact rRNA from *petite* mitochondria was confirmed by KELLERMAN et al. (1971) and by YU (1971), the former using chemostat cultures limited for glucose so that the loss of activity could not be ascribed to catabolite repression, and the latter using glutaraldehyde prefixation before isolation of the mitochondria to limit losses from the organelles during isolation. *Petite* mitochondria *in situ* or isolated from cells do not have protein synthesizing activity (SCHATZ and SALTZGABER, 1969; KUZELA and GRECNA, 1969), confirming the findings on the absence of RNA.

In *petite* mutants in which mitochondrial DNA has been altered to varying degrees, hybridization studies with mitochondrial leucine tRNA and valine tRNA show that genes for these tRNA species may be retained and replicated in the altered DNA of subsequent generations (COHEN et al., 1972). The conservation of small functional cistrons of this sort in *petite* mitochondrial DNA that may be otherwise grossly altered in base composition and sequence may provide linkage data that will be most useful for the genetic mapping of mitochondrial DNA (COHEN et al., 1972).

The *poky* mutation in *Neurospora* is expressed differently compared with the *petite* mutation in yeast. While the latter mutation is irreversible, the phenotypic expression of *poky* to a considerable extent is dependent on the physiological state of the culture. Thus, during logarithmic phase of growth, cytochromes *a* and *b* are formed in unusually small amounts and at the same time the rate of synthesis of the small subunits of mitochondrial ribosomes is very much less than that of the large subunits. Consequently normal levels of ribosomes in the organelle are not maintained in log-phase cells (RIFKIN and LUCK, 1971). In stationary phase, cytochrome content and the proportion of normal ribosomes in the mitochondria rise; the rate of synthesis of small subunits is thus adequate when generation of new mitochondrial mass is low. These results could be explained in terms of altered gene-dosage in the mutant for the mitochondrial ribosomal RNAs, or by changes in the rate of transcription of these cistrons on mutant mitochondrial DNA. Other explanations clearly are also possible, for instance involving the synthesis of ribosomal proteins. Nevertheless, this area of organelle development is only beginning to be explored and information from *poky* and *petite* mutants should provide important insight into the involvement of mitochondrial RNA in developmental processes.

H. Concluding Remarks

It is clear that very significant progress has been made in our understanding of the properties of mitochondrial RNA in the period covered by this review. Nevertheless, the field still contains contradictions, inconsistencies and unresolved questions. Most of our knowledge in this field, with important exceptions, is derived from organisms from extreme ends of the phylogenetic scale— mammals on the one hand, yeasts and fungi on the other. The evolutionary significance of the differences in the properties and size of mitochondrial nucleic acids between the mammals and fungi may not be clear until organisms representing intermediate stages of evolution of eukaryotes have been examined. But the tempo of investigation in the field is high, and this information should soon be available.

References

Reviews

ASHWELL, M., WORK, T.S.: Ann. Rev. Biochem. **39**, 251 (1970).
BORST, P., GRIVELL, L.A.: FEBS Lett. **13**, 73 (1971).
NASS, S.: Int. Rev. Cytol. **27**, 55 (1970).
RABINOWITZ, M., SWIFT, H.: Physiol. Rev. **50**, 376 (1970).

Other References

Allen, N.E., Suyama, Y.: Biochim. Biophys. Acta **259**, 369 (1972).
Aloni, Y., Attardi, G.: Proc. Nat. Acad. Sci. U.S. **68**, 1757 (1971a).
Aloni, Y., Attardi, G.: J. Mol. Biol. **55**, 251 (1971b).
Aloni, Y., Attardi, G.: J. Mol. Biol. **55**, 271 (1971c).
Attardi, B., Attardi, G.: Proc. Nat. Acad. Sci. U.S. **61**, 261 (1968).
Attardi, B., Attardi, G.: J. Mol. Biol. **55**, 231 (1971).
Borst, P., Aaij, C.: Biochem. Biophys. Res. Commun. **34**, 205 (1969).
Bosmann, H.B.: J. Biol. Chem. **246**, 3817 (1971).
Buck, C.A., Nass, M.M.K.: J. Mol. Biol. **41**, 67 (1969).
Casey, J., Cohen, M., Rabinowitz, M., Fukuhara, H., Getz, G.S.: J. Mol. Biol. **63**, 431 (1972).
Chi, J.C.H., Suyama, Y.: J. Mol. Biol. **53**, 531 (1970).
Clayton, D.A., Vinograd, J.: Nature **216**, 652 (1967).
Cohen, M., Casey, J., Rabinowitz, M., Getz, G.S.: J. Mol. Biol. **63**, 441 (1972).
Curgy, J.J.: Exp. Cell Res. **62**, 359 (1970).
Davey, P., Yu, R.S., Linnane, A.W.: Biochem. Biophys. Res. Commun. **36**, 30 (1969).
Dawid, I.B.: J. Mol. Biol. **63**, 201 (1972).
Dawid, I.B., Chase, J.W.: J. Mol. Biol. **63**, 217 (1972).
Dubin, D.T., Czaplicki, S.M.: Biochim. Biophys. Acta **224**, 663 (1971).
Dubin, D.T., Montenecourt, B.S.: J. Mol. Biol. **48**, 279 (1970).
Edelman, M., Verma, I.M., Herzog, R., Galun, E., Littauer, U.Z.: Eur. J. Biochem. **19**, 372 (1971).
Edelman, M., Verma, I.M., Littauer, U.Z.: J. Mol. Biol. **49**, 67 (1970).
Epler, J.L., Shugart, L.R., Barnett, W.E.: Biochemistry **9**, 3575 (1970).
Fairfield, S.A., Barnett, W.E.: Proc. Nat. Acad. Sci. U.S. **68**, 2972 (1971).
Forrester, I.T., Nagley, P., Linnane, A.W.: FEBS Lett. **11**, 59 (1970).
Fukuhara, H.: Biochim. Biophys. Acta **134**, 143 (1967).
Fukuhara, H.: Eur. J. Biochem. **11**, 135 (1969).
Fukuhara, H.: Mol. Gen. Genetics **107**, 58 (1970).
Fukuhara, H., Kujawa, C.: Biochem. Biophys. Res. Commun. **41**, 1002 (1970).
Galper, J.B., Darnell, J.E.: Biochem. Biophys. Res. Commun. **34**, 205 (1969).
Gamble, J.G., McCluer, R.H.: J. Mol. Biol. **53**, 557 (1970).
Grivell, L.A., Reijnders, L., Borst, P.: Eur. J. Biochem. **19**, 64 (1971a).
Grivell, L.A., Reijnders, L., Borst, P.: Biochim. Biophys. Acta **247**, 91 (1971b).
Groot, G.S.P., Rouslin, W., Schatz, G.: J. Biol. Chem. **247**, 1735 (1972).
Groot, P.H.E., Aaij, C., Borst, P.: Biochem. Biophys. Res. Commun. **41**, 1321 (1970).
Kellerman, G.M., Griffiths, D.E., Hansby, J.E., Lamb, A.J., Linnane, A.W.: In: Autonomy and biogenesis of mitochondria and chloroplasts (Boardman, N.K., Linnane, A.W., and Smillie, R.M., eds.), p. 346. Amsterdam: North Holland 1971.
Kleinow, W., Neupert, W., Bucher, T.: FEBS Lett. **12**, 129 (1971).
Krawiec, S., Eisenstadt, J.M.: Biochim. Biophys. Acta **217**, 132 (1970).
Kuntzel, H.: Nature **222**, 142 (1969).
Kuntzel, H., Noll, H.: Nature **215**, 1340 (1967).
Kuntzel, H., Schafer, K.P.: Nature New Biol. **231**, 265 (1971).
Kuzela, S., Grecna, E.: Experientia **25**, 776 (1969).
Lederman, M., Attardi, G.: Biochem. Biophys. Res. Commun. **40**, 1492 (1970).
Lizardi, P.M., Luck, D.J.L.: Nature New Biol. **229**, 140 (1971).
Loening, U.E.: Biochem. J. **113**, 131 (1969).
Mager, J.: Biochim. Biophys. Acta **38**, 150 (1960).
Mansour, A.M., Nass, S.: Nature **228**, 665 (1970).
Millis, A.J.T., Suyama, Y.: J. Biol. Chem. **247**, 4063 (1972).
Montenecourt, B.S., Langsam, M.E., Dubin, D.T.: J. Cell Biol. **46**, 245 (1970).
Morimoto, H., Halvorson, H.O.: Proc. Nat. Acad. Sci. U.S. **68**, 324 (1971).
Morimoto, H., Scragg, A.H., Nekhorocheff, J., Villa, V., Halvorson, H.O.: In Autonomy and biogenesis of mitochondria and chloroplasts (Boardman, N.K., Linnane, A.W., and Smillie, R.M., eds.), p. 282. Amsterdam: North Holland 1971.
Nass, S., Nass, M.M.K., Hennix, U.: Biochim. Biophys. Acta **95**, 426 (1965).
Neupert, W., Sebald, W., Schwab, A.J., Massinger, P., Bucher, T.: Eur. J. Biochem. **10**, 589 (1969).

NOLL, H.: In Control of organelle development (MILLER, P. L., ed.), p. 419. Cambridge: Cambridge University Press 1970.

NUSSDORFER, G. G., MAZZOCHI, G.: Z. Zellforsch. **118**, 35 (1971).

O'BRIEN, T. W.: J. Biol. Chem. **246**, 3409 (1971).

O'BRIEN, T. W., KALF, G. F.: J. Biol. Chem. **242**, 2172 (1967).

PENMAN, S., FAN, H., PERLMAN, S., ROSBASH, M., WEINBERG, R., ZYLBER, E.: Cold Spring Harbor Symp. Quant. Biol. **35**, 561 (1970).

PICA-MATTOCCIA, L., ATTARDI, G.: J. Mol. Biol. **57**, 615 (1971).

RABBITS, T. H., WORK, T. S.: FEBS Lett. **14**, 214 (1971).

REGER, B. J., FAIRFIELD, S. A., EPLER, J. L., BARNETT, W. E.: Proc. Nat. Acad. Sci. U.S. **67**, 1207 (1970).

REID, B. D., PARSONS, P.: Proc. Nat. Acad. Sci. U.S. **68**, 2830 (1971).

RIFKIN, M. R., LUCK, D. J. L.: Proc. Nat. Acad. Sci. U.S. **68**, 287 (1971).

RIFKIN, M. R., WOOD, D. D., LUCK, D. J. L.: Proc. Nat Acad. Sci. U.S. **58**, 1052 (1967).

ROBBERSON, D., ALONI, Y., ATTARDI, G., DAVIDSON, N.: J. Mol. Biol. **60**, 473 (1971).

ROBBERSON, D., ALONI, Y., ATTARDI, G., DAVIDSON, N.: J. Mol. Biol. **64**, 313 (1972).

ROGERS, P. J., PRESTON, B. N., TITCHENER, E. B., LINNANE, A. W.: Biochem. Biophys. Res. Commun. **27**, 405 (1967).

ROODYN, D. B., WILKIE, D.: The biogenesis of mitochondria. London: Methuen 1968.

SACCONE, C., CADALETA, N. N., QUAGLIARELLO, E.: Biochim. Biophys. Acta **138**, 474 (1967).

SCHAFER, K. P., KUNZEL, H.: Biochem. Biophys. Res. Commun. **46**, 1312 (1972).

SCHATZ, G., SALTZGABER, J.: Biochem. Biophys. Res. Commun. **37**, 996 (1969).

SCHIMMELPFENNIG, K., SAUERBERG, M., NEUBERT, D.: FEBS Lett. **10**, 269 (1970).

SCHMITT, H.: Eur. J. Biochem. **17**, 278 (1970).

SMITH, A. E., MARCKER, K. A.: J. Mol. Biol. **38**, 241 (1969).

SOKOLOFF, L., ROBERTS, P. A., JANUSKA, M. M., KLINE, J. E.: Proc. Nat. Acad. Sci. U.S. **60**, 652 (1968).

SOUTH, D. J., MAHLER, H. R.: Nature **218**, 1226 (1968).

SUYAMA, Y., MIURA, K.: Proc. Nat. Acad. Sci. U.S. **60**, 235 (1968).

SWANSON, R. F., DAWID, I. B.: Proc. Nat. Acad. Sci. U.S. **66**, 117 (1970).

TSAI, M. J., MICHAELIS, G., CRIDDLE, R. S.: Proc. Nat. Acad. Sci. U.S. **68**, 473 (1971).

VAN TUYLE, G. C., KALF, G. F.: Arch. Biochem. Biophys. **149**, 425 (1972).

VERMA, I. M., EDELMAN, M., HERZBERG, M., LITTAUER, U. Z.: J. Mol. Biol. **52**, 137 (1970).

VESCO, C., PENMAN, S.: Proc. Nat. Acad. Sci. U.S. **62**, 218 (1969).

VIGNAIS, P. V., HUET, J., ANDRE, J.: FEBS Lett. **4**, 234 (1969).

WEISLOGEL, P. O., BUTOW, R. A.: J. Biol. Chem. **246**, 5113 (1971).

WERNER, S., NEUPERT, W.: Eur. J. Biochem. **25**, 379 (1971).

WINTERSBERGER, E.: Hoppe-Seylers Z. Physiol. Chem. **348**, 1701 (1967).

WINTERSBERGER, E.: Biochem. Biophys. Res. Commun. **40**, 1179 (1970).

YU, F. L., FEIGELSON, P.: Biochim. Biophys. Acta **213**, 134 (1970).

YU, R. S.: Ph. D. Thesis, Australian National University, Canberra, Australia 1971.

YU, R. S., POULSON, R., STEWART, P. R.: Mol. Gen. Genetics **114**, 339 (1972a).

YU, R. S., POULSON, R., STEWART, P. R.: Mol. Gen. Genetics **114**, 325 (1972b).

ZYLBER, E., PERLMAN, S., PENMAN, S.: Biochim. Biophys. Acta **240**, 588 (1971).

CHAPTER 9

Chloroplast RNA

P. R. WHITFELD

Introduction

The first convincing demonstration that chloroplasts contain RNA came from the observation by LYTTLETON (1962) that one of the two species of ribosomes which can be distinguished in leaf cells is actually localized within the chloroplasts. Confirmation of the presence of RNA in these organelles resulted from the direct analysis of purified chloroplasts which showed that as much as 25% to 35% of the leaf cell RNA could be in the chloroplasts (HEBER, 1963). The identification of chloroplast-specific ribosomal RNAs (rRNA) was followed by evidence for the existence of chloroplast-specific transfer RNAs (tRNA) and messenger RNAs (mRNA). Chloroplasts are thus analogous to mitochondria in that they contain ribosomal and transfer RNA species which can be clearly differentiated from the corresponding cytoplasmic RNA species. That the RNA plays an active role in the functioning of chloroplasts follows from the observation that transfer of dark-grown plants to the light brings about a significant change in the level of certain of the chloroplast RNA species, and from the fact that isolated chloroplasts can synthesize protein. In this chapter an attempt will be made to provide a general yet practical introduction to the subject of chloroplast RNA (both algal and higher plant) but an exhaustive review of the literature will not be undertaken. Recent articles which contain sections dealing with chloroplast RNA are: KIRK, 1970; KIRK and TILNEY-BASSETT, 1967; LOENING, 1968; SMILLIE and SCOTT, 1969; TEWARI, 1971; WOODCOCK and BOGORAD, 1971.

A. Isolation of Chloroplasts

Although chloroplast ribosomal RNA can be prepared by fractionation of a total leaf RNA extract, its preparation is greatly facilitated by the prior isolation of the organelle. For the preparation of other chloroplast RNA species such a preparatory step is usually essential. Methods for the isolation of chloroplasts are numerous and detailed descriptions can be found in articles by SPENCER (1967), GRAHAM and SMILLIE (1971), STOCKING (1971) and WALKER (1971). The method of choice in any situation will depend partly on the precise aim of the experiment, and partly on the nature of the plant species being used as source material. The major concern is to select conditions which will minimize the possibility of degradation of the RNA during purification of the organelles but which will yield a highly enriched fraction of chloroplasts free of cytoplasmic contamination. Two basically different procedures are available.

1. Nonaqueous Method

This involves freeze-drying the leaf material, blending or grinding the dried residue in hexane-carbon tetrachloride and separating the chloroplasts from other subcellular components by means of density gradient fractionation in organic solvents (STOCKING, 1971). The virtue of this method is that water-soluble, readily-diffusible compounds are not lost from the organelles during the purification procedure. On the other hand contamination of the chloroplasts with fragments of cytoplasm can be a problem unless the tissue-disruption step is carefully controlled. This procedure is usually only applied to the isolation of soluble RNA species.

2. Aqueous Method

Disruption of the plant tissue is a crucial step in this method because plant cell walls are tough and it is difficult to select conditions which are drastic enough to break the cell walls but which are not so severe that they also fragment the chloroplasts. This is especially the case with the unicellular algae, *Chlamydomonas* and *Chlorella*, which contain a single chloroplast almost as large as the cell itself. Chopping with razor blades or blending in a commercial blender are the two most satisfactory methods for macerating higher plants (SPENCER, 1967). Blending for approximately 4 seconds has the distinct advantage of speed but it can result in considerable shearing of nuclei which may cause difficulties in the subsequent purification steps. Razor-chopping is not satisfactory for large quantities (>50 g) of tissue unless a semi-automatic machine is available but it leaves the nuclei more or less intact. *Euglena, Chlorella* and *Chlamydomonas* can be disrupted by passage through a French pressure cell (GRAHAM and SMILLIE, 1971). In the case of *Euglena*, the cells can be broken by homogenizing in a Waring blendor for 50 seconds if they have first been exposed to trypsin (RAWSON and STUTZ, 1969).

The medium used for cell disruption contains, in addition to buffer and an SH-reducing agent, an osmoticum, usually sucrose or salt but it may be sorbitol or mannitol or sugar polymers such as Ficoll or dextran. Magnesium ions and ethylenediaminetetraacetate, individually or together, and isoascorbate are frequently included constituents (WALKER, 1971).

Purification of the chloroplasts from the filtered, cell-free extract may involve simply one or two very brief centrifugations in which case the pellet is considered to be an "enriched" chloroplast fraction, contaminated by nuclear material. The initial centrifugation should be carried out as rapidly as possible so as to reduce to a minimum the time during which the chloroplasts are exposed to the vacuolar sap and to other deleterious substances resulting from the disruption of peroxisomes, spherosomes etc. Further purification of the "enriched" chloroplast pellet can be achieved by zonal centrifugation through continuous or discontinuous density gradients of glycerol or sucrose (e.g. TEWARI and WILDMAN, 1966), or by flotation from dense glycerol solution (e.g. JAGENDORF, 1955; KUNG and WILLIAMS, 1969). These additional purification steps tend to increase the likelihood of the outer chloroplast membrane being damaged with concomitant loss of stromal material, particularly in situations where there is a lot of starch in the chloroplasts. Also, the relative effectiveness of the fractionation procedure is very much a function of the physiological state of the leaf. In young, rapidly expanding leaves the size range of chloroplasts and nuclei can overlap to a marked degree and, to achieve satisfactory separation of the two types of organelle, considerable numbers of the chloroplasts may have to be sacrificed.

B. Extraction of Leaf and Chloroplast RNAs

Methods for the extraction of leaf and chloroplast nucleic acids are basically the same as those which are used for the extraction of RNA from other tissues and which are described in detail in Chapter 11. Leaf tissue can be conveniently reduced to a very fine powder by grinding in a mortar with liquid nitrogen. Before the powdered leaf thaws out, buffer, detergent and phenol are added and the usual extraction routine is followed. Another very effective method for tissue disintegration is to blend the leaves at high speed directly in a mixture of buffer, detergent and phenol.

For whole-leaf RNA preparations an alkaline buffer such as Tris-glycine, pH 9.5, may be used in the initial extraction in order to counteract the acids released from the cell vacuole (CLICK and HACKETT, 1966) and to maintain the pH of the extract well above pH 5.5, which is the optimum pH for the ribonuclease most prevalent in plants. Where purified chloroplasts or ribosomes are the source of RNA, a more neutral buffer will suffice. A divalent cation such as Mg^{2+} is frequently included in the buffer because it helps to stabilize the large ribosomal RNA component of the chloroplast (LEAVER and INGLE, 1971).

Of the many detergent-type compounds which can be included in the extracting buffer, sodium dodecyl sulphate, 4-aminosalicylate, naphthalene-1,5-disulphonate and tri-isopropylnaphthalene sulphonate have all been used with success (LOENING and INGLE, 1967; DYER and LEECH, 1968; LEAVER and INGLE, 1971). Phenol, m-cresol—phenol—8-hydroxyquinoline, or phenol-chloroform-isoamylalcohol are the preferred protein denaturants because chlorophyll partitions readily into the organic phase. It should be noted, however, that with certain plant tissues the use of phenol results in very low recoveries of ribosomal RNAs and it is necessary to substitute chloroform as the denaturing agent (INGLE and BURNS, 1968).

Leaf and chloroplast nucleic acids may also be successfully isolated using the diethylpyrocarbonate-sodium dodecyl sulphate method of SOLYMOSY et al. (1970).

Further purification of the RNA can be achieved by extraction with methoxyethanol to remove polysaccharides, followed by precipitation with cetyltrimethylammonium bromide (BELLAMY and RALPH, 1968), by digestion with deoxyribonuclease, by banding in cesium sulphate density gradients, by filtration through

Sephadex G50 or G100 columns, or, of course, by various combinations of these procedures.

The RNA is then fractionated into the different molecular species by chromatography on methylated serum albumin-kieselguhr (MAK) columns (DYER and LEECH, 1968), by DEAE-cellulose chromatography (PAYNE and DYER, 1971a), by sucrose density gradient centrifugation (STUTZ and NOLL, 1967) or by polyacrylamide gel electrophoresis (LOENING and INGLE, 1967). If separation into high and low molecular weight species is all that is desired, then gel filtration or high-speed centrifugation (30,000 g for 17 hours) is satisfactory (PAYNE and DYER, 1971a). Alternatively, high molecular weight ribosomal RNA may be precipitated from a total nucleic acid extract by adjusting the solution to 2M sodium chloride, or low molecular weight RNA may be selectively extracted from a pellet of total nucleic acid by extracting with 3M sodium acetate, pH 6 (KIRBY, 1965).

C. Transfer RNA

Evidence for the existence of low molecular weight RNAs in chloroplasts came initially from the analysis of chloroplast RNA preparations on MAK columns (WOLLGIEHN et al., 1966; RUPPEL, 1967; DYER and LEECH, 1968). The possibility that the chloroplast soluble RNAs were different from plant cytoplasmic soluble RNAs was verified by comparison of the MAK elution profiles of total leaf RNA, of chloroplast RNA and of root RNA from *Vicia faba* (DYER and LEECH, 1968). Subsequent examination of these RNAs using the greater resolving power of polyacrylamide gel electrophoresis (DYER et al., 1971) established that the mean size of chloroplast tRNAs was slightly greater than that of the cytoplasmic tRNAs and, in addition, chloroplasts contained several small components intermediate in size between 4S and 5S RNAs which were absent from the cytoplasm. It is now generally accepted that chloroplasts contain many of their own specific tRNAs and it is likely that there is a full complement of tRNAs to support the protein synthesis machinery of chloroplasts.

1. Formylmethionine tRNA

Interest in the general question of the origin of chloroplasts and particularly in the possible relationship between chloroplasts and prokaryotic organisms was no doubt responsible for the deliberate search for the presence of formylmethionine tRNA in chloroplasts. This tRNA is the initiating tRNA for protein synthesis in prokaryotes and is a useful characteristic for distinguishing protein synthesis in prokaryotic systems from that in eukaryotic systems.

Formylmethionine tRNA has been identified as a component of chloroplasts of bean (*Phaseolus*), wheat and cotton. If the tRNA fraction from *Phaseolus* chloroplasts is incubated with ^{35}S-methionine, formyltetrahydrofolate and chloroplast extract and the RNA subsequently recovered and digested with pancreatic ribonuclease, one of the products can be identified as ^{35}S-formylmethionyl-adenosine (BURKARD et al., 1969). In the case of wheat, there is a methionyl-tRNA in the chloroplasts which can be formylated by a transformylase present in wheat extracts and which has been shown to function as formylmethionyl-tRNA in chain initiation in chloroplast protein synthesis (LEIS and KELLER, 1970). Wheat cytoplasm, on the other hand, contains a methionyl-tRNA which is not formylated by the endogenous transformylase and functions in peptide chain initiation as methionyl-tRNA. Likewise in chloroplasts isolated from cotyledons of germinating cotton there are two methionyl-tRNAs, only one of which can be formylated either by an endogenous transformylase or by an E. coli formylase (MERRICK and DURE III, 1971). The cytoplasmic methionyl-tRNA cannot be formylated.

It is clear that chloroplasts contain a methionyl-tRNA which, by virtue of the fact that it can be formylated, is distinguishable from plant cytoplasmic methionyl-tRNAs. This observation provides a further argument in support of the proposition that chloroplasts have much in common with prokaryotic organisms.

2. Other Species of tRNAs in Chloroplasts

In comparing the tRNAs and amino acid-tRNA ligases of chloroplasts and cytoplasm of Phaseolus vulgaris, BURKARD et al. (1970) found that three of the five leucine tRNAs present in chloroplasts are specific to the organelle and can be acylated only by the chloroplast-localized leucine-tRNA ligase. The remaining two chloroplast leucine tRNAs are similar to cytoplasmic leucine tRNAs. It was also established that, of the three valine tRNAs present in chloroplasts, one species is unique to the organelle and can be charged only by the chloroplast valine-tRNA ligase. The other two species of valine tRNA are common to both cytoplasm and chloroplast and can be charged by either the chloroplast or cytoplasmic ligases.

A similar situation exists in tobacco leaves where, of the six resolvable leucine tRNAs, two are localized within the chloroplasts and are aminoacylated exclusively by the chloroplast ligases (GUDERIAN et al., 1972).

BARNETT and his colleagues (BARNETT et al., 1969; REGER et al., 1970) in a study of the effect of light on the pattern of tRNAs present in dark-grown, wild-type Euglena gracilis observed the appearance of new species of phenylalanine tRNA, glutamine tRNA and isoleucine tRNA following transfer of the culture to photosynthetic growth conditions. Light does not induce the synthesis of these new species of tRNA in the UV-bleached mutant, W$_3$BUL, which lacks chloroplast DNA and identifiable chloroplast structure. The inference that the new tRNA species which appear in response to light might be chloroplast localized was borne out by reverse-phase chromatography of the tRNAs from isolated chloroplasts. It was further established that the chloroplast isoleucine tRNA and phenylalanine tRNAs can be acylated only by the chloroplast-localized isoleucine-tRNA and phenylalanine-tRNA ligases and not by the cytoplasmic enzymes. The chloroplast isoleucine-tRNA ligase, but not the phenylalanine-tRNA ligase is light induced and cannot be detected in dark-grown or in W_3BUL mutant cells.

Extrapolating from this kind of evidence one can conclude that there are likely to be chloroplast specific tRNAs for all twenty amino acids and that these tRNAs are acylated by chloroplast-localized amino acid-tRNA ligases. Chloroplasts may also contain tRNAs similar to the analogous cytoplasmic tRNAs, but whether or not they are identical will only be established when their nucleotide sequences are known.

3. Coding Information for Chloroplast tRNAs

With the recognition of the fact that chloroplasts contain their own DNA came the possibility that certain of the chloroplast-localized macromolecules might be coded for by this DNA and not by nuclear DNA. This possibility was considered to be a very real one in the case of the chloroplast tRNAs and a number of experiments have been carried out to test it. By saturation-hybridization of a total chloroplast tRNA preparation from tobacco with DNA from tobacco chloroplasts, TEWARI and WILDMAN (1970) established that 0.4%–0.7% of the chloroplast DNA is involved in coding for chloroplast tRNA. Assuming a molecular weight of 10^8 daltons for tobacco chloroplast DNA [kinetic complexity data indicate a molecular weight of 1.14×10^8 (TEWARI and WILDMAN, 1970) but electron micrographs of isolated circular chloroplast DNA from *Euglena* (MANNING et al., 1971) and spinach (MANNING et al., 1972) indicate a size closer to 9.0×10^7 daltons], 0.4% to 0.7% represents a size of 400,000 to 700,000 daltons. This is equivalent to 20 molecules of 25,000 molecular weight, which suggests that each chloroplast DNA molecule (and there may be 15 to 30 such molecules per chloroplast) codes for 20 tRNA species.

A similar type of study involving hybridization of purified leucyl-tRNA from bean leaves, acylated with ³H-leucine, to *Phaseolus* chloroplast DNA yielded the information that 0.025% of the DNA codes for this tRNA (WILLIAMS and WILLIAMS, 1970). Assuming again that the bean chloroplast DNA has a molecular weight of 10^8 daltons, 0.025% represents a size of 25,000 daltons or the equivalent of one molecule of leucine tRNA. From the data of BURKARD et al. (1970) we have already learnt that there are 3 species of leucine tRNA in *Phaseolus* chloroplasts, so it might be concluded that the chloroplast DNA contains a cistron for only one of the species of chloroplast leucine tRNA.

The existence of cistrons for chloroplast tRNA in chloroplast DNA does not in itself prove that chloroplast DNA is actually responsible for the synthesis of these RNAs. Duplicate cistrons within the nuclear DNA may exist and these could be the functional elements. However the fact that enucleated *Acetabularia* are capable of incorporating ¹⁴C-uracil into tRNA (SCHWEIGER et al., 1967) shows that the cistrons in the chloroplast DNA can operate where circumstances demand it and they may well be the only repository of genetic information for the chloroplast specific tRNAs.

D. Ribosomes

1. Properties and Distribution within the Chloroplast

Brief mention only will be made of chloroplast ribosomes because our main concern is with chloroplast RNA. As was pointed out earlier, chloroplast ribosomes are clearly distinguishable in size from plant cytoplasmic ribosomes (see, for example, LYTTLETON, 1962; STUTZ and NOLL, 1967; HOOBER and BLOBEL, 1969). In common with the prokaryotes, chloroplasts from algae and higher plants contain ribosomes belonging to the 70S class, whereas plant cytoplasmic ribosomes, like most other eukaryotic cytoplasmic ribosomes, belong to the 80S class (see SMILLIE and SCOTT, 1969, for a comprehensive tabulation of sedimentation values for ribosomes and ribosomal RNAs of chloroplasts and cytoplasm from algae and higher plants). Sedimentation analysis in the ultracentrifuge of ribosome preparations from green leaves and from roots provides a classic demonstration of the relative size difference of the two classes of ribosome and also very suggestive evidence of the localization of the 70S particles within chloroplasts (Fig. 1).

Like bacterial ribosomes, chloroplast ribosomes require at least $5\,mM\ Mg^{2+}$ for stability, dissociating into 50S and 30S subunits if the Mg^{2+} concentration is lowered to $1\,mM$. They are also sensitive to the same antibiotic inhibitors of protein synthesis as are bacterial ribosomes. Plant cytoplasmic ribosomes, on the other hand, are 80S in size, they are relatively stable at low Mg^{2+} concentrations [with the exception perhaps of *Chlamydomonas* cytoplasmic ribosomes (SAGER and HAMILTON, 1967)] and, like animal cytoplasmic ribosomes, are in-

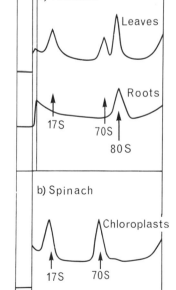

Fig. 1. Analytical sedimentation patterns of ribosomal preparations from (a) tobacco leaves and roots (MATSUDA et al., 1970) and (b) spinach chloroplasts (SPENCER, 1965). Line drawings were traced from Schlieren photographs. The 17S component in the leaf and chloroplast patterns is Fraction I protein of chloroplasts, which, because of its size, is often a contaminant of ribosomal preparations from leaves

sensitive to chloramphenicol and lincomycin but sensitive to cycloheximide (for references, see BOULTER et al., 1972).

In a total ribosomal preparation from leaves, 30% to 40% of the ribosomes may be of the 70S type and a variable proportion of these will be in polysome form. Chloroplast ribosomes can be visualized in electron micrographs of sections of leaves and algae (BARTELS and WEIER, 1967; GIBBS, 1968), and, if the picture is of good quality, direct measurement confirms that their diameter is slightly less than that of cytoplasmic ribosomes. Both free and membrane-bound ribosomes occur in chloroplasts. Electron micrographs of *Phaseolus vulgaris* leaf-sections reveal whorl-like polysomes on the outermost thylakoid membranes of the grana stacks, as well as on the single, stromal thylakoids (FALK, 1969). Thylakoid-associated cyclic polysomes have also been detected in pea chloroplasts and the suggestion has been advanced that they may play an important role in the formation of chloroplast lamellae (PHILIPPOVICH et al., 1970). A similar conclusion concerning the distribution of ribosomes within chloroplasts comes from the direct measurement of ribosomes released from isolated chloroplasts by different treatments. If intact chloroplasts are osmotically shocked a proportion (approximately 40% in the case of tobacco and spinach) of their ribosomes will no longer sediment with the lamellar structure at low centrifugal forces. These presumably are the monosomes and polysomes which initially are free in the stroma phase of the organelle. The remainder of the chloroplast ribosomes are membrane-bound and can only be detached by solubilizing the lamellae with deoxycholate or a non-ionic detergent such as Triton X-100 (CHEN and WILDMAN, 1970; SPENCER, unpublished data).

2. Isolation of Chloroplast Ribosomes

Isolation of the total complement of the cell's ribosomes usually involves disruption of leaf or algal cells in an appropriate Mg^{2+}-containing buffer, followed by the addition of deoxycholate or of a non-ionic detergent such as Triton X-100 (STUTZ and NOLL, 1967). For optimal recovery of membrane-bound ribosomes of chloroplasts, the membrane-solubilizing agent should be added prior to the first low-speed spin. A series of differential low and high speed centrifugations is then carried out, followed by a sucrose density gradient purification step if desired. Ribonuclease inhibitors such as bentonite (final concentration 1 mg/ml) or diethylpyrocarbonate (final concentration 0.1%) may be added at the initial cell disruption stage but excess should be avoided. In the case of bentonite, adsorption of ribosomes to the clay can lead to significant losses (TESTER and DURE III, 1966) and in the case of diethylpyrocarbonate dissociation of ribosomes into subunits may occur (ANDERSON and KEY, 1971). Such a ribosome preparation will of course contain both 70S and 80S particles and these can be resolved on a preparative scale by isokinetic sucrose density gradient centrifugation as demonstrated so elegantly by STUTZ and NOLL (1967).

The direct isolation of chloroplast ribosomes completely free of contaminating cytoplasmic ribosomes is not without problems. Chloroplasts must first be purified and, in order to avoid losing the stroma ribosomes, they should be kept as intact as possible during the purification procedure. The chloroplasts, or a chloroplast-enriched fraction, are lysed with deoxycholate or Triton X-100 and the ribosomes purified by differential centrifugation as already described for total cell ribosomes. Some contamination with cytoplasmic ribosomes is usually apparent but, as these are only a minor component, they are relatively easy to remove by centrifugation through a sucrose density gradient.

E. Ribosomal RNA

1. Molecular Size of Ribosomal RNA from Chloroplasts

Chloroplast ribosomes contain three species of RNA: 23S (1.1×10^6 daltons), 16S (0.56×10^6 daltons) and 5S RNA (38,000 daltons). As can be seen in Fig. 2

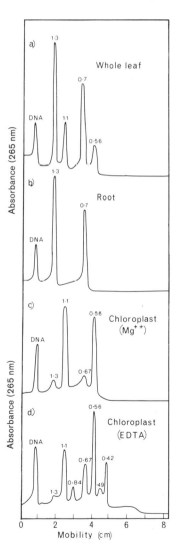

Fig. 2. Polyacrylamide gel electrophoresis of spinach RNAs. (a) Whole leaf; (b) roots; (c) and (d) chloroplasts. Chloroplasts were prepared as described by ELLIS and HARTLEY (1971). RNAs were extracted by a phenol-sodium dodecyl sulphate procedure. The buffer used for the extraction of chloroplast RNA, shown in pattern (d), contained 0.1 mM EDTA; buffers used for the other three RNA preparations contained 10 mM Mg^{2+}. 2.4% gels in a Tris-phosphate-EDTA buffer containing 0.1% sodium dodecyl sulphate (LOENING and INGLE, 1967) were run for 3 hours at 5 mA per gel and scanned at 265 nm in a Joyce-Loebl Chromoscan. Transfer RNAs are not shown in the patterns because they migrate as broad bands and usually run off the bottom of the gel under these conditions. DNA was not removed from the RNA preparations; the relatively large amount of DNA in the chloroplast RNA samples reflects the level of contamination with nuclear DNA

the two larger chloroplast species are clearly distinguishable from the corresponding two cytoplasmic RNA species of plants (25S or 1.3×10^6 daltons and 18S or 0.7×10^6 daltons). The 23S and 5S RNAs are derived from the 50S subunit of the chloroplast ribosome whereas the 16S RNA comes from the 30S subunit. In the case of *Acetabularia* it is still not clear what the size of the chloroplast RNA is. SCHWEIGER (1970) has reported the presence of 23S and 16S RNA in enucleated *Acetabularia* cells but WOODCOCK and BOGORAD (1970) detected only 25S and 18S RNA, with trace amounts of 16S and 14S RNA, in both whole cell and enucleated cell extracts. *Acetabularia* presents a particular experimental problem due to the presence of powerful ribonucleases within the cell and because of the possibility of contamination of organelle preparations by bacteria. However, it should be borne in mind that the electrophoretic migration of RNAs in polyacrylamide gels can give rise to anomalous results because of the effect of molecular conformation on the mobility. RNA molecules having a low guanine plus cytosine $(G + C)$ content can assume different shapes as the degree of intramolecular base-pairing varies with temperature. It would be interesting to know whether the electrophoretic pattern of whole-cell *Acetabularia* RNA remains the same if carried out at different temperatures or whether the apparently single band of 25S RNA splits into two bands, corresponding to a chloroplast and cytoplasmic species.

2. Isolation of Chloroplast Ribosomal RNA

There are four possible approaches to the isolation of chloroplast rRNA.

1) Extraction of total RNA from whole cells or from a ribosome preparation from whole cells, followed by fractionation of the RNA into its various molecular species by sucrose density gradient centrifugation or polyacrylamide gel electrophoresis.

2) Isolation of the total cell ribosomes, fractionation into 70S and 80S species as described in Section D, and then extraction of the RNA from the purified 70S ribosomes.

3) Preparation of a chloroplast fraction, purification of the 70S ribosomes followed by extraction of the rRNA.

4) Extraction of total RNA from a purified chloroplast sample and then fractionation into the various molecular species.

The method chosen will depend upon many factors such as type of plant species being studied, whether the RNA is simply to be analyzed in terms of the relative proportion of the various molecular sizes, or whether it is to be a large-scale preparation of rRNA. Choice of procedure will also depend on whether the possibility of contamination with mitochondrial RNA is of importance or not, or whether ribonucleases are particularly active in the tissue under examination. Isolation of ribosomes from chloroplasts is a relatively lengthy procedure and the possibility of limited degradation of the RNA taking place during the course of centrifugation and other manipulations is a real one. If it is critical to minimize degradation then the best approach is to extract total cell RNA, or at least total chloroplast RNA, and then fractionate it into the various species.

Methods for the extraction of RNA have already been outlined (Section B) and will not be discussed again. However, it should be noted that, where the starting material for the preparation of ribosomal RNA is purified ribosomes, a relatively simple isolation procedure can be adopted. Addition of EDTA and sodium dodecyl sulphate results in the dissociation of ribosomes into RNA and protein, and the 23S and 16S RNA species can be recovered by centrifuging the sample through a 5%–20% sucrose density gradient (STUTZ and NOLL, 1967). Alternatively the detergent-dissociated ribosomes can be fractionated directly by electrophoresis on a 2.4% polyacrylamide gel in the presence of a buffer containing sodium dodecyl sulphate (GOODENOUGH and LEVINE, 1970).

In our laboratory we have extended this type of approach to permit the direct analysis of whole chloroplasts. Chloroplasts containing the equivalent of 50 to 100 µg chlorophyll are treated briefly with EDTA and sodium dodecyl sulphate (ratio of detergent to chlorophyll at least 20:1) and then layered directly onto a 2.4% polyacrylamide gel. Chlorophyll and chlorophyll-protein complexes migrate close to the front along with 4S RNA and are well clear of the large ribosomal RNAs (Fig. 3). This procedure is useful not only for monitoring of chloroplast RNAs but also for analyzing the distribution of radioactivity incorporated into chloroplast RNA species, either *in vivo* or *in vitro*.

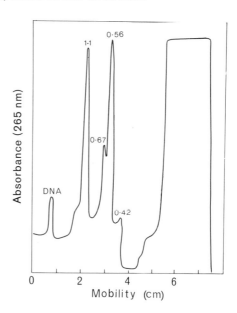

Fig. 3. Polyacrylamide gel electrophoresis of lysed spinach chloroplasts. Chloroplasts were prepared as described by ELLIS and HARTLEY (1971). A sample containing the equivalent of approximately 50 µg chlorophyll was centrifuged for 1 min at 2,500 g and the chloroplast pellet resuspended in 20 µl of gel buffer containing 0.05 M EDTA and 1 mg sodium dodecyl sulphate (ratio of detergent to chlorophyll should be greater than 20:1). Complete lysis of the chloroplasts was achieved, by warming the sample to 35° for 2 min. Sucrose was added and the mixture layered onto a 2.4% polyacrylamide gel. Other conditions are as described in Fig. 2. The chlorophyll and chlorophyll-protein complexes migrate close to, and obscure the tRNA. RNAs of mol. wt. greater than 2×10^5 are detectable and resolution of the large ribosomal RNA species is usually as good as it is for the fractionation of purified chloroplast RNA

3. Characteristics of Chloroplast Ribosomal RNAs

a) Unstable Nature of Chloroplast 23S RNA (1.1×10^6 mol. wt.). For the purposes of this discussion it is necessary to refer to all RNA species in terms of their approximate molecular weights rather than in terms of sedimentation values. Since each ribosome subunit contains one molecule of high molecular weight RNA, and because the large rRNA has a molecular weight of 1.1×10^6 daltons and the small rRNA has one of 0.56×10^6 daltons, it would be expected that, on a mass basis, the amount of the large rRNA would be twice that of the small rRNA in chloroplast ribosomes. However, a survey of the published electrophoretic patterns of chloroplast RNAs derived from a whole range of plant species shows that the expected ratio of 2:1 is very rarely achieved. The 1.1×10^6 mol. wt. RNA is usually considerably depleted and in species such as swisschard it may be almost non-existent.

The reason for this is simply that the chloroplast 1.1×10^6 mol. wt. rRNA is particularly susceptible to degradation. Degradation is not a random process, however, but proceeds via a number of relatively specific breaks to produce a family of RNA molecules of discrete sizes. This phenomenon has been examined in great detail by INGLE and his colleagues (LOENING and INGLE, 1967; INGLE, 1968a; INGLE et al., 1970; LEAVER and INGLE, 1971). The pattern of breakdown is not precisely the same for all species of chloroplast 1.1×10^6 mol. wt. RNA and, depending on the actual size of the primary cleavage products, the extent of degradation may or may not be immediately apparent from gel electrophoretic analysis. For instance, in the case of *Phaseolus*, the major split of the 1.1×10^6 mol. wt. rRNA occurs at the exact centre of the molecule yielding RNA of approximately 0.56×10^6 daltons. The electrophoretic mobility of this product coincides with that of the 0.56×10^6 mol. wt. RNA from the 30S ribosomal subunit, so the only observable change is in the relative amounts of the two basic rRNA species. On the other hand, the 1.1×10^6 mol. wt. RNA of spinach chloroplasts first splits to give 0.67×10^6 and 0.45×10^6 mol. wt. RNAs and the degradation is clearly discernible in a gel electrophoretic pattern of the RNA. It should be noted however, that the 0.67×10^6 mol. wt. RNA more or less coincides with the 0.7×10^6 mol. wt. RNA from the small subunit of the 80S cytoplasmic ribosomes (Fig. 2). Thus a gel pattern of spinach whole leaf RNA usually shows a depleted 1.1×10^6 mol. wt. RNA peak and an enhanced 0.7×10^6 mol. wt. RNA peak.

Fragmentation of 1.1×10^6 mol. wt. rRNA into two products is very common but more extensive breakdown can occur to yield a variety of low molecular weight RNA species. The cleavage products of 1.1×10^6 mol. wt. rRNA from a number of plant species have been tabulated by LEAVER and INGLE (1971) and their data is reproduced in Fig. 4.

Tissue	Preparation	Molecular weights of chloroplast rRNA species (mol.wt.×10⁻⁶)		Cleavage products of 1.1×10^6-mol.wt. rRNA
Radish	Mg	1.04	0.56	
	EDTA	1.03	0.57	
Spinach	Mg	1.01	0.56	
	EDTA	1.03	0.56	
Broad bean	Mg	1.08	0.57	
• French bean	Mg	1.06	0.56	
• Pea	Mg	1.06	0.57	
• Maize	Mg	1.12	0.56	
• Chlamydomonas	Mg	1.05	0.57	

Fig. 4. Cleavage products of the chloroplast 1.1×10^6 mol. wt. rRNA. A summary showing the range of cleavage products from a variety of plant species. ▮ major fragments; ▯ minor components. (From LEAVER and INGLE, 1971)

The actual pattern of degradation of the large chloroplast rRNA species, as observed on polyacrylamide gels, is a function not only of the species of plant but also of the age of the leaves being analyzed and of the conditions used for extraction and fractionation of the RNA. During the development of radish cotyledons the ratio of the large to the small chloroplast ribosomal RNAs decreases from a value close to 1 for 3-day old seedlings to a value of less than 0.5 for 5-day old seedlings (INGLE, 1968a). Similarly during development of the French bean leaf, the ratio drops from slightly less than 1 for a 5-day old plant to approximately 0.3 for a 10-day old plant (LOENING and INGLE, 1967).

If a divalent cation such as Mg^{2+} or Ca^{2+} is present in the buffer used for extracting the RNA and if Mg^{2+} is a component of the gel electrophoresis buffer, then the amount of 1.1×10^6 mol. wt. RNA relative to the amount of 0.56×10^6 mol. wt. RNA will be much greater than is the case if EDTA is present in the buffer (LEAVER and INGLE, 1971). Presumably, the RNA fragments are held together and behave as an intact 1.1×10^6 mol. wt. RNA provided Mg^{2+} or Ca^{2+} is present from the moment the ribosome is disrupted. In the absence of Mg^{2+}, or if Mg^{2+} is removed prior to electrophoresis of the RNA, or if the RNA is heated briefly at 50° to 60°, then the break in the RNA molecule is revealed and the fragments migrate on polyacrylamide gels according to their true size (Fig. 5).

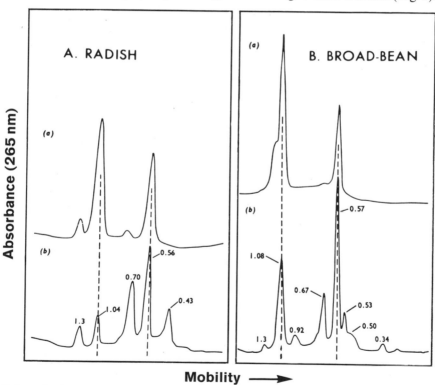

Fig. 5. Polyacrylamide gel fractionation of chloroplast RNAs prepared in the presence of Mg^{2+} and electrophoresed in either Mg^{2+} or EDTA-containing buffers. A. Radish cotyledon chloroplast RNA electrophoresed in (a) Mg^{2+} buffer and (b) EDTA buffer. B. Broad-bean chloroplast RNA electrophoresed in (a) Mg^{2+} buffer and (b) EDTA buffer. (From LEAVER and INGLE, 1971)

For this reason Mg^{2+} (10 mM) is frequently included in the buffer used for chloroplast RNA extraction. However, total leaf RNA prepared in the presence of Mg^{2+} contains more non-nucleic acid material than that prepared in EDTA and polyacrylamide gel electrophoresis in Mg^{2+}-containing buffers gives poorer resolution than does electrophoresis in EDTA-containing buffers (Leaver and Ingle, 1971).

Breaks in chloroplast 1.1×10^6 mol. wt. RNA could be the result of ribonuclease activity during extraction of the RNA but the extreme specificity of the break would argue against this explanation. Moreover, there is no obvious reason why only the larger of the chloroplast ribosomal RNAs should be attacked while the smaller chloroplast rRNA and both cytoplasmic ribosomal RNAs are resistant. The most plausible explanation of the phenomenon is that a break is introduced into the RNA during the lifetime of the ribosome. Ingle (1968a) has clearly shown that the large rRNA of chloroplasts is synthesized as an intact molecule of 1.1×10^6 daltons. Chloroplast RNA extracted from radish cotyledons which have been labelled with ^{32}P-orthophosphate for 6 hours shows a 2:1 distribution of radioactivity between the 1.1×10^6 and 0.56×10^6 mol. wt. RNAs, despite the fact that the unlabelled RNA species show a ratio of 1 or less. As the ribosome ages there is a finite chance that the 1.1×10^6 mol. wt. RNA will be nicked, presumably at a point which is rendered vulnerable to ribonuclease by virtue of the conformation of the RNA within the ribosomal structure. The

Table 1. Base composition (moles %) of chloroplast and cytoplasmic rRNAs

			UMP	GMP	CMP	AMP	CMP/ AMP	Reference [a]
Ivy	Chlor.	total	21.1	34.1	21.6	23.4	0.92	1
	Cyt.	total	22.0	31.7	24.1	22.2	1.08	1
Lettuce	Chlor.	total	19.6	33.3	21.9	24.2	0.90	1
	Cyt.	total	23.7	30.9	23.1	22.3	1.04	1
Cabbage	Chlor.	total	19.8	32.6	23.0	24.5	0.94	1
	Cyt.	total	21.5	32.0	24.4	22.1	1.10	1
Turnip	Chlor.	total	19.8	33.1	22.7	24.4	0.93	1
	Cyt.	total	22.7	32.2	23.6	21.5	1.10	1
Spinach	Chlor.	23S	19.0	34.8	21.7	24.5	0.89	1
		16S	22.4	30.6	24.0	23.0	1.04	1
		total	20.0	32.4	23.4	24.2	0.97	1
	Cyt.	total	23.3	31.7	23.5	21.6	1.09	1
Beet	Chlor.	23S	18.7	35.2	21.5	24.6	0.87	1
		16S	22.9	30.9	23.2	22.8	1.02	1
		total	20.3	33.6	22.3	23.7	0.94	1
	Cyt.	total	23.4	31.4	23.2	22.0	1.05	1
Broad bean	Chlor.	5S	22.2	28.7	22.6	26.5	0.85	2
	Cyt.	5S	23.5	28.6	23.8	24.1	0.99	2
Euglena	Chlor.	total	23.0	31.1	20.4	25.5	0.80	3
	Cyt.	total	22.3	32.0	24.1	21.6	1.11	3

[a] 1, Rossi and Gualerzi (1970); 2, Payne and Dyer (1971b); 3, Rawson and Stutz (1969).

point of attack appears to be so specific that one might predict it relates to a normal function of the ribosome such as a site for attachment of the 50S subunit to a membrane.

b) Base Composition of Chloroplast RNA. The most extensive survey on the base composition of higher plant chloroplast ribosomal RNA is that done by Rossi and Gualerzi (1970). From their data and from that on *Euglena* (Rawson and Stutz, 1969) and that on *Chlorella* (Oshio and Hase, 1968), the generalizations can be made that guanylic acid is the nucleotide present in the greatest amount and that the G+C content is approximately 56%. In the few instances where comparisons have been made (spinach and beet by Rossi and Gualerzi, 1970; *Antirrhinum majus* by Ruppel, 1967) the base composition of the 23S RNA differs somewhat from that of the 16S RNA.

Data on the base composition of chloroplast rRNA is of interest primarily in terms of its similarity or dissimilarity to that of the corresponding cytoplasmic RNAs. Differences are not very striking but significance can be attached to them because they show up fairly consistently in all the published analyses (Table 1). The proportion of the pyrimidine nucleotides is usually slightly less in chloroplast rRNA than it is in cytoplasmic rRNA. Rossi and Gualerzi (1970) point out that the parameter which reflects the greatest difference between chloroplast and cytoplasmic ribosomal RNAs is the ratio of cytidylic to adenylic acid. The value for most chloroplast ribosomal RNAs is in the region of 0.90–0.97 for higher plants (0.80 for *Euglena* and *Chlorella*) whereas that for plant (including *Euglena*) cytoplasmic RNAs is 1.04–1.10.

4. Synthesis of Chloroplast Ribosomal RNA

Synthesis of cytoplasmic ribosomal RNAs in animal and plant systems involves the initial transcription of a precursor RNA molecule considerably larger than the sum of the two individual rRNAs (see Chapter 6). These precursor molecules are processed through various intermediate stages to yield finally the large and the small ribosomal RNAs. Despite deliberate efforts to locate similar precursor molecules in the chloroplast system, no evidence for their existence as yet has been forthcoming. If synthesis of chloroplast rRNAs is analogous to the synthesis of bacterial rRNAs (a quite real possibility in view of the other known points of similarity existing between chloroplast and bacterial systems) then the likelihood of identifying precursor rRNA molecules is not very great. In the first place, RNA molecules of the same size as bacterial precursor rRNAs (24S and 17S) would be largely obscured by the plant cytoplasmic rRNAs (25S and 18S) whose synthesis usually dominates that of the chloroplast rRNAs. Secondly, chloroplast rRNA synthesis takes place during a very limited period in the growth of a leaf and the labelling of chloroplasts adequately with radioactive precursors presents a major technical obstacle to detecting transitory molecules such as precursor rRNA.

Synthesis of chloroplast rRNA in higher plants is largely restricted to the phase of chloroplast development and once formed it is relatively stable metabolically. By following the incorporation of ^{32}P-orthophosphate into 23S and 16S RNA of developing radish cotyledons, Ingle (1968a) established that chloroplast rRNA attained its highest specific activity in 3-day old seedlings although the absolute level of the RNA did not reach a maximum until the fifth day. At no stage did the level of incorporation of label into chloroplast rRNA approach that into the cytoplasmic rRNA. Similar conclusions concerning the association of chloroplast rRNA synthesis with chloroplast division and development can be drawn from the results of experiments on wheat (INGLE et al., 1970; PATTERSON and SMILLIE, 1971) and spinach (DETCHON and POSSINGHAM, 1972).

A direct estimate of the half-life of chloroplast rRNA has been made using duck-weed, *Lemna minor* (TREWAVAS, 1970). The value for chloroplast rRNA is 15 days which is in marked contrast to the 4-day half-life of the corresponding cytoplasmic rRNA. This provides a solid basis for the inference suggested by the data of Ingle's experiments on the labelling patterns of chloroplast and cytoplasmic RNAs in developing radish cotyledons. What this means in practical terms is that, if one needs to obtain labelled chloroplast RNA, radioactive precursors should be supplied to the plant or leaves at the proplastid stage, or at least at a stage when chloroplast numbers per cell are still increasing.

One interesting difference which helps to differentiate chloroplast rRNA synthesis from cytoplasmic rRNA synthesis in *Chlorella* is that chloroplast ribosomes are preferentially labelled if ^{3}H-uridine is supplied to the cells whereas cytoplasmic ribosomes are preferentially labelled if ^{32}P-orthophosphate is supplied (GALLING and SSYMANK, 1970). Whether this phenomenon occurs in other plants is not known.

5. Coding Information for Chloroplast Ribosomal RNA

The presence in chloroplast DNA of cistrons coding for chloroplast ribosomal RNA is now well established. Quantitative saturation-hybridization of *Euglena* chloroplast rRNA to chloroplast DNA showed that approximately 2% of the DNA is involved in coding for chloroplast rRNA (SCOTT and SMILLIE, 1967; SCOTT et al., 1971 a). A higher value (6%) has recently been obtained by STUTZ and VANDREY (1971) who isolated *Euglena* chloroplast DNA by a method which relied on the exhaustive purification of the organelles to remove nuclear contamination rather than on the fractionation of extracted DNA by preparative CsCl buoyant density centrifugation. In the latter procedure the chloroplast DNA is recovered from the 1.685 g/cm^3 density region of the gradient and, according to STUTZ and VANDREY, is thereby automatically depleted of DNA which is enriched in rRNA cistrons and which bands in the 1.701 g/cm^3 region.

As might be expected, the 23S RNA and 16S RNA do not compete with each other in the hybridization reaction and thus must be coded for by separate cistrons (SCOTT et al., 1971 a). *Euglena* chloroplast DNA can be separated into heavy and light strands by alkaline CsCl equilibrium density gradient centrifugation (STUTZ and RAWSON, 1970). Hybridization of chloroplast rRNA to each of the strands revealed that the cistrons are localized in the heavy strand, saturation occurring at a level of 2.2%.

Accepting that the molecular weight of *Euglena* chloroplast DNA is approximately 10^8 daltons (MANNING et al., 1971) then 2% of this is equivalent to

2×10^6 daltons which is sufficient to code for both the large and small chloroplast ribosomal RNAs (size 1.1×10^6 and 0.56×10^6 daltons). The value of 6% (STUTZ and VANDREY, 1971) would allow for three copies of each of the rRNAs per chloroplast DNA molecule.

Chloroplast RNA-chloroplast DNA hybridization experiments with tobacco (TEWARI and WILDMAN, 1970) and with swisschard (INGLE et al., 1970) have given results comparable to those with *Euglena*. Hybridization-saturation in the case of tobacco chloroplast DNA occurred at a level of 1.6%, so again it would appear that each chloroplast DNA molecule has a cistron for each of the chloroplast ribosomal RNAs.

In a number of instances chloroplast ribosomal RNA has been found to hybridize to nuclear DNA as well as to chloroplast DNA (TEWARI and WILDMAN, 1968; MATSUDA et al., 1970; INGLE et al., 1970). Although the extent of hybridization is far less with nuclear DNA (0.13% for *Euglena* and for tobacco), the potential number of chloroplast rRNA cistrons in the nucleus is roughly equivalent to the sum of those located in all the chloroplasts of the cell because there is far more nuclear DNA per cell than there is chloroplast DNA. Attempts have been made to assess the significance of these nuclear DNA sites which react with chloroplast rRNA. By centrifuging swisschard nuclear DNA ($\rho = 1.694 \text{ g/cm}^3$) on a preparative CsCl equilibrium density gradient and then hybridizing the individual fractions to cytoplasmic rRNA and to chloroplast rRNA, INGLE et al. (1970) observed that DNA from exactly the same density region (1.705 g/cm^3) of the gradient hybridized with both species of rRNA to a similar extent. Furthermore, hybridization of the chloroplast and cytoplasmic rRNAs to this DNA fraction was competitive not additive, suggesting that only one type of cistron was involved in the reaction. However, the fidelity of hybridizations involving the rRNA cistrons of plant nuclear DNA may not be too exacting because it is known that cytoplasmic rRNAs from a whole range of plant species will all hybridize with approximately equal efficiency to the nuclear DNA of a single species (MATSUDA and SIEGEL, 1967). Therefore the fact that chloroplast rRNA also hybridizes to nuclear DNA is not too surprising. Actually the problem is more complex than this would suggest, because the hybridization of cytoplasmic rRNA to chloroplast DNA is only 20% as efficient as the chloroplast rRNA-chloroplast DNA reaction (INGLE et al., 1970). Thus, while accepting that the hybridization of chloroplast rRNA to nuclear DNA is real, it does not necessarily follow that there are cistrons for chloroplast rRNA in the nucleus.

There are other good reasons for believing that chloroplast, not nuclear DNA, is actively responsible for coding for chloroplast ribosomal RNA. In the first place, it is known that enucleated *Acetabularia* cells can incorporate ^{3}H-uridine into 23S and 16S RNA species (SCHWEIGER, 1970) and that this incorporation is inhibited by rifampicin, an antibiotic which, in algae, appears to block chloroplast RNA polymerase but not nuclear RNA polymerase (BRÄNDLE and ZETSCHE, 1971). Secondly, UV-induced mutants of *Euglena* which have lost the ability to form normal chloroplasts and which lack chloroplast DNA, do not contain chloroplast rRNA (SCOTT and SMILLIE, 1967). Thirdly, when *Ochromonas danica* is supplied with ^{3}H-uridine for 30 minutes and the location of the RNA synthesized within the cells determined by autoradiography of electron micrograph sections, the chloroplasts are seen to become labelled immediately whereas there is a lag before cytoplasmic RNA becomes labelled (Gibbs, 1968). This clearly indicates that chloroplast RNA is transcribed *in situ* from chloroplast DNA. And finally, if rifampin, an inhibitor of RNA synthesis in *Chlamydomonas* chloroplasts, is added to a culture of *Chlamydomonas* then inhibition of chloroplast rRNA results (Surzycki, 1969).

Taken together then, all the evidence points to the conclusion that chloroplast ribosomal RNA is transcribed from chloroplast DNA. It is interesting to note, however, that the synthesis of rRNA in isolated chloroplasts from higher plants has not yet been unequivocally demonstrated.

6. Chloroplast 5S Ribosomal RNA

There have been reports on the occurrence of 5S RNA in chloroplast ribosomes of *Vicia faba* (Dyer and Leech, 1968), *Allium porrum* and *Zea mays* (Ruppel, 1969) and *Chlamydomonas* (Bourque et al., 1971). Initial observations indicated that the chloroplast 5S RNA was distinguishable from the plant cytoplasmic ribosomal 5S RNA by virtue of its slightly different behaviour on MAK columns. A higher salt concentration was necessary to elute the chloroplast species than the cytoplasmic species.

More recently a very detailed comparison of the chloroplast and cytoplasmic ribosomal 5S RNA species from *Vicia faba* has been made by Payne and Dyer (1971 b). Polyacrylamide gel electrophoresis of low molecular weight RNAs from whole leaf tissue reveals the presence of two species of 5S RNA whereas that from chloroplasts reveals a single species, corresponding to the slower moving component of the whole leaf pattern. 5S RNA from root tissue also moves as a single band but it coincides with the faster moving component of the whole leaf pattern. The electrophoretic mobility of *E. coli* 5S RNA lies between that of chloroplast and that of plant cytoplasmic 5S RNA and, from this observation, Payne and Dyer calculate that *Vicia faba* chloroplast 5S RNA contains 122 nucleotides and the cytoplasmic 5S RNA, 118 nucleotides. When chloroplast ribosomes are fractionated into 50S and 30S subunits and the RNAs derived therefrom subjected to polyacrylamide gel electrophoresis, 5S RNA is found to be associated only with the 50S subunit. Analysis of 5S RNAs from *Vicia* chloroplast and cytoplasmic ribosomes shows that their base compositions are very similar and quite unlike that of *E. coli* 5S RNA (Table 1).

Although there is as yet no evidence to support such a claim, it would seem reasonable to assume that chloroplast ribosomal 5S RNA is coded for by chloroplast DNA just as are the large chloroplast ribosomal RNAs.

F. Messenger RNA (mRNA)

Most of the evidence pertaining to the presence of mRNA in chloroplasts is of an indirect nature. One exception to this was the demonstration by Brawerman and EISENSTADT (1964) that *Euglena* chloroplast RNA could stimulate protein synthesis in an *in vitro*, RNA-dependent system from *E. coli*. RNA extracted from *Euglena* chloroplasts was fractionated on a sucrose density gradient and the ability of the various fractions to stimulate protein synthesis in a cell-free *E. coli* system was determined. The fractions which elicited the greatest response came from the 12S region of the gradient, a region intermediate between the small ribosomal RNA and transfer RNA.

Somewhat less direct evidence comes from analysis of the rapidly labelled RNA fraction which can be extracted from chloroplasts of young spinach plants exposed to ^{3}H-uridine for 30 minutes. Fractionation of the RNA by sucrose density gradient centrifugation or polyacrylamide gel electrophoresis reveals a polydisperse RNA ranging in size from 6S to greater than 25S and generally similar to the type of pattern considered to be representative of mRNA in bacteria (SPENCER and WHITFELD, 1967).

A compelling argument for the existence of mRNA in chloroplasts comes from the knowledge that isolated chloroplasts can synthesize proteins (for detailed references see BOULTER et al., 1972). If a partially disrupted chloroplast system is used, then incorporation of amino acids into protein is extremely sensitive to inhibition by ribonuclease. Sucrose density gradient fractionation of the ribosome population released from *in vitro* labelled tobacco chloroplasts shows that the radioactive proteins are associated with both the 70S monosome and the polysome regions (CHEN and WILDMAN, 1967). Direct demonstration of the presence of polysomes in chloroplasts has come from a number of studies including the analytical ultracentrifugation of ribosomes recovered from chinese cabbage chloroplasts (CLARK, 1964) and the sucrose density gradient analysis of ribosome preparations from pinto bean chloroplasts (STUTZ and NOLL, 1967). In the latter case confirmation of the presence of 70S ribosomes in the polysomes was provided by showing that the extracted RNA belonged to the 23S-16S chloroplast type.

Protein synthesis in isolated chloroplasts is not inhibited by deoxyribonuclease or actinomycin D. Nor is it stimulated by the addition of the four nucleoside triphosphates, although synthesis of RNA in the chloroplasts results from their addition. Thus synthesis of protein in this system is not coupled to transcription of DNA and it is reasonable to conclude that protein synthesis in isolated chloroplasts is programmed by a long-lived messenger RNA.

Messenger RNA species in chloroplasts could conceivably be the products of transcription of chloroplast DNA or of nuclear DNA, or of both. At present evidence in support of the first alternative is good but there is nothing for or against the other two possibilities. Isolated chloroplasts synthesize RNA and such RNA must necessarily be chloroplast-DNA coded. This RNA moves as a heterogeneous population of molecules in sucrose density gradient centrifugation or polyacrylamide gel electrophoresis and the pattern resembles closely that of the rapidly labelled, presumed messenger RNA which is synthesized in chloroplasts *in vivo* (SPENCER and WHITFELD, 1967). Hybridization of RNA synthesized in isolated tobacco chloroplasts to chloroplast DNA indicates that 21% of the chloroplast DNA nucleotide sequences are being transcribed (TEWARI and WILDMAN, 1970). This represents a much larger proportion of the chloroplast genome than is required to code only for ribosomal and transfer RNAs, the implication being that messenger RNA species are also synthesized on chloroplast DNA.

Recently it has been established by genetic crosses of *Nicotiana* species that the large protein subunit of ribulose diphosphate carboxylase (Fraction I protein) is coded for by chloroplast DNA because the pattern of inheritance is of the maternal type (WILDMAN, 1972). It has also been shown that isolated pea chloroplasts synthesize the large subunit of Fraction I protein (BLAIR and ELLIS, 1972). Thus the messenger RNA for this particular protein must be transcribed and translated within the chloroplast. It should be remembered, however, that very many of the chloroplast proteins are coded for by the nuclear DNA (KIRK, 1971) so that the possibility of mRNA species transcribed from nuclear DNA appearing in the chloroplasts is by no means remote.

G. RNA Synthesis in Isolated Chloroplasts

1. General Features

The ability of isolated chloroplasts to incorporate labelled ribonucleoside triphosphates into RNA is a well-characterized property of these organelles from a variety of plant species (KIRK, 1964; SEMAL et al., 1964; SPENCER and WHITFELD, 1967; TEWARI and WILDMAN, 1969; SURZYCKI, 1969). Incorporation will usually proceed for 30 to 40 minutes and requires the presence of all four of the nucleoside triphosphates; it is sensitive to deoxyribonuclease and actinomycin D and the product is degraded by alkali or by ribonuclease.

Chloroplasts which have been disrupted by treatment with a Mg^{2+}-containing, hypotonic buffer still retain the capacity to synthesize RNA. In fact, the active component can be sedimented at low speeds even after solubilization of the chloroplast lipids with the detergent Triton X-100 (BOTTOMLEY, 1970b). Thus it appears that RNA synthesis in chloroplasts takes place in a DNA-RNA polymerase-membrane protein complex.

There is reason to believe that the incorporation of nucleotides into RNA of chloroplasts *in vitro* reflects the continuing synthesis of polynucleotide chains, already in existence at the time the chloroplasts were isolated from the leaf, and not the initiation of new RNA chains. Addition of calf thymus DNA to the system causes some increase in nucleotide incorporation but this appears to be due to a non-specific, polyanionic stimulation of the chloroplast DNA-RNA polymerase complex and not to the initiation of a new round of RNA synthesis by chloroplast polymerase on the exogenous DNA template (BOTTOMLEY et al., 1972). The absence of unattached RNA polymerase in chloroplasts is further attested to by the difficulty one has in isolating the enzyme in a form free of DNA. Thus the apparent failure of the isolated chloroplast system to initiate the synthesis of new RNA chains is probably a result of the enzyme failing to dissociate from the DNA when transcription of the template terminates.

2. Product of the Reaction

The major portion of the labelled RNA product remains associated with the DNA-RNA polymerase-membrane complex and sediments at low centrifugal forces. If the product is isolated by a routine RNA extraction procedure and analysed by sucrose density gradient centrifugation or polyacrylamide gel electrophoresis, it is found to be polydisperse, ranging in size from 2×10^4 to 2×10^6 daltons (SPENCER et al., 1971; TEWARI and WILDMAN, 1969). Maximum radioactivity is usually located in the region of 0.5 to 1.0×10^6 daltons and there are no discrete species which stand out from the rest of the product (Fig. 6). As

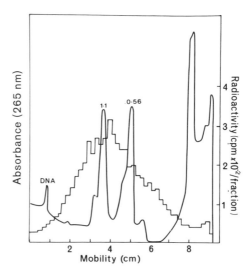

Fig. 6. Polyacrylamide gel electrophoresis of RNA synthesized by isolated spinach chloroplasts. Chloroplasts were incubated for 20 min at 25° in a buffer containing ATP, CTP, GTP and ^{3}H-UTP as described by SPENCER et al. (1971). They were then sedimented at 10,000 g for 10 min, lysed with sodium dodecyl sulphate and electrophoresed as detailed in Fig. 3. The gel was scanned at 265 nm, frozen, sliced and the radioactivity measured (BOTTOMLEY et al., 1971b). Radioactivity in each slice is shown in the histogram

mentioned before, the sedimentation pattern of RNA species labelled in isolated chloroplasts resembles closely the pattern of rapidly-labelled chloroplast RNA from young spinach plants supplied with ^{3}H-uridine for a short period. In other words, the pattern exhibits the characteristics expected of a messenger RNA fraction.

One of the established functions of chloroplast DNA is to code for chloroplast ribosomal and transfer RNAs and it is reasonable to ask whether synthesis of these particular species occurs in isolated chloroplasts. To date, the answer is that although RNA species in the same size range as the ribosomal and transfer RNAs are synthesized, there is no specific coincidence of incorporated radioactivity with these stable RNA species (SPENCER et al., 1971). If the chloroplast rRNA species are synthesized by way of a larger, precursor molecule, and if the isolated chloroplasts have lost the ability to process such precursor molecules down to their final size, one might expect to see the synthesis of large molecules. However, as already indicated above, polyacrylamide gel electrophoresis has not as yet revealed evidence of the synthesis of *any* discrete RNA species by isolated chloroplasts. It should be noted, though, that rRNA and tRNA cistrons account for only a few percent of the total information content of the chloroplast genome and, unless controls operate to promote specific transcription of these cistrons, detection of the synthesis of their RNA transcripts against a background of hundreds of other RNA molecules is rather unlikely. The question of whether low levels of rRNA are being synthesized in isolated chloroplasts might best be tackled by competition-hybridization experiments but there is one serious drawback to this approach. The bulk of the RNA synthesized by chloroplasts *in vitro* remains associated with the membrane-DNA-enzyme complex which also has a significant number of chloroplast ribosomes attached to it. Isolation of the highly radioactive, but quantitatively insignificant, product RNA also releases the relatively large amount of unlabelled ribosomal RNA. Therefore, in terms of the ribosomal RNA species, the product is of very low specific activity and the possibility of measuring a small reduction in the hybridization of the product to chloroplast DNA by competing with additional unlabelled ribosomal RNA is remote.

3. Chloroplast RNA Polymerase

One of the more interesting aspects of RNA synthesis in chloroplasts is the question of whether the RNA polymerase involved in transcribing the chloroplast DNA is the same as that involved in transcribing nuclear DNA. Rifampicin, an antibiotic which specifically inhibits many prokaryotic RNA polymerases but not eukaryotic nuclear RNA polymerases, blocks RNA synthesis in chloroplasts, but not in nuclei, from algae (SURZYCKI, 1969; BRÄNDLE and ZETSCHE, 1971; GALLING, 1971). This differential inhibitory effect on RNA synthesis is observed for *in vivo* systems as well as *in vitro* ones. Thus in algae it is obvious that the nuclear and chloroplast RNA polymerases are different. However, there is no such clear distinction between the nuclear and chloroplast RNA polymerases from higher plants. RNA synthesis by isolated chloroplasts from peas and spinach is not affected by rifampicin (BOTTOMLEY et al., 1971 b). This could be attributed to the fact that initiation of RNA chains is not going on in isolated chloroplasts, but parallel studies on the effect of rifampicin on RNA synthesis in whole plants support the conclusion that the RNA polymerase of higher plant chloroplasts is not sensitive to this inhibitor. A range of inhibitors, which were selected because they can differentiate between prokaryotic and eukaryotic RNA polymerases, also failed to distinguish between the RNA polymerase activities of isolated chloroplasts and nuclei from peas and spinach (BOTTOMLEY et al., 1971 b). There is, however, one way to show a difference between the activities from the two types of organelle and that is by their relative sensitivity to sonication (BOTTOMLEY, 1970 a). Polymerase activity of plastids can be inhibited more than 90 % by conditions of sonication which have virtually no effect on the activity of nuclei. This differential sensitivity does not establish that the RNA polymerases of the two organelles are actually different because it could be the association of the enzyme with its DNA template which is the sonication-sensitive site.

A more exhaustive comparison of the properties of the RNA polymerase of chloroplasts with those of plant nuclei is possible with the help of purified enzymes.

There have been only two reports on the solubilization and partial purification of RNA polymerase from chloroplasts. One is on the enzyme from maize chloroplasts (BOTTOMLEY et al., 1971a) and one is on the enzyme from wheat chloroplasts (POLYA and JAGENDORF, 1971). The properties of the chloroplast polymerases are very similar to those of the maize and wheat nuclear polymerases but there are some differences, particularly with respect to template specificity, which suggest that the enzyme from the chloroplast is not the same as those from the nucleus. Neither the wheat nor the maize chloroplast enzyme is sensitive to rifampicin, an observation which supports the conclusion drawn above on the basis of *in vivo* studies with these inhibitors.

H. RNA Synthesis in Developing Chloroplasts

Illumination of dark-grown plants results in the development of proplastids and etioplasts into mature chloroplasts, and synthesis of RNA is one of the earliest events associated with these changes, usually preceding the formation of chloroplast pigments, lipids and proteins. This aspect of chloroplast RNA metabolism has attracted a great deal of attention and the reasons for this are at least twofold. One is the expectation that such studies will help to establish the nature of the primary events triggered by light and how they are elaborated into the multitude of responses which ultimately constitute the development and maturation of the chloroplast. The other reason is concerned with the question of the informational role of chloroplast DNA. Very little is known about the function of the major portion of chloroplast DNA but it would seem a reasonable proposition to assume that analysis of changes in the RNA which take place when immature plastids are exposed to light might provide some insight into this problem.

There is ample evidence that the leaves of dark-grown higher plants contain plastid-specific ribosomal and transfer RNAs. In very young plants, the absolute amount of plastid ribosomal RNA is low, and in fact in many situations it cannot be detected at all (SMITH, 1970). This is in marked contrast to the cytoplasmic rRNA which is present even in ungerminated seeds. During growth of the plant in the dark the amount of plastid ribosomal RNA gradually increases and it can reach a level almost equivalent to that which exists in chloroplasts from light-grown plants. Such is the case in bean (*Phaseolus*), cucumber, wheat and corn (BOARDMAN, 1966; VEDEL and D'AOUST, 1970; PATTERSON and SMILLIE, 1971; DYER et al., 1971). Other species of plants, for example *Vicia faba*, do not accumulate as much plastid RNA in the dark as in the light (DYER et al., 1971). There is some evidence that when such plants are exposed to light, they green-up more slowly than do plants like *Zea mays* which are able to synthesize their full complement of plastid RNA in the dark. This implies that the rapid maturation of etioplasts into fully developed chloroplasts depends on the presence of the maximum number of plastid ribosomes (DYER et al., 1971).

From the above it is obvious that light is not essential for synthesis of plastid rRNA. However, light does have a significant effect on the rate of plastid rRNA synthesis, particularly if plants are exposed to the light at a very early stage of their development (INGLE 1968 b; SMITH et al., 1970). In radish cotyledons, synthesis of plastid rRNA in dark-grown seedlings occurs at only one third the rate that it occurs in light-grown plants (INGLE, 1968 b). The stimulation of RNA synthesis following illumination can be observed either by measuring the incorporation of labelled precursors into RNA or by direct determination of absolute levels of plastid RNA. Transfer of dark-grown radish seedlings into the light also has some effect on the level of cytoplasmic RNA, but the percent stimulation is only a quarter of that which occurs with the plastid RNA (INGLE, 1968 b). Increased synthesis of cytoplasmic RNAs following exposure of plants to light is not a universal phenomenon. Illumination of two-week old, dark-grown *Vicia faba* plants causes hardly any change in the cytoplasmic rRNA level, although there is a four-fold increase in the level of plastid rRNA (DYER et al., 1971).

Plastid-specific tRNAs also occur in etiolated higher plants. For instance, the levels of chloroplast valine tRNA and isoleucine tRNAs in germinating cotton seedlings are as high in five-day old etiolated cotyledons as in cotyledons which have been allowed to green (DURE III and MERRICK, 1971). However, as is the case with plastid ribosomal RNAs, synthesis of plastid tRNAs can be stimulated by light (DYER et al., 1971). In *Phaseolus vulgaris* all six of the chloroplast leucine tRNA species, and all five of the valine tRNA species are present in the etioplasts of dark-grown plants (BURKARD et al., 1972). The level of two of the leucine tRNAs and two of the valine tRNAs is relatively higher in mature chloroplasts than it is in etioplasts, which suggests that their synthesis may be specifically stimulated by light.

The light-induced stimulation of plastid RNA synthesis is a phytochrome-mediated response. Irradiation of four-day old dark-grown pea plants with red light results in an increased synthesis of plastid rRNA, similar to that which results from illumination of plants with white light (SCOTT et al., 1971b). Exposure of red-light treated plants to far-red light negates the effect, suggesting that maximum levels of plastid RNA can be attained if leaf phytochrome is maintained in the active form.

Unlike the situation in higher plants, where light can stimulate, but is not obligatory for plastid RNA accumulation, dark-grown *Euglena gracilis* cells apparently lack plastid ribosomal RNA (BROWN and HASELKORN, 1971). Transfer of dark-grown cells to the light results in the gradual accumulation of the plastid RNA. Its synthesis can first be detected by radioactive labelling experiments 2 to 3 hours after the cells have been placed in the light but the RNA is not quantitatively detectable until 10 to 12 hours in the light. After 16 to 24 hours in the light, the plastid rRNA levels are comparable to those of cells grown continuously in the light. The period of most rapid incorporation of precursors into chloroplast RNA occurs during the first 8 hours of illumination and, in this early phase, the specific activity of chloroplast RNA exceeds that of cytoplasmic RNA (MUNNS et al., 1972). If culture conditions are selected so that cell division does not take place during the greening process, the total amount of RNA per cell does not change although the proportion of plastid RNA increases from zero to 7% in 16 to 24 hours (MUNNS et al., 1972). This suggests that some mobilization of the cytoplasmic RNA is taking place, providing nucleotide precursors for use in plastid RNA synthesis. The notion is supported by the observation that transfer of cells to the light causes stimulation of incorporation of labelled orotic acid into cytoplasmic RNA which presumably is turning over at a significant rate.

Transfer of dark-grown *Euglena* cells to photosynthetic conditions of growth results in the appearance of new species of phenylalanine tRNA, glutamine tRNA and isoleucine tRNA localized within the chloroplasts (BARNETT et al., 1969; REGER et al., 1970). Cells of the *Euglena* mutant, W$_3$BUL, which lack chloroplast DNA do not synthesize these tRNA species when exposed to light.

From these experiments it would appear that light is obligatory for plastid RNA synthesis in *Euglena*. However, this is not the case for all unicellular algae. In *Chlamydomonas reinhardi*, levels of chloroplast ribosomes remain constant whether the cells are grown phototropically on a minimal salt medium or hetero-trophically in the dark on an acetate-supplemented medium (GOODENOUGH et al., 1971).

The possibility that new messenger RNA species might play a role in the transformation of etioplasts to chloroplasts would seem a reasonable one. Looking at this problem in *Euglena*, BROWN and HASELKORN (1971) could not detect any species of messenger RNA in photosynthetic cells which was not also present in dark-grown cells. In other words, transferring cells from dark to light does not seem to generate any new chloroplast DNA-coded messenger RNAs. However the situation could be different in higher plants. Transfer of chinese cabbage plants into the light shortly before isolation of the chloroplasts results in most of the chloroplast ribosomes being transformed into polysomes (CLARK, 1964). This of course does not prove that synthesis of new species of messenger RNA has resulted from the light treatment but it is indicative of an increase in the level of messenger RNA.

In general, however, the potential which these studies seem to offer in the way of contributing to our knowledge on the role of chloroplast DNA and of elucidating the mechanism of light-induced development of chloroplasts, has not, so far, been realized.

I. Concluding Remarks

There is still much to be learnt with respect to the chloroplast specific tRNAs which are coded for by chloroplast DNA. The question of whether those tRNA species, whose synthesis is induced by exposure of dark-grown plants to light, play a role in the control of the light-induced synthesis of chloroplast proteins is of obvious interest. As far as the ribosomal RNA species are concerned, it has yet to be determined whether their synthesis involves precursor RNA molecules and whether those rRNA cistrons which apparently reside in the nuclear DNA are ever transcribed. Nothing definite is known about the site of chloroplast ribosome assembly and very little is known about the synthesis of chloroplast ribosomal proteins. At present there is evidence that some of the ribosomal proteins may be synthesized in the cytoplasm whereas others may be synthesized within the chloroplasts.

In general, however, the purely descriptive phase concerned with the charac-terization of the stable chloroplast RNA species is drawing to a close and attention is currently being directed to the question of what is the role of the major portion of the chloroplast DNA. The stable chloroplast RNA species account for only 5% of the total potential genetic content and the few proteins whose amino acid sequence is known to be specified by chloroplast DNA account for another 5% at most. How much, then, of the remaining information encoded in the chloroplast DNA is actually transcribed into RNA, and what function might such RNA have in the development and biochemical activities of the chloroplast? Are the RNA species mainly acting as messengers, coding for structural and enzymic proteins of the chloroplast, or are they involved in the elaborate inter-organelle control mechanisms which must exist in plant cells? Does the chloroplast RNA polymerase exercise a control function or does it transcribe all available cistrons in chloroplast DNA without discrimination? Do any of the RNA transcripts of chloroplast DNA appear in the cytoplasm of the cell? And what of the nuclear-coded chloroplast proteins? Are any of them actually translated in the chloroplast on 70S ribosomes or are they all synthesized in the cytoplasm and then transported to the chloroplast? These are the types of question which are now attracting attention and no doubt many of the answers will soon be forthcoming.

References

Reviews

KIRK, J.T.O.: Ann. Rev. Plant Physiol. **21**, 11 (1970).
KIRK, J.T.O., TILNEY-BASSETT, R.A.E.: The plastids. London and San Francisco: Freeman 1967.
LOENING, U.E.: Ann. Rev. Plant Physiol. **19**, 37 (1968).
SMILLIE, R.M., SCOTT, N.S.: Progr. Mol. Subcell. Biol. **1**, 136 (1969).
TEWARI, K.K.: Ann. Rev. Plant Physiol. **22**, 141 (1971).
WOODCOCK, C.L.F., BOGORAD, L.: In Structure and function of chloroplasts (GIBBS, M., ed.), p. 89. Berlin-Heidelberg-New York: Springer 1971.

Other References

ANDERSON, J.M., KEY, J.L.: Plant Physiol. **48**, 801 (1971).
BARNETT, W.E., PENNINGTON, C.J., FAIRFIELD, S.A.: Proc. Nat. Acad. Sci. U.S. **63**, 1261 (1969).
BARTELS, P.G., WEIER, T.E.: J. Cell Biol. **33**, 243 (1967).
BELLAMY, A.R., RALPH, R.K.: In Methods in enzymology (GROSSMAN, L., and MOLDAVE, K., eds.), vol. 12B, p. 156. New York and London: Academic Press 1968.
BLAIR, G.E., ELLIS, R.J.: Biochem. J. **127**, 42p. (1972).
BOARDMAN, N.K.: Exp. Cell Res. **43**, 474 (1966).
BOTTOMLEY, W.: Plant Physiol. **45**, 608 (1970a).
BOTTOMLEY, W.: Plant Physiol. **46**, 437 (1970b).
BOTTOMLEY, W., SMITH, H.J., BOGORAD, L.: Proc. Nat. Acad. Sci. U.S. **68**, 2412 (1971a).
BOTTOMLEY, W., SPENCER, D., WHEELER, A., WHITFELD, P.R.: Arch. Biochem. Biophys. **143**, 269 (1971b).
BOTTOMLEY, W., WHITFELD, P.R., SPENCER, D.: Arch. Biochem. Biophys. **151**, 35 (1972).
BOULTER, D., ELLIS, R.J., YARWOOD, A.: Biol. Rev. **47**, 113 (1972).
BOURQUE, D.P., BOYNTON, J.E., GILLHAM, N.W.: J. Cell Sci. **8**, 153 (1971).
BRÄNDLE, E., ZETSCHE, K.: Planta **99**, 46 (1971).
BRAWERMAN, G., EISENSTADT, J.: J. Mol. Biol. **10**, 403 (1964).
BROWN, R.D., HASELKORN, R.: Proc. Nat. Acad. Sci. U.S. **68**, 2536 (1971).
BURKARD, G., ECLANCHER, B., WEIL, J.H.: FEBS Lett. **4**, 285 (1969).
BURKARD, G., GUILLEMAUT, P., WEIL, J.H.: Biochim. Biophys. Acta **224**, 184 (1970).
BURKARD, G., VAULTIER, J.P., WEIL, J.H.: Phytochemistry **11**, 1351 (1972).

CHEN, J. L., WILDMAN, S. G.: Science **155**, 1271 (1967).

CHEN, J. L., WILDMAN, S. G.: Biochim. Biophys. Acta **209**, 207 (1970).

CLARK, M. F.: Biochim. Biophys. Acta **91**, 671 (1964).

CLICK, R. E., HACKETT, D. P.: Biochim. Biophys. Acta **129**, 74 (1966).

DETCHON, P., POSSINGHAM, J. V.: Phytochemistry **11**, 943 (1972).

DURE III, L. S., MERRICK, W. C.: In Autonomy and biogenesis of mitochondria and chloroplasts (BOARDMAN, N. K., LINNANE, A. W., and SMILLIE, R. M., eds.), p. 413. Amsterdam: North-Holland 1971.

DYER, T. A., LEECH, R. M.: Biochem. J. **106**, 689 (1968).

DYER, T. A., MILLER, R. H., GREENWOOD, A. D.: J. Exp. Bot. **22**, 125 (1971).

ELLIS, R. J., HARTLEY, M. R.: Nature New Biol. **233**, 193 (1971).

FALK, H.: J. Cell Biol. **42**, 582 (1969).

GALLING, G.: Planta **98**, 50 (1971).

GALLING, G., SSYMANK, V.: Planta **94**, 203 (1970).

GIBBS, S. P.: J. Cell Sci. **3**, 327 (1968).

GOODENOUGH, U. W., LEVINE, R. P.: J. Cell Biol. **44**, 547 (1970).

GOODENOUGH, U. W., TOGASAKI, R. K., PASZEWSKI, A., LEVINE, R. P.: In Autonomy and biogenesis of mitochondria and chloroplasts (BOARDMAN, N. K., LINNANE, A. W., and SMILLIE, R. M., eds.), p. 224. Amsterdam: North-Holland 1971.

GRAHAM, D., SMILLIE, R. M.: In Methods in enzymology (SAN PIETRO, A., ed.), vol. 23 A, p. 228. New York and London: Academic Press 1971.

GUDERIAN, R. H., PULLIAM, R. L., GORDON, M. P.: Biochim. Biophys. Acta **262**, 50 (1972).

HEBER, U.: Planta **59**, 600 (1963).

HOOBER, J. K., BLOBEL, G.: J. Mol. Biol. **41**, 121 (1969).

INGLE, J.: Plant Physiol. **43**, 1448 (1968 a).

INGLE, J.: Plant Physiol. **43**, 1850 (1968 b).

INGLE, J., BURNS, R. G.: Biochem. J. **110**, 605 (1968).

INGLE, J., POSSINGHAM, J. V., WELLS, R., LEAVER, C. J., LOENING, U. E.: Symp. Soc. Exptl. Biol. **24**, 303 (1970).

JAGENDORF, A. T.: Plant Physiol. **30**, 138 (1955).

KIRBY, K. S.: Biochem. J. **96**, 266 (1965).

KIRK, J. T. O.: Biochem. Biophys. Res. Commun. **14**, 393 (1964).

KIRK, J. T. O.: Ann. Rev. Biochem. **40**, 161 (1971).

KUNG, S. D., WILLIAMS, J. P.: Biochim. Biophys. Acta **194**, 434 (1969).

LEAVER, C. J., INGLE, J.: Biochem. J. **123**, 235 (1971).

LEIS, J. P., KELLER, E. B.: Proc. Nat. Acad. Sci. U.S. **67**, 1593 (1970).

LOENING, U. E., INGLE, J.: Nature **215**, 363 (1967).

LYTTLETON, J. W.: Exp. Cell Res. **26**, 312 (1962).

MANNING, J. E., WOLSTENHOLME, D. R., RICHARDS, O. C.: J. Cell Biol. **53**, 594 (1972).

MANNING, J. E., WOLSTENHOLME, D. R., RYAN, R. S., HUNTER, J. A., RICHARDS, O. C.: Proc. Nat. Acad. Sci. U.S. **68**, 1169 (1971).

MATSUDA, K., SIEGEL, A.: Proc. Nat. Acad. Sci. U.S. **58**, 673 (1967).

MATSUDA, K., SIEGEL, A., LIGHTFOOT, D.: Plant Physiol. **46**, 6 (1970).

MERRICK, W. C., DURE III, L. S.: Proc. Nat. Acad. Sci. U.S. **68**, 641 (1971).

MUNNS, R., SCOTT, N. S., SMILLIE, R. M.: Phytochemistry **11**, 45 (1972).

OSHIO, Y., HASE, F.: Plant Cell Physiol. **9**, 69 (1968).

PATTERSON, B. D., SMILLIE, R. M.: Plant Physiol. **47**, 196 (1971).

PAYNE, P. I., DYER, T. A.: Biochim. Biophys. Acta **228**, 167 (1971 a).

PAYNE, P. I., DYER, T. A.: Biochem. J. **124**, 83 (1971 b).

PHILIPPOVICH, I. I., TONGUR, A. M., ALINA, B. A., OPARIN, A. I.: Exp. Cell Res. **62**, 399 (1970).

POLYA, G. M., JAGENDORF, A. T.: Arch. Biochem. Biophys. **146**, 649 (1971).

RAWSON, J. R., STUTZ, E.: Biochim. Biophys. Acta **190**, 368 (1969).

REGER, B. J., FAIRFIELD, S. A., EPLER, J. L., BARNETT, W. E.: Proc. Nat. Acad. Sci. U.S. **67**, 1207 (1970).

ROSSI, L., GUALERZI, C.: Life Sci. **9**, 1401 (1970).

RUPPEL, H. G.: Z. Naturforsch. B **22**, 1068 (1967).

RUPPEL, H. G.: Z. Naturforsch. B **24**, 1467 (1969).

SAGER, R., HAMILTON, M. G.: Science **157**, 709 (1967).

Schweiger, H. G.: Symp. Soc. Exptl. Biol. **24**, 327 (1970).
Schweiger, H. G., Dillard, W. L., Gibor, A., Berger, S.: Protoplasma **64**, 1 (1967).
Scott, N. S., Munns, R., Graham, D., Smillie, R. M.: In Autonomy and biogenesis of mitochondria and chloroplasts (Boardman, N. K., Linnane, A. W., and Smillie, R. M., eds.), p. 383. Amsterdam: North-Holland 1971 a.
Scott, N. S., Nair, H., Smillie, R. M.: Plant Physiol. **47**, 385 (1971 b).
Scott, N. S., Smillie, R. M.: Biochem. Biophys. Res. Commun. **28**, 598 (1967).
Semal, J., Spencer, D., Kim, Y. T., Wildman, S. G.: Biochim. Biophys. Acta **91**, 205 (1964).
Smith, H.: Phytochemistry **9**, 965 (1970).
Smith, H., Stewart, G. R., Berry, D. R.: Phytochemistry **9**, 977 (1970).
Solymosy, F., Lazar, G., Bagi, G.: Anal. Biochem. **38**, 40 (1970).
Spencer, D.: Arch. Biochem. Biophys. **111**, 381 (1965).
Spencer, D.: In Methods in developmental biology (Wilt, F. H., and Wessells, N. K., eds.), p. 645. New York: T. Y. Crowell Co. 1967.
Spencer, D., Whitfeld, P. R.: Arch. Biochem. Biophys. **121**, 336 (1967).
Spencer, D., Whitfeld, P. R., Bottomley, W., Wheeler, A. W.: In Autonomy and biogenesis of mitochondria and chloroplasts (Boardman, N. K., Linnane, A. W., and Smillie, R. M., eds.), p. 372. Amsterdam: North-Holland 1971.
Stocking, C. R.: In Methods in enzymology (San Pietro A., ed.), vol. 23A, p. 221. New York and London: Academic Press 1971.
Stutz, E., Noll, H.: Proc. Nat. Acad. Sci. U.S. **57**, 774 (1967).
Stutz, E., Rawson, J. R.: Biochim. Biophys. Acta **209**, 16 (1970).
Stutz, E., Vandrey, J. P.: FEBS Lett. **17**, 277 (1971).
Surzycki, S. J.: Proc. Nat. Acad. Sci. U.S. **63**, 1327 (1969).
Tester, C. F., Dure III, L.: Biochem. Biophys. Res. Commun. **23**, 287 (1966).
Tewari, K. K., Wildman, S. G.: Science **153**, 1269 (1966).
Tewari, K. K., Wildman, S. G.: Proc. Nat. Acad. Sci. U.S. **59**, 569 (1968).
Tewari, K. K., Wildman, S. G.: Biochim. Biophys. Acta **186**, 358 (1969).
Tewari, K. K., Wildman, S. G.: Symp. Soc. Exptl. Biol. **24**, 147 (1970).
Trewavas, A.: Plant Physiol. **45**, 742 (1970).
Vedel, F., D'Aoust, M. J.: Plant Physiol. **46**, 81 (1970).
Walker, D. A.: In Methods in enzymology (San Pietro, A., ed.), vol. 23A, p. 211. New York and London: Academic Press 1971.
Wildman, S. G.: In The biochemistry of gene expression in higher organisms (Pollak, J. K., and Lee, J. W., eds.). Sydney: Australia and New Zealand Book Co., 1972 (in press).
Williams, G. R., Williams, A. S.: Biochem. Biophys. Res. Commun. **39**, 858 (1970).
Wollgiehn, R., Ruess, M., Munsche, D.: Flora **157**, 92 (1966).
Woodcock, C. L. F., Bogorad, L.: Biochim. Biophys. Acta **224**, 639 (1970).

CHAPTER 10

Viral RNA

A. J. GIBBS and J. J. SKEHEL

Introduction

Nowadays such a wide range of biological entities are considered to be viruses that most recent definitions of viruses have been outdated within a year or so, and it is increasingly difficult to devise a general definition that would be accepted by most virologists. In this chapter we will only discuss viruses that are *transmissible parasites, whose nucleic acid genome is less than 3×10^8 daltons in weight, and that need ribosomes and other components of their host cells for replication.*

The genome of some viruses is DNA; the DNA of some is single stranded, of others double, of some circular, of others linear. It is probable that the genetic information of all types of virus DNA is expressed via messenger RNA in the same way as that of the DNA of pro- and eukaryotes. By contrast, the genome of other viruses is RNA; the RNA of some is single stranded, of others double, however, it seems that all RNA genomes are linear but some are divided into two or more segments. The genetic information in some of these viral RNAs is expressed directly, in others it is expressed via complementary messenger RNA, and in one interesting group of viruses, it is perhaps expressed via DNA and messenger RNA.

There are many types of viral RNAs. The largest are those of the paramyxoviruses, and are about 7×10^6 daltons in weight, but most viral genomes are around $2-4 \times 10^6$ daltons in weight. The smallest well characterized viral RNA is that in the particles of the satellite virus of tobacco necrosis virus, which is only 4×10^5 daltons in weight (KASSANIS, 1970). Even smaller RNAs are found in preparations of the virus-like agent or viroid that causes potato spindle tuber disease (DIENER, 1972). The smallest individual viral RNAs are viral monocistronic mRNAs, though the smallest functional viral RNAs are the few ribonucleotides found attached to the replicative form DNA of M13 and ϕX174 bacteriophages and which are needed to prime their synthesis (SCHEKMAN et al., 1972).

Within the restricted length of a single chapter we cannot review exhaustively all that is known of viral RNAs and therefore we will merely outline the major areas of interest and describe in more detail particular illustrative examples. The chapter is divided into two principal sections, the first of which describes the functions and synthesis of different viral RNAs and their place in the life cycles of different viruses; the second section describes the structure of viral RNA at different stages of the life cycles.

The names of viruses and virus groups that we use come from the collation of Wildy (1971), and a glossary at the end of this chapter gives brief notes on all the viruses mentioned.

A. Synthesis and Function

RNA is indispensable to the life cycle of all viruses. Some of the simplest viruses have RNA that serves both as genome and as mRNA, whereas the most complex have a DNA genome, and mRNA and sometimes tRNA are synthesized at other parts of their life cycle.

In most viruses the transcription of the genome and the translation of its message are so closely interdependent that they are best discussed together as in this chapter.

The ways in which the genomes of viruses are expressed in infected cells have received a great deal of attention (see, for example, the review of Sugiyama et al., 1972) and, as a result, most viruses can be conveniently placed into one or other of four basic groups.

1. Viruses Whose Particles Contain Single Stranded Infectious RNA

a) Viruses Whose Genome Is a Single RNA Species. Much is known about this type of virus since the group includes the RNA bacteriophages R17, Qβ and MS2, the animal picornaviruses, polio and encephalomyocarditis virus (EMC) and the plant virus, tobacco mosaic virus (TMV) all of which have been extensively examined from several aspects. The virion RNA of all these viruses can be purified free of virus protein and shown to be a single infectious polynucleotide. The fact that these RNAs are infective is now thought to imply that they serve as mRNAs in infected cells and are translated directly to produce proteins some of which are required for virus replication.

1) The ribophages. R17, MS2 and Qβ are typical of the small bacteriophages; they have isometric particles (Fig. 1) about 25 nm in diameter, which contain the RNA genome of about 10^6 daltons molecular weight. In infected cells the genome is translated into virus specific proteins; a RNA replicase subunit, a coat protein and an A protein or maturation protein, the latter two of which together with the RNA are the sole components of infectious virus particles. Once the replicase subunit has been synthesized it combines with three host proteins to form a virus specific replicase, whose properties are discussed below. Transcription of the viral RNA can then start.

Although there is still some debate about the identity of certain intermediates in replication it is generally agreed that the polymerase enzyme attaches first to the 3' end of the virus RNA, and moves along the RNA reading it from 3' to 5' and synthesizing a complementary RNA molecule as an antiparallel strand, i.e. in a 5' $\rightarrow$ 3' direction. When this molecule is complete the replicase attaches to its 3' end and in turn copies it to produce a polynucleotide identical in sequence

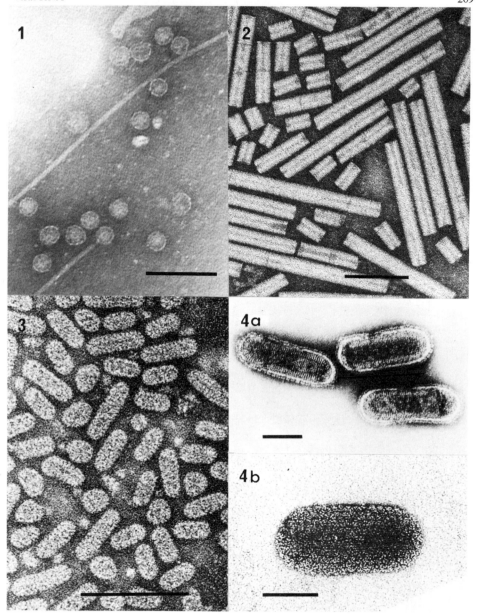

Fig. 1. The icosahedrally constructed particles of Qβ bacteriophage, a ribophage. The bar markers in Figs. 1–8 are 100 nm long

Fig. 2. The helically constructed tubular particles of tobacco rattle virus, note that the particles are mostly of two lengths; the genome is divided between these two types of particles (courtesy of B.D. HARRISON and I. M. ROBERTS)

Fig. 3. Particles of alfalfa mosaic virus, in structure they are tubular variants of icosahedra. They are mostly of four lengths; the genome of the virus is divided among these four types of particles

Fig. 4. Particles of sowthistle yellow vein virus, a rhabdovirus. The particles, fixed with gluteraldehyde, were negatively stained either with a) sodium phosphotungstate to show the helically constructed core and 'edge view' of the envelope subunits, or b) uranyl acetate to show the arrangement of the envelope subunits (courtesy of D. PETERS)

to the initial virus RNA (Stavis and August, 1970). The major product of prolonged RNA synthesis *in vitro* is viral RNA as shown by assays for infectivity and the results of hybridization experiments. Thus it is clear that after the initial synthesis of RNA complementary to virion RNA, the replicase preferentially copies the complementary RNA and ensures the predominant synthesis of infectious viral RNA. This selection process will be mentioned later in connection with the subunit composition of the replicase molecule.

It has been suggested that the viral RNA is synthesized either using a double stranded RNA template or a free complementary single stranded RNA as template. The evidence most often cited in support of the first idea is that RNAase-resistant RNA (i.e. double stranded RNA) is readily isolated from polymerase reaction mixtures and from virus infected cells. This RNA can be fractionated into completely double stranded RNA, called replicative form (RF) (Amman et al., 1964), and partially RNAase-resistant RNA, called replicative intermediates (RI) (Franklin, 1966). The RF is thought to be the product of the first stage of RNA synthesis; and is the template plus complementary RNA. The RI is thought to be produced later and to consist of complexes of several nascent viral RNA molecules hydrogen-bonded for parts of their sequence to a single complementary RNA template. Pulse-chase experiments with radioactive RNA precursors support this idea as the radioactivity in RI decreases with concomitant increase in the radioactivity of free viral RNA. These observations suggest that the mechanism of RNA replication is similar in outline to that previously shown for the replication of ϕX174 single stranded DNA molecules (Sinsheimer et al., 1962). However, although the involvement of double stranded structures in RNA synthesis has not been disproved, much doubt has been cast upon their significance (Weissman et al., 1968). Thus, contrary to prediction neither the RF nor RI will act as templates for RNA synthesis using purified Qβ replicase. Moreover, several reports indicate that little RNAase-resistant RNA is obtained unless polymerase reaction mixtures are deproteinized using either detergent or phenol, and it is suggested that the double stranded RNAs are artefacts and are not present in the infected cell. Therefore, the second scheme has been proposed, and this simply suggests that the RNA template, the complementary strand mentioned earlier, is free and is not extensively base-paired with progeny RNA. This scheme predicts that before deproteinization of the reaction mixtures both the template and the product will be sensitive to RNAase digestion, and that RNA complementary to viral RNA will function as template for RNA synthesis *in vitro*. Both these predictions have proved true.

Thus it is now thought that the first product of polymerase action, complementary RNA, is subsequently used as a single stranded template in the production of virus RNA. Both complementary and viral RNA molecules are transcribed in a 3' to 5' direction and are synthesized from 5' to 3', and RNA molecules while being transcribed are not necessarily base paired with much of the nascent progeny.

Studies of Qβ RNA synthesis both *in vivo* and *in vitro* have contributed greatly to our knowledge of RNA replication, and Qβ replicase is the best known RNA-dependent RNA polymerase. It has been purified in various ways (Eikhom and Spiegelman, 1967; Kamen, 1970; Stavis and August, 1970). In most methods extracts of virus-infected cells are fractionated either by precipitating differentially with protamine sulphate or ammonium sulphate, or by polyethylene glycol/ Dextran phase separation. The enzyme is then further purified by DEAE-cellulose and phosphocellulose column chromatography and density gradient centrifugation.

During purification the specificity of the enzyme is modified. Before purification the enzyme will replicate either $Q\beta$, or a small virus-specific RNA present in $Q\beta$ infected cells, or even synthetic polymers containing cytidylic acid. After purification, however, viral RNA cannot serve as template, perhaps because host components have been removed from the polymerase complex; two such components have been resolved (STAVIS and AUGUST, 1970). These host factors are specifically required for viral RNA to function as template, perhaps at some stage after the replicase attaches to the viral RNA.

The purest functional replicase so far obtained consists of four different polypeptides of molecular weights 70, 65, 45, and 35 thousand daltons (KAMEN, 1970; KONDO et al., 1970). Of these, the second largest subunit is the product of the virus replicase gene and the other three are from the host and can be found in uninfected bacteria. The role of these host components in uninfected cells has provoked much interest. It was initially suggested that the two smallest polypeptides stimulated transcription of host DNA by DNA-dependent RNA polymerase and they appeared to be involved in ribosomal RNA synthesis (TRAVERS et al., 1970). However, this now seems unlikely (PETTIJOHN, 1972), and there now is evidence that all are concerned with protein biosynthesis; the two smaller replicase proteins are elongation factors (BLUMENTHAL et al., 1972), and the largest replicase component is the interference factor i which appears to interact directly with the protein synthesis initiation factor IF3 and affect the specificity of ribosome binding to messenger RNAs (GRONER et al., 1972 b).

Thus several bacterial factors are required for $Q\beta$ replicase. Two of these allow the replicase to synthesize complementary RNA using virus RNA as template and three others appear to be required independently of template for replicase action.

Work on the translation of the ribophage genome has also yielded many interesting results. During infection the amount and time of appearance of each of the virus proteins varies. Thus the replicase is mostly synthesized early in infection, the maturation protein is synthesized at a constant rate throughout infection, and the coat protein is made throughout infection but at an exponentially increasing rate and in much greater amounts than either of the other two proteins (NATHANS et al., 1969). The most likely explanation for these differences is that the translation of the three proteins is specifically and independently controlled, and that the genome probably has separate sites for initiating and terminating the translation of the gene for each protein. This subject is further discussed in Chapter 4.

Experiments have also shown that the three genes in the genome of ribophages are arranged in the order 5' terminus—maturation (A protein)—coat protein— replicase subunit—3' terminus. The way in which independent translation of the ribophage genes is controlled, and the solution to the dilemma that the genome is translated in a 5' to 3' sense but is transcribed in the opposite direction (WEISSMAN et al., 1968) is turning out to be a fascinating story (ANON., 1971) even though it is not yet complete. However it is clear that most of the molecules involved have dual functions. For example the coat protein molecules not only form the coat of the virus particle, but they also repress translation of the replicase subunit gene (Bernardi and Spahr, 1972), so that during infection, as coat protein accumulates, a decreasing amount of replicase is produced. Similarly the replicase subunit has dual functions for not only does it combine with three host enzyme subunits to form the virus specific replicase (Blumenthal et al., 1972), but it also competes with ribosomes for mRNA, as it is impossible for mRNA to be translated and transcribed at the same time since the ribosomes and replicase move in opposite directions. When the viral replicase is added to complexes of the viral mRNA and ribosomes, no more ribosomes attach and those already attached complete translation before transcription starts. It is likely that the replicase, which begins transcription at the 3' terminus, stops ribosomes attaching to the coat protein initiation site by attaching to both the initiation site and the 3' terminus, as when mixed with viral RNA it protects the coat protein initiation site from nuclease digestion. Thus, early in infection both the coat protein and replicase genes are translated, but later, when coat protein inhibits translation of the replicase gene, only coat protein is produced. However an additional complexity is that the largest replicase subunit, factor i, can modify the specificity of ribosome attachment to the different initiation sites on the viral RNA, so that initiation at the coat protein cistron is inhibited and initiation at the replicase cistron stimulated (Groner et al., 1972 a). As mentioned before the replicase preferentially transcribes complementary RNA to produce progeny viral RNA, and it seems probable that the A protein gene is translated only on nascent progeny RNA and thus only one A protein molecule will be produced for each progeny RNA molecule.

These mechanisms ensure that the viral progeny molecules are produced at the right stage of infection and in the stoichiometric proportions required for viral replication: one viral mRNA molecule, one A protein molecule, 180 coat protein molecules (all of which combine to form progeny virus particles), and less than one replicase subunit required in the replication process.

2) Poliovirus. By contrast with the ribophages the replication of poliovirus seems to be controlled, if at all, in quite a different manner. Its replication has probably received more attention than that of any other animal virus because its genome is only about 2.7×10^6 daltons in weight and because of the rapidity with which large numbers of its isometric virus particles are produced in infected cells. Like the ribophages discussed above the genome of poliovirus is a single stranded infectious RNA molecule which can direct the synthesis of virus proteins both *in vivo* and *in vitro*. Unlike the ribophages, however, the number of distinct functional gene products synthesized and hence the number and order of the poliovirus genes are not known with certainty. This is largely because it seems that mammalian cells do not have the ability to terminate and re-initiate polypeptide synthesis within a polycistronic mRNA. Thus poliovirus RNA is translated into one large polypeptide chain which is equivalent in size to the complete virus genome (Summers and Maizel, 1968; Jacobson et al., 1970; Öberg and Shatkin, 1972). Subsequently, by processes which are not understood, this large precursor molecule is rapidly divided within the infected cell to yield the functional

viral proteins. Analyses of the proteins produced at various times during infection (SUMMERS et al., 1965) indicate that the same polypeptide molecules are synthesized at all times, a seemingly wasteful process.

The order of the genes in the poliovirus genome, and hence the order of poly-peptides in the large precursor protein molecule, has been determined by genetical recombination experiments (COOPER et al., 1971), and by use of the inhibitor pactamycin (TABER et al., 1971; SUMMERS and MAIZEL, 1971). These experiments show that the coat protein gene is at the 5' end of the genome with the gene(s) associated with polymerase function at the 3' end.

The role of double stranded RNAs in poliovirus replication has also caused much debate, though it now seems probable that poliovirus intermediates are similar to those of the ribophages (ÖBERG and PHILIPSON, 1971).

Work with other virus specific replicases has so far been much less productive than that with Qβ replicase, perhaps because it now seems that many are bound to host cytoplasmic membranes. However the replicase of EMC, which is an enterovirus related to poliovirus, has been purified and closely resembles that of Qβ; it consists of five polypeptides, four of which are almost identical in size to the four polypeptides of Qβ replicase (ROSENBERG et al., 1972).

The togaviruses are viruses that infect both vertebrates and their invertebrate vectors, and are perhaps the only animal viruses with infectious segmented RNA genomes. In cells infected with togaviruses there is found not only the viral genome RNA, which has a molecular weight of 4×10^6 daltons, but also several smaller pieces of viral RNA that range in size from $2-3 \times 10^6$ daltons (SIMMONS and STRAUSS, 1972). It is possible that these components are viral mRNA, since RNAs of similar size have been isolated from the polysomes of cells infected with Semliki Forest virus (KENNEDY, 1972). It is thus possible that togavirus mRNAs smaller than the genome are involved in replication.

3) Other viruses. There is at present no evidence on how plant viruses replicate and whether their replication is similar to that of the ribophages, or poliovirus or some other system. Several plant viruses including tobacco mosaic virus (TMV) and turnip yellow mosaic virus (TYMV) have RNA genomes which are infectious when removed from the particles, and it has been assumed, by analogy with animal and bacterial viruses, that their virion RNA is therefore their mRNA, however recent evidence from alfalfa mosaic virus (discussed later in this Chapter) suggested that this assumption may not always be correct.

The virion RNAs of several plant viruses have been shown to resemble tRNAs in having a 3' terminal sequence of —CCCA with the terminal adenosine perhaps added by a tRNA nucleotidyltransferase, though it is missing from the RNA of some viruses (e.g. tobacco rattle virus). The similarity to tRNAs is even closer as RNAs of these viruses specifically accept and bind amino acids from yeast amino acid-tRNA ligase preparations, for example TYMV RNA specifically accepts valine (PINCK et al., 1970) and TMV RNA accepts histidine (LITVAK et al., 1973). Furthermore the amino acids are released from the virus RNAs by N-acylamino-acyl-tRNA hydrolase. TYMV RNA does not have to be intact to accept valine; PROCHIANTZ and HAENNI (1973) have found that when the 23S viral RNA is treated with tRNA maturation endonuclease it yields a 4.5S 3' terminal fragment which can be acylated. The significance of these findings is not known, but it is possible that this tRNA function may be involved in transcription as Litvak et al. (1973) have shown that aminoacylated TYMV or TMV RNAs will bind to wheat embryo protein elongation factor (EF 1) whereas unacylated RNAs will not, and it is known that the analogous factor in *E. coli* (Tu or EF Tu) is one of the components of Qβ replicase.

There have been claims that the RNA from the particles of some plant viruses can direct the synthesis of proteins in cell-free protein synthesizing systems. Most of these reports are unconvincing, except that of van Ravenswaay Claasen et al. (1967), who showed that the top *a* RNA of alfalfa mosaic virus could be translated into a protein closely resembling the coat protein of the virus, a result confirmed by genetical complementation experiments.

Little is known of the position of genes in the RNA of these viruses as complementation experiments like those done with divided genome viruses are difficult or impossible (Kado and Knight, 1968). However Kado and Knight (1966) estimated that the coat protein gene of TMV is about one quarter of the distance from the 5′ end of the genome. They combined the fact that when TMV particles are treated with the detergent sodium dodecyl sulphate, they disassemble from that end of the particle which contains the 3′ end of the genome, with the fact that exposed TMV RNA is more susceptible to the mutagenic action of nitrous acid than RNA in intact particles. Using particles stripped to differing extents with detergent they measured the frequency of mutation in the coat protein gene.

Another virus that should yield information on the synthesis and function of its RNA, but is in need of further study is satellite virus (Kassanis, 1970). Its genome is a single stranded RNA molecule of molecular weight 4×10^5 daltons, and it replicates only in plants already infected with certain strains of tobacco necrosis virus, whose genome is also a single stranded RNA molecule but of 1.5×10^6 daltons molecular weight.

b) Viruses With Segmented Genomes. Certain plant viruses that have infectious RNA in their particles have the RNA divided into several distinct separable molecules. One of the best studied of these is tobacco rattle virus (Harrison, 1970). The purified particles of this virus are helically-constructed tubes about 25 nm in diameter and most are either about 180 nm long or, depending upon the isolate, between 50–90 nm long (Fig. 2). Particles of the two lengths are readily separated by centrifugation. Mixtures of the two or of the RNA they contain are infectious and plants infected with the mixtures yield virus preparations containing particles of both lengths and indistinguishable from the parent isolate. By contrast, although the long particles, or their RNA, can infect plants, the infected plants apparently contain only infectious viral RNA but no virus particles. The short particles, or the RNA they contain, are not infective, but when plants infected with long particles are rubbed with the short particles, or their RNA, they yield virus particles of both lengths. These and other experiments, which involve mixing long and short particles of distinguishable strains of the virus, show that the RNA of the short particles contains the gene for the coat protein from which both long and short particles are made (Lister and Bracker, 1969; Sänger, 1969).

There are other plant viruses with divided genomes. In most of these the segments of the genomes are completely interdependent, and only tobacco rattle virus and other tobraviruses have their genomes divided so that one of the segments is able to replicate alone, albeit defectively. Cowpea mosaic virus is an example of the other type; preparations of it contain two ribonucleoprotein components, one component containing 23% RNA, the other 32% RNA. The components when separated are not infectious but mixtures of them, or of the RNA molecules extracted from them, are (VAN KAMMEN, 1968).

The genomes of some other plant viruses are divided into more than two segments. For example the genomes of the bromoviruses are divided into at least three segments, though even more species of RNA are always found in particles of these viruses and the function, if any, of the additional RNA species is not known.

The isometric particles of the bromoviruses were once thought to be all of one type, each containing the viral genome of about 1.1×10^6 daltons molecular weight; the smallest autonomous plant virus. However recent work on brome mosaic virus (BANCROFT, 1970) and other bromoviruses (LANE and KAESBERG, 1971) has shown that the relatively homogenous preparations of particles of these viruses contain at least three closely similar types of particle, which contain four different RNA species. One type of particle contains a single RNA molecule of molecular weight 1.1×10^6 daltons, the second another of 1.0×10^6 daltons, and the third two RNA molecules of molecular weights 0.7×10^6 and 0.3×10^6 daltons. Only mixtures containing the three largest RNA species are infective, and it is tempting to suggest that the smallest RNA merely provides packing for the 0.7×10^6 dalton RNA segment. However this is not known nor has it been reported whether the smallest RNA has a unique sequence distinct from those of the other three RNAs.

The different particles of alfalfa mosaic virus (BOS and JASPERS, 1971; Fig. 3) are much more readily distinguished than those of the bromoviruses, even though they are not easily separated (VAN VLOTEN-DOTING and JASPERS, 1967). Most of the particles obtained from alfalfa mosaic infected plants are of four types and these are called, in order of decreasing sedimentation coefficient, the bottom, middle, top b and top a components. Despite earlier reports to the contrary (BANCROFT and KAESBERG, 1960) most now agree that alone none is infectious. The extremely interesting results of work on this virus obtained by VAN VLOTEN-DOTING and her colleagues will be discussed in the next section of this chapter, but we note here that much detailed work has been done on the initiation of translation of top a RNA in an *in vitro* system by BOSCH and his colleagues (VERHOEF and BOSCH, 1971; VERHOEF et al., 1971).

2. Viruses Whose Particles Contain Single Stranded Noninfectious RNA

a) Viruses Whose Particles Contain a Single RNA Species. This type of viral RNA has so far only been found in particles of paramyxoviruses and rhabdoviruses (Fig. 4), which include viruses of mammals, insects and plants. The molecular weights of the genomes of viruses of these two groups are about 7×10^6 and 4×10^6 daltons respectively, and the RNAs from these viruses are not infectious. During the initial stages of virus replication a RNA-dependent RNA polymerase present in the virus particles transcribes the virion RNA; the resulting polynucleotides which are complementary in sequence to the genome are the viral mRNAs (HUANG et al., 1970, 1971). Moreover, these mRNAs are smaller than the virion RNA and range in molecular weight from about 3×10^5 to 2×10^6 daltons and may, therefore, be monocistronic mRNA molecules (KINGSBURY, 1966; BRATT and ROBINSON, 1967).

b) Viruses With Segmented Genomes.

1) Orthomyxoviruses and leukoviruses. These viruses have large pleomorphic particles that contain lipid and an incompletely defined number of single stranded RNA molecules. In the particles of orthomyxoviruses the genome probably consists of eight or nine distinct segments of RNA which together make up a total molecular weight of approximately 5×10^6 daltons (SKEHEL, 1972). The particles of leukoviruses (Fig. 6) contain four or five pieces of RNA of total molecular weight approximately 10^7 daltons (CANAANI and DUESBERG, 1972). Apart from possessing segmented genomes the two groups of viruses differ from each other in the way in which their genetic information is translated. Like those of the rhabdoviruses and paramyxoviruses, orthomyxovirus particles contain a RNA polymerase, but, by contrast with Qβ polymerase, it is unlikely that there are host components in this polymerase (SKEHEL, 1971). In infected cells the polymerase transcribes the virion RNA into mRNA molecules of equivalent size but of complementary sequence (BISHOP et al., 1972). Since these molecules can be isolated from polysomes (PONS, 1972) they are probably virus mRNAs, although how many of the virion gene segments are copied in this way and whether or not some segments are translated directly, remains to be determined. It is, however, clear that different orthomyxovirus genes are transcribed and translated into proteins independently and it is therefore possible that this independence allows independent control of their synthesis.

Less is known about the way leukovirus genomes are translated despite a great deal of research. There has been much interest in these viruses since it was shown, and repeatedly confirmed, that their particles contain a RNA-dependent DNA polymerase otherwise called a reverse transcriptase (BALTIMORE, 1970; TEMIN and MIZUTANI, 1970). This enzyme is of particular interest as it probably transcribes the RNA genome of the virus into DNA, which in turn is transcribed into progeny virus molecules but may also be incorporated into, and alter, the genome of the host cell. This latter propensity probably accounts for the oncogenicity (i.e. ability to cause cancer) of these viruses, and for the occurrence of proteins of these viruses in cells isolated from tumours of animals. Reverse transcriptase is also of interest as the transcription it catalyses is contrary to what is popularly assumed to be one of the basic tenets of molecular biology. As stated above, the transcriptase copies the virion RNA to produce complementary DNA. The enzyme can be extracted from virus particles in a template dependent form, purified and shown to contain at least two of the polypeptide components of the virus (KACIAN et al., 1971). When pure it will accept a variety of natural and synthetic polynucleotides as template and, like other DNA polymerase enzymes, a RNA primer is needed to start synthesis of complementary DNA; the virus particles contain a small polyribonucleotide primer (VERMA et al., 1972).

Experiments on the effects of anti-metabolites on the replication of leukoviruses show that DNA synthesis is required early in infection and the reverse

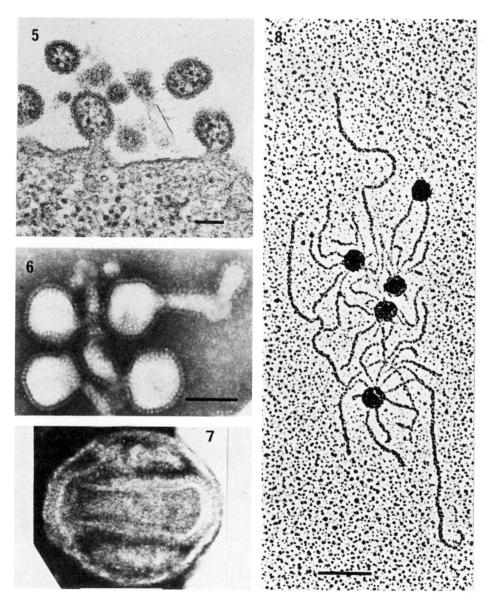

Fig. 5. A thin section of L cell infected with lymphocytic choriomeningitis virus, an arenavirus. Particles of the virus form at the cell membrane, note the ribosomes (electron-dense granules) both in the cytoplasm of the cell and within the particles (courtesy of F. LEHMANN-GRUBE)

Fig. 6. The pleomorphic particles of mouse mammary tumour virus, a leukovirus (courtesy of D. H. MOORE)

Fig. 7. A particle of vaccinia virus treated with detergent and negatively stained to show the outer membranes, lateral bodies and dumbell-shaped core (courtesy of K. B. EASTERBROOK)

Fig. 8. Cores of reovirus particles that have synthesized mRNA *in vitro* (GILLIES et al., 1972) and been prepared for electron microscopy by the Kleinschmidt technique (courtesy of A. R. BELLAMY)

transcriptase is presumably involved in this synthesis (Vigier, 1970). Other experiments show that DNA-dependent RNA synthesis is also essential throughout infection (Temin, 1963) and it is probable that newly formed viral DNA is transcribed to virus mRNA molecules. The relationship of these molecules to the virion RNAs is, however, not known.

The complementary DNA is itself copied by another polymerase component of virus particles to produce double stranded DNA molecules and it is thought that these might integrate with the host genome (Mizutani et al., 1970). The integrated DNA molecules are presumably transcribed by host DNA-dependent RNA polymerase to form virus mRNA and also RNA for encapsidation in progeny virus particles. Many of the steps involved in RNA production in leuko-virus infected cells are not known, but it is clear that genome replication involves the synthesis and transcription of virus specific DNA probably by both virus and host specific polymerase enzymes.

2) Other viruses. Another virus with a segmented genome which may have a life cycle similar to that of the orthomyxoviruses and leukoviruses is alfalfa mosaic virus (Fig. 3). The genome, isolated from particles, is infectious, suggesting that it may replicate like the viruses described in section b) of this chapter. However recent interesting results obtained by Bol et al. (1971), van Vloten-Doting et al. (1971) and Bol and van Vloten-Doting (1973) suggest that its life cycle may be different from these. Van Vloten-Doting and her colleagues purified the four major types of nucleoprotein particles of the virus and also the nucleic acids from these particles. They found that only the following mixtures were infective:

1. bottom + middle + top *b* ribonucleoproteins,
2. bottom RNA + middle RNA + top *b* RNA + isolated coat protein,
3. bottom RNA + middle RNA + top *b* RNA + top *a* RNA.

Thus a mixture of bottom + middle + top *b* RNAs is infective only when either coat protein or top *a* RNA is present too. The amount of coat protein required for infection is small; a few molecules per RNA molecule. These results suggest that coat protein helps initiate infection, and that top *a* RNA carries the gene for coat protein, which has been confirmed directly (van Ravenswaay Claasen et al., 1967), and is translated at infection to yield coat protein. However, other experiments show that top *b* RNA also carries the coat protein gene, and the progeny of mixtures of RNAs from different strains of the virus has the coat protein of the top *b* parent irrespective of the top *a* RNA or protein used to initiate infection.

Although other explanations for these results exist we suggest that the most likely is that top *a* RNA, or the coat protein it specifies, is a RNA polymerase needed to transcribe the three larger RNAs into complementary mRNAs. Thus the life cycle of alfalfa mosaic perhaps resembles those of the orthomyxoviruses and leukoviruses. Furthermore, cucumber mosaic virus, which is very similar to alfalfa mosaic virus, and also the bromoviruses may replicate in the same way.

All viruses in this class, therefore, probably present their genetic information to the cellular translation machinery in the form of distinct, possibly mono-cistronic, mRNAs and as a consequence they may share with the rhabdo- and paramyxoviruses similar mechanisms for controlling the synthesis of virus proteins.

3. Viruses Whose Particles Contain Double Stranded RNA

The viruses of this group which have received most attention are the reoviruses (JOKLIK et al., 1970). Not only is the genome of these viruses double stranded RNA but it is also segmented. Particles of these viruses contain ten pieces of RNA which are readily separated by sucrose density gradient centrifugation or polyacrylamide gel electrophoresis into three size classes of mean molecular weights approximately 2.5×10^6, 1.6×10^6 and 0.8×10^6 daltons, with a total molecular weight of 15×10^6 daltons. In infected cells each of these ten segments of double stranded RNA is transcribed, again by a RNA-dependent RNA polymerase which is a component of the virus particles, into single stranded mRNA molecules. These mRNA species can also be separated into three size classes. The mRNAs direct the synthesis of virus-specific polypeptides and again these fall into similar size classes (SMITH et al., 1969). Particles of the reoviruses thus contain ten genes of double stranded RNA which are transcribed in infected cells into ten monocistronic mRNA molecules, and these are translated into ten virus polypeptides.

As mentioned above the infective particles of the reoviruses, like those of some other viruses already discussed, contain the transcriptase of the virus. However the mechanism of synthesis of their mRNA is unlike that of the single stranded RNA viruses, but in contrast, is entirely analogous to that of the DNA genome viruses in that the double stranded RNA template is completely conserved (JOKLIK et al., 1970). In reovirus, which has been examined in most detail, all ten double stranded genome segments can be transcribed *in vitro* to produce ten species of messenger RNA. *In vivo*, however, transcription of certain of the segments may be restricted early in infection (WATANABE et al., 1968).

The RNA produced acts not only as mRNA but also as template upon which the complementary strand is synthesized to produce the RNA duplex of the genome (SAKUMA and WATANABE, 1972). The enzyme involved in this latter process is unknown.

The outer parts of reovirus particles must be removed before mRNA synthesis can be detected *in vitro* (Fig. 8). These resulting cores contain only three types of polypeptide, all of which are virus specified (JOKLIK et al., 1970), and it is probable that these are the transcriptase, although they have not yet been purified free of template. However as they are active *in vitro* it is likely that the enzyme does not depend on host factors.

It is probable that other viruses such as clover wound tumour virus and the cytoplasmic polyhedrosis viruses of insects (KALMAKOFF et al., 1969) whose particles contain similar segmented double stranded RNA genomes also replicate in the same way.

4. Viruses Whose Particles Contain DNA

Many viruses have DNA genomes, but none have their genomes segmented like those of the RNA viruses described above. The smallest viruses of this group have DNA genomes of around $1-2 \times 10^6$ daltons molecular weight and include

Kilham's rat and minute mouse viruses, whose genomes are single stranded linear molecules, and ϕX174 and the unrelated bacteriophage fd, whose genomes are single stranded circular molecules. In addition, the group also includes viruses like the adeno-associated (adeno-satellite) viruses and *Galleria* densonucleosis virus, all of which have isometric particles that contain single linear DNA molecules; however half of the particles contain single stranded DNA with one sequence and the other half contain DNA with the complementary sequence, so that when deproteinized they combine to form double stranded molecules (Barwise and Walker, 1970). There are also viruses whose genomes are circular double stranded DNA. These include the papovaviruses and cauliflower mosaic virus, which have isometric particles and genomes with molecular weights around 4×10^6 daltons. There is also evidence that at least one baculovirus has a genome of circular double stranded DNA of molecular weight 100×10^6 daltons (Summers and Anderson, 1972).

The other major group of DNA viruses have linear double stranded genomes. These are the largest viruses and range in size and complexity from the adenoviruses (25×10^6 daltons) to the iridoviruses and T-even phages (130×10^6 daltons), and the poxviruses (160–200×10^6 daltons) (Fig. 7). The DNA genomes of all these viruses are transcribed by DNA-dependent RNA polymerases in infected cells to produce mRNAs, which are translated into virus proteins.

a) The Bacteriophages.

1) ϕX174. The replication of ϕX174 and of the filamentous single stranded DNA-containing bacteriophages is mostly similar though it differs in some details. ϕX174 particles contain single-stranded circular DNA molecules of molecular weight 1.7×10^6 daltons (Sinsheimer, 1959). On infection, the phage DNA is converted by the action of host enzymes to a double stranded circular form, RF (replicative form), from which template, single stranded progeny DNA molecules are subsequently copied (Sinsheimer et al., 1962). The phage-specific messenger RNA required for replication is also transcribed by the host polymerase from the RF template. In distinction from the larger DNA-containing bacteriophages, all of the messenger RNA is complementary to one strand of the template and since it is not complementary in sequence to virion DNA appears to be transcribed from the complementary strand (Hayashi et al., 1963). Again unlike transcription of the larger virus genomes, the same messenger RNA species are made throughout infection and there appear to be no distinct "early" or "late" species.

2) The T-phages. T4 and T7 bacteriophages have provided much information on the mode of synthesis and function of RNA from the viruses with DNA genomes. The genome of these viruses is transcribed, initially at least, by the DNA-dependent RNA-polymerase (transcriptase) of the host. This enzyme has been well studied and consists of four major polypeptides termed α, β, β' and σ with approximate molecular weights of 40, 150, 160 and 90 thousand daltons respectively (Burgess, 1971). These are present in the active molecule in the molar ratio 2α: 1β: $1\beta'$: 1σ. Two different forms of the enzyme have been examined *in vitro* one of which, called the holoenzyme, contains all four polypeptides, whereas the other, called the core polymerase, lacks σ. *In vitro* both core polymerase and holoenzyme can use either dAT copolymer or fragmented DNA as

template but only the holoenzyme can efficiently transcribe intact T4 DNA. There is some evidence on the function of three of the individual polypeptides; β is the catalytic subunit (ZILLIG et al., 1970), β' is involved in the primary binding of polymerase to specific sites on the double stranded DNA, and σ recognises specific initiation sites (BURGESS et al., 1969). In addition to these polypeptides, other protein factors are needed for the enzyme to transcribe efficiently, principal among which is a factor called ρ which is involved in the accurate termination of some RNA transcripts (ROBERTS, 1969). Further information on bacterial RNA polymerase is given in Chapter 2.

The synthesis of RNA at various stages in the life cycles of T4 and T7 illustrates how the transcriptase functions during virus replication. RNAs produced during transcription are usually classified by the time after infection at which they appear; three main classes of virus RNA are recognized and are called immediate early, delayed early and late. In cells infected with either virus, the bacterial holoenzyme synthesizes the immediate early class of mRNA mentioned before. After this stage, however, the method of transcription employed by the two viruses differs. The product of one of the early genes of T7 is a DNA-dependent RNA polymerase and this enzyme rather than the host polymerase synthesizes all later virus mRNA (CHAMBERLIN et al., 1970). T7 polymerase contains only one type of polypeptide of molecular weight 110,000 daltons and is specific for T7 DNA as template, but it is not yet known how T7 RNA is synthesized nor how the host and virus polymerases are distinguished. In contrast, T4 mRNA is synthesized throughout infection by the bacterial polymerase, which becomes progressively modified; early in infection σ is altered (TRAVERS, 1970, 1971) and α is adenylated (WALTER et al., 1968); and later in infection both the β and β' polypeptides are also modified and a new σ factor may be produced (ZILLIG et al., 1970). How these changes affect the enzyme is unknown, but host RNA synthesis is inhibited shortly after infection, and polymerase obtained from T4 infected cells differs from that from uninfected cells. Another difference between the early and late RNAs of both bacteriophages is that the early RNAs are transcribed from part of one strand of the DNA whereas the late are transcribed from another part of the DNA, but from the other strand.

In addition to virus specified mRNAs several tRNA molecules are transcribed from the virus genome in T4 and T5 infected cells. Their role is not yet known but possibly they are needed to efficiently translate virus RNAs and may also help regulate the synthesis of particular virus proteins (LITTAUER et al., 1971; see also Chapter 5).

   b) *Animal Viruses.* The animal viruses whose particles contain DNA can be divided into two groups according to whether DNA isolated from their particles is infectious or not. The first group includes SV40, polyoma and the adenoviruses. Their isolated DNA genomes, like those of the small bacteriophages, are infectious. These viruses multiply in the nucleus and their DNA is transcribed by host polymerase enzymes. This transcription, again like that involved in bacteriophage replication, is controlled and early and late classes of virus specific mRNA can be isolated from infected cells.

   1) Adenovirus. The transcription of the DNA of adenovirus seems typical of the group. Human cells infected with type 2 adenoviruses yield early and late

classes of RNA (Green, 1970). The early RNA is transcribed even in the absence of protein synthesis and about 10–20% of the virus genome is involved in early RNA production. At least two size classes of mRNA with molecular weights about 6.0×10^5 and about 1.2×10^6 daltons are produced and are found both in nuclei, where they are synthesized, and in polysomes. Later in infection, after virus DNA is synthesized, late virus mRNA is formed. Late mRNA isolated from the nucleus consists of large molecules of about 4×10^6 daltons molecular weight, which is about 40% of the size of the virus genome. However, mRNA from polysomes isolated late in infection contain much smaller polynucleotides that are the size of monocistronic mRNAs. It is likely, therefore, that late adenovirus mRNA is transcribed in the nucleus as large polycistronic RNA molecules and that these are cleaved to produce the smaller mRNAs found in the cytoplasm. Furthermore, at the time of cleavage sequences of about 150–250 residues of adenylic acid are coupled to the 3' ends of the RNAs and it is these A-rich mRNAs which are found in virus specific polysomes. The possible significance of the A-rich segments will be discussed later. Early and late adenovirus mRNAs can also be partially separated since, although most of the late RNA will hybridize to the one strand of the genome, approximately 40% of the early RNA hybridizes to the other. Thus as with T4, mRNAs are transcribed from both strands of the genome.

Late in infection, adenovirus infected cells contain not only mRNA but also large amounts of a small RNA species (molecular weight about 5×10^4 daltons) which has been termed viral-associated RNA since it is also found in the virion.

In comparison with bacterial and bacteriophage systems, little is known about the transcription machinery of mammalian cells. Recent work with sea urchin embryos, *Xenopus laevis* cells and rat liver cells has shown that there are several DNA-dependent RNA-polymerases in the nuclei of higher organisms (Roeder and Rutter, 1969; Roeder et al., 1970). They are distinguished by their chromatographic properties, by their differential stimulation by Mg^{2+} and Mn^{2+} ions and by their sensitivity to the toxic polypeptide from mushrooms called α-amanitin. Purified nucleoli contain most of an enzyme named polymerase I whereas the other principal enzyme, polymerase II, is nucleoplasmic (Roeder and Rutter, 1970). At present it can only be assumed that all DNA viruses which multiply in the nuclei of eukaryotes use these enzymes for their transcription and the sensitivity of, for example, adenovirus transcription to α-amanitin is because adenovirus uses polymerase II (Price and Penman, 1972 b). By contrast, it is possible that the viral-associated RNA of adenovirus is synthesized by polymerase I (Price and Penman, 1972 a).

Certain strains of adenovirus are also able to transform particular host cells to cancer-like cells under conditions where no virus particles are produced. In cells transformed by adenovirus type 2 the only virus specific RNA found is homologous to half the early viral RNA sequences detected during lytic infections, and it therefore represents no more than 5–10% of the genome. However, in both the nucleus and the cytoplasm of transformed cells the viral RNAs are covalently linked to host RNA and are consequently part of larger molecules than the early mRNAs produced during a lytic infection. It appears, therefore, that the viral DNA has been integrated into the host genome and together with neighbouring sections of host DNA is transcribed by the host polymerase.

2) Vaccinia virus. The larger DNA viruses whose isolated genomes are not infectious are typified by vaccinia virus (Fig. 7) whose particles contain a single DNA molecule of about 160×10^6 daltons molecular weight. In contrast with adenoviruses, vaccinia replicates in the host cell cytoplasm and during the early stages of infection mRNA is transcribed from the still partially encapsidated virion DNA by a DNA-dependent RNA polymerase from the virus particles (KATES, 1970). As a consequence, transcription can occur in the absence of protein synthesis, and the early mRNAs produced have molecular weights of about $2–4 \times 10^5$ daltons and represent 14% of the viral genome. Again as with adenovirus, the vaccinia mRNA molecules are terminated by a sequence of adenylic acid residues which are in this case polymerized by the virion transcriptase, at least early in infection.

After early mRNA synthesis, the viral DNA is completely uncoated. DNA is then synthesized and late mRNA sequences are transcribed. DNA-RNA hybridization data indicate, however, that some species of early mRNA are synthesized even late in infection (ODA and JOKLIK, 1967).

It is clear that vaccinia, like other DNA viruses, synthesizes its mRNA in a controlled way but that vaccinia mRNAs unlike those of adenovirus may be monocistronic.

It is possible that the entomopoxviruses of insects replicate in the same way as the poxviruses of mammals, as their particles contain similar enzymes (POGO et al., 1971).

c) *Plant Viruses.* Cauliflower mosaic virus and other caulimoviruses have a double stranded circular DNA genome in their isometric particles, which are about 50 nm in diameter. This genome when extracted from the particles is infective (SHEPHERD et al., 1968), and in all these properties the caulimoviruses resemble most closely the papovaviruses of animals. However they differ in one important respect for whereas papovaviruses replicate in the host nucleus (perhaps using host DNA polymerases) the caulimoviruses seem to replicate in the host cytoplasm where they are found in dense inclusion bodies (KAMEI et al., 1969); the development of these inclusions has not been studied so it is not known whether they originate from a cell organelle (with its own DNA polymerase), though when mature they are not bounded by a membrane.

B. Structure of Viral RNAs

1. RNAs Extracted from Virus Particles

Most is known of the structure of the viral genome RNAs, those that occur in virus particles. This is because virus particles can often be readily separated from host cell constituents and, once purified, the virus particles can be disrupted in various ways (RALPH and BERGQUIST, 1967) and the genome RNAs extracted.

Viral RNAs are composed of the usual four nucleotides and vary greatly in composition [see collations in papers by BELLETT (1967) and GIBBS (1969)]. Most of those analyzed so far contain less cytosine than uracil and around

45–50% G+C. There are, of course, exceptions, for example the RNA genomes of the tymoviruses contain about 40% cytidylic acid, and those of the potexviruses about 32% adenylic acid.

To determine the sequence of nucleotides of even the smallest viral RNA is a daunting problem. This is because there are few specific ribonucleases and RNA molecules are built from only four different nucleotides, which occur in the sequence in approximately equal amounts. Despite these basic difficulties, progress has been made because it has been found that when viral RNAs are in conditions which allow folding and maintenance of secondary structure, or when they react specifically with ribosomes or viral coat proteins, parts of the RNA resist ribonuclease digestion and large and distinct polynucleotide fragments are obtained. The fact that these fragments are characteristic non-random parts of the RNA indicates that the protein binding sites and the parts of the sequence involved in secondary structure are specific and are determined by the sequence of nucleotides in particular parts of the RNA.

Using this approach several parts of the genome of different ribophages, especially Qβ and MS2, have been isolated. Each fragment has then been degraded further using fractionation methods pioneered by Sanger and his colleagues (Adams et al., 1969) and its sequence determined. For example, Fig. 9 shows the sequence and possible secondary structure of the part of the MS2 genome that includes the gene for the coat protein (Min Jou et al., 1972). Further aspects of this approach to the sequencing of ribophage RNA are discussed in Chapter 4.

An alternative method has been used by Billeter et al., (1969) to determine the sequence of part of the Qβ genome. They used Qβ replicase to synthesize particular parts of the Qβ genome using as template the complementary Qβ RNA. Under defined conditions the genome is synthesized at a constant rate, and thus when the start of synthesis is synchronized it is possible to obtain defined, radioactively labelled parts of the molecule, whose sequence can then be determined by standard fractionation methods. In this way the 330 nucleotides at the 5′ end of Qβ RNA have been sequenced.

From the known ribophage nucleotide sequences together with other data (Anon., 1972; Min Jou et al., 1972; Steitz, 1972; Staples and Hindley, 1971; Staples et al., 1971) various deductions can be made:

a) Each ribophage genome has a unique and specific nucleotide sequence.

b) The three genes in the ribophage genome are in the order 5′—A protein (maturation protein)—coat protein—replicase—3′.

c) Before and after each of the genes there are seemingly untranslated sequences of different lengths some even several hundred nucleotides long; that at the 5′ terminus of Qβ is 61 nucleotides long.

d) Ribosomes, coat protein and replicase bind to particular parts of the genome and make them resistant to ribonucleases. These binding sites are usually about 40 nucleotides long, and less than half of each such sequence corresponds to the terminal part of a known phage protein; most of the nucleotides are from seemingly untranslated sequences.

e) There is no obvious similarity of sequence between the three ribosome binding sites of each phage, nor between analogous binding sites of unrelated phages. However homologous binding sites of related phages of the MS2 group

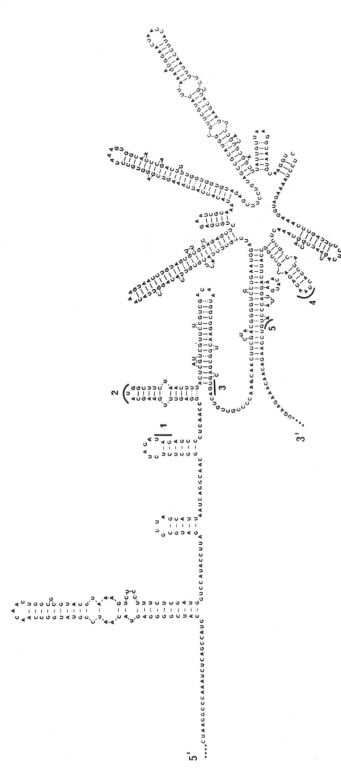

Fig. 9. The nucleotide sequence of the central part of the MS 2 genome from near the end of the A-protein gene to the beginning of the polymerase gene (modified from MIN JOU et al., 1972; CONTRERAS et al., 1973). The codon marked 1 normally terminates the end of the A-protein, though sometimes translation continues to codon 3. Codon 2 is the start of the coat protein gene and codon 4 is its end. Codon 5 is the start of the polymerase gene

have almost identical nucleotide sequences, even though the nucleotide sequences of their coat protein genes are not identical and, in fact, are less similar than the amino acid sequence of their coat proteins would indicate. This suggests that while evolving from a common ancestor, the MS2 phages conserved the sequence of their inter-cistronic regions most rigidly, the amino acid sequence of their coat protein less rigidly, and the nucleotide sequence of their coat protein genes least rigidly (Robertson and Jeppesen, 1972; Bernardi and Spahr, 1972).

f) The sequences so far determined can be examined by mathematical techniques such as that of Tinoco et al. (1971) to search for regions where stable secondary structures might form by base pairing within loops of the sequence. No satisfactory unequivocal folding pattern can be computed for the intergene sequences, whereas the gene sequences suggest folding patterns that would be stable and would have few or no alternatives. Furthermore, other properties (such as the relative nuclease sensitivity of different parts of the sequence) predicted from the computed folding patterns coincide closely with observed properties.

Much less is known about other viral genomes, but what is known confirms and extends the impressive results obtained by those working with ribophages. For example, analyses of oligonucleotide mixtures obtained by treating the RNAs of various viruses with specific nucleases confirm that each viral RNA has a specific sequence of nucleotides and that the nucleotide sequences of related viruses are similar (Horst et al., 1972; Rushizky and Knight, 1960; Symons et al., 1963).

The terminal nucleotides of several viral genomes are known. The 5'-terminus of all the ribophages has the sequence pppGpGpGp..., whereas the 5'-terminal nucleotide of at least three plant viruses and also poliovirus is adenylic acid, which for at least two of the plant viruses is not phosphorylated (Fraenkel-Conrat, 1969; Suzuki and Haselkorn, 1968; Wimmer, 1972). The 3'-terminus of all the ribophages, of several strains of tobacco mosaic virus and of all segments of the brome mosaic virus genome is .. CpCpCpA (Glitz and Eichler, 1971; Fraenkel-Conrat, 1969). However, not all viral RNAs terminate in A, thus, for example, the 3'-terminus of tobacco necrosis and its satellite virus, and also both segments of the tobacco rattle virus genome have a 3'-terminal cytosine (Wimmer et al., 1968; Darby and Minson, 1972). It is probable that the 3'-terminal adenosine of the ribophages, and perhaps the plant viruses, is added by a cellular enzyme after transcription as the ribophages have 5'-terminal guanylic acid rather than uridylic acid, and the 3'-terminal adenosine can be removed without loss of infectivity (Kamen, 1970). Those viral RNAs with the 3'-terminal sequence .. CCCA behave in many ways like the transfer RNAs, as has been discussed earlier in this chapter [see Section A.1.a)3)].

Ribophage genome RNAs are not alone in having a defined non-random secondary structure, and it is probable that all single stranded viral RNAs have a definite secondary structure. All such RNAs when protein-free and in normal solvents become hypochromic to an extent which indicates that more than half their bases are paired. Furthermore many, though not all, single stranded viral RNAs in virus particles are hypochromic to a similar extent. It is possible that this secondary structure is important in the control of transcription or translation

(MIN JOU et al., 1972), and, where it occurs in particles, is important in the assembly or stability of the particles. However the precise role is probably different for different viruses. For example, the secondary structure of the mRNAs of T4 bacteriophage may control their translation or their stability (SALSER and RICARD, 1970). There may be a similar explanation for the base-paired structures found in free tobacco mosaic virus RNA (MUNDRY, 1969), because when arranged in virus particles, the RNA of this virus has no paired bases (DOBROV et al., 1972). Other viruses, such as foot and mouth disease virus have hypochromic RNA in their particles, and it is likely that some of the folding is concerned with the assembly or stability of the particles because the secondary structure of the RNA when in the particles differs, at least in amount, from that when in solution (BACHRACH, 1964). However, it is clear that whatever is the function of secondary structure, it will considerably influence the selection of viable nucleotide sequences during evolution, and may conflict with the messenger function of the nucleotide sequence (BALL, 1972).

Recently it has been found that some of the mRNA of eukaryotic cells contains segments, 100–250 nucleotides long of which 90 % are adenylic acid. It was thought that these A-rich segments might be involved in the transport of mRNA from the nucleus to the cytoplasm. However, they are also found in the RNA transcripts of adenovirus, which has a DNA genome that replicates in the nucleus (PHILLIPSON et al., 1971), and more significantly in the mRNA of vaccinia virus, which has a DNA genome that replicates in the cytoplasm (KATES, 1970). Furthermore, poly A tracts are found in several viruses with RNA genomes including Sendai, Sindbis, polio, tobacco mosaic and various leukoviruses (GILLESPIE et al., 1972; GREEN and CARTAS, 1972; JOHNSTON and BOSE, 1972; MUNDRY, 1969; PRIDGEN and KINGSBURY, 1972; and various others cited in these papers). It is possible that the poly A tracts are concerned with translation of the RNA as they are found in viral genome RNAs that act as messenger but not in vesicular stomatitis virus whose genome is the complement of the message (JOHNSTON and BOSE, 1972).

A few viruses have genomes that consist of fully base-paired double stranded segments of RNA. These viruses include several, such as reovirus, that infect mammals, others, such as wound tumour and rice dwarf viruses, that infect plants, and also the cytoplasmic polyhedrosis viruses of insects. The first of this interesting group was discovered by GOMATOS et al. (1962), who found that inclusions in L cells infected with reovirus contained DNAase-resistant material that fluoresced green when stained with acridine orange; this is characteristic of double stranded but not single stranded nucleic acid. Many other characteristics confirmed that this RNA was double stranded. These included its composition, its density and its resistance to ribonuclease, to formaldehyde and to heat denaturation (reviewed by SHATKIN, 1968), all of which showed that the RNA had a structure similar to double stranded DNA (ARNOTT et al., 1966; LANGRIDGE and GOMATOS, 1963).

Earlier in this chapter we mentioned that some viruses have single stranded genomes that are divided into two or more segments. These segments may either be packaged in separate particles as are those of alfalfa mosaic virus (VAN VLOTEN-DOTING and JASPERS, 1967), or may be grouped in different particles as, for example, are those of brome mosaic virus (LANE and KAESBERG, 1971), alternatively, they may be all grouped in one type of particle as are those of reovirus.

Virus particles may contain nucleic acid other than that of the virus genome. This may be viral RNA that is apparently inessential and non-infectious as, for example, is part of the RNA in one type of particle in preparations of brome mosaic virus (LANE and KAESBERG, 1971), or the small A-rich oligonucleotide in reovirus particles (NICHOLS et al., 1972). Alternatively it may be host RNA, such as, for example, the host ribosomal RNA in the ribosomes found in arena-virus particles (PEDERSEN, 1971) (Fig. 5) or the host RNA, presumably mRNA, found in up to 2% of tobacco mosaic virus particles (SIEGEL, 1972).

2. RNA within Virus Particles

The study of the structure of RNA within virus particles is of great interest because, amongst other rewards, it gives us information on specific protein/nucleic acid interactions.

There has been much research on the structure and assembly of virus particles; some details are discussed here but general recent reviews and important reports include those by CASPAR and KLUG (1962), CRICK and WATSON (1957), EISERLING and DICKSON (1972), KAPER (1968), KUSHNER (1969) and LEBERMAN (1968).

One of the best studied viruses is tobacco mosaic virus. More is known about its particles (FRAENKEL-CONRAT, 1969; FRANKLIN et al., 1956; KAPER, 1968) than those of any other virus. They are helically constructed tubes about 300 nm long with inner and outer diameters of 4 nm and 18 nm respectively. The basic single-start helix of each particle is right-handed, has a pitch of 2.3 nm and there are 49 protein subunits in every 3 turns of the helix; each subunit is a single poly-peptide of molecular weight 17,800 daltons, and all the subunits of the particles of one strain of the virus are of a single protein species. Each particle contains a single RNA molecule which is held between the protein subunits and follows the basic helix at a radius of 4 nm. Each protein subunit is closely associated with 3 nucleotides of the RNA molecule. Each particle thus contains about 2,100 protein subunits, and its RNA is about 6,300 nucleotides long.

Measurements of the optical properties of tobacco mosaic virus preparations treated in various ways (DOBROV et al., 1972; SCHACHTER et al., 1966) show that the RNA in the particles is not base-paired, but is held with each base perpendicular to the sugar/phosphate backbone of the RNA in such a way that formaldehyde cannot react with its amino groups; the last property contrasts with the sus-ceptibility of the RNA in the particles to various chemical mutagens (SINGER and FRAENKEL-CONRAT, 1969). The lack of effect of formaldehyde seems to be correlated with the fact that when TMV particles, but not the free RNA, are irradiated with ultraviolet light no photoreactivable dimers are formed (CAR-PENTER and KLECZKOWSKI, 1969), a phenomenon that also suggests that the bases of adjacent nucleotides are isolated from one another in intact particles.

The particles of potato virus X resemble those of TMV in their basic archi-tecture. They consist of a helix of protein subunits protecting a single RNA molecule of 2×10^6 daltons; however the particles are nearly twice as long, are flexuous and their basic helix has a pitch of 3.3–3.6 nm (VARMA et al., 1968). The RNA in potato virus X particles like that in particles of TMV is fully hyperchromic, but unlike that in TMV reacts with formaldehyde and can suffer photoreactivable damage when irradiated with ultraviolet light, indicating that though adjacent bases are not paired, their amino groups are free and they can interact to form pyrimidine dimers (DOBROV et al., 1972). When potato virus X RNA is assembled with TMV protein to form TMV-like particles, it becomes unable to suffer photoreactivable damage, suggesting that it is the protein which stops pyrimidine dimerization (BRECK and GORDON, 1970).

No viruses of mammals have tubular or filamentous particles that are naked like those of plant viruses. Some however, including the paramyxoviruses, orthomyxoviruses, leukoviruses and rhabdoviruses, have particles with a RNA-containing helically constructed core, or nucleocapsid, which acquires a lipid-containing envelope when 'maturing' at the surface of the infected cell (Fig. 6). The structure of the RNA is these enveloped particles is not known, but, after removal of the envelope, the nucleocapsids of paramyxoviruses seem to have a structure similar to that of TMV (FINCH and GIBBS, 1970).

Isometric virus particles, such as those of turnip yellow mosaic virus (KAPER, 1968), tomato bushy stunt virus (WEBER et al., 1970), the bromoviruses (BANCROFT, 1970), and the picornaviruses of mammals (JOHNSTON and MARTIN, 1971), are constructed of an outer shell of 60 or 180 or more identical polypeptides or groups of different polypeptides (as perhaps in picornavirus particles) arranged in icosahedral symmetry. An icosahedron is a geometric solid with 20 equilaterally triangular faces, thus the number of asymmetric units, such as polypeptides, from which an icosahedral shell can be constructed fits definite rules (GOLDBERG, 1937; WRIGLEY, 1969). The nucleic acid is packed loosely or tightly inside the icosahedral shell of these particles and may, as in tymovirus particles, evaginate into clusters of the subunits of the shell to give 32 large morphological subunits (KAPER, 1968).

At one time it was thought that the particles of some viruses such as tomato bushy stunt and turnip crinkle contained a protein core as well as shell, and had the structure of a 'spherical sandwich'; this now seems unlikely (MARTELLI et al., 1971).

The unusual bacilliform particles of alfalfa mosaic virus are related in structure to icosahedra. Those of alfalfa mosaic virus are short round-ended rods of 18 nm diameter and various definite lengths between 18 nm and 60 nm (Fig. 3). The round ends of these particles are half icosahedra and the central tubular part is constructed of the same subunits arranged without vertices of 5-fold symmetry; the 5-fold vertices curve the surface of the shell.

3. Particle Reassembly Experiments

Tobacco mosaic virus particles are readily disassembled into their constituent protein and nucleic acid (BAWDEN and PIRIE, 1940). They may be experimentally reassembled (FRAENKEL-CONRAT and WILLIAMS, 1955) and such experiments have given much information on the morphogenesis of the particles, but whether the conditions required for assembly *in vivo* are similar to those required *in vitro* is not known, though they are probably similar. In assembly experiments particles form most efficiently and quickly when TMV RNA is mixed with a suspension of two-turn disc aggregates of the viral protein (34 subunits arranged as two 17 subunit rings). Virus-like particles can also be made by mixing tobacco mosaic virus protein with other RNAs or with 5′ exonuclease-treated tobacco mosaic RNA, but they form more slowly, indicating that there are specific interactions between TMV RNA and protein during assembly. The particles grow in length, until the RNA is coated, by the addition of more protein. Whether growth is fastest when the protein is in discs or some other polymer is at present being discussed (BUTLER and KLUG, 1972; OHNO et al., 1972) as also is the mode of growth (HARRIS, 1972). Assembly experiments using protein from different strains of tobacco mosaic virus show that the protein species first added to the RNA determines the protein species of which the rest of the particle can be built; when protein of a particular strain is used to start the assembly of the particle then only the same protein will assemble subsequently, even in mixtures of proteins from different isolates (OKADA et al., 1970). Experiments with RNA treated with specific exonucleases (GUILLEY et al., 1971; BUTLER and KLUG, 1971), or RNA labelled at one end (THOUVENEL et al., 1971) show that the particles assemble starting at the 5′ end of the RNA, and that a length of about 50 nucleotides (one turn of the basic helix) at the 5′ end of the RNA is specifically involved in the recognition of the first disc.

Thus, to summarize, the specificity of assembly of TMV RNA and its protein *in vitro* seems to depend firstly on the sequence of the 50 or so nucleotides at the 5′ end of the RNA, with the later addition of protein to the growing TMV particle depending on a protein to protein specificity (OKADA et al., 1970). The initial RNA/protein specificity seems to be at the viral 'group' level as the RNA of one TMV strain will apparently readily accept the proteins from any normal TMV strain, whereas the protein/protein interaction is much more specific and distinguishes between strains of virus.

Experiments with bromoviruses also suggest that protein/protein interactions are more specific than RNA/protein interactions in particle reassembly. BANCROFT and his colleagues (BANCROFT, 1970) have found that various nucleic acids and polyanions seem equally efficient as nucleating agents for the reassembly of particle-like shells of bromovirus proteins. However, in experiments with proteins from different bromoviruses, they found that mixtures of proteins of brome mosaic and cowpea chlorotic mottle viruses formed hybrid particles more readily than did mixtures of either protein with that of broad bean mottle virus (WAGNER and BANCROFT, 1971).

Reassembly experiments and studies with mutants of TMV suggest that proteins interact with proteins by a combination of electrovalent and hydrophobic forces, and also presumably by surface shape factors. Much less is known about how RNA and protein interact, though presumably the interactions involve similar factors. The way in which the TMV RNA specifically interacts with the first protein disc to start the growth of the particle is not known. It is unlikely that it is determined by the secondary structure of the 5′-terminus of the free TMV RNA, as the reactivity of TMV RNA with its protein when treated with exonuclease decreases at a linear rate (BUTLER and KLUG, 1971), suggesting that the primary sequence of the 5′-terminus not its secondary structure determines its specificity of behaviour.

Specificity of recognition and interaction between the constitutents of virus particles is likely to be a general phenomenon for it would help to ensure the correct assembly of the particles, but few examples of such specificity, other than those listed above, are known. However, one unusual example of this phenomenon is the ability of isolated RNA of alfalfa mosaic virus, and no other, to disrupt intact particles of the virus, a result which suggests that the isolated RNA can specifically compete for the protein in intact particles (VAN VLOTEN-DOTING and JASPERS, 1972).

It is not known whether the results of *in vitro* reassembly experiments give insight into what happens *in vivo*. It is possible that the seeming unspecificity of some nucleic acid/protein interactions *in vitro*, if similarly unspecific *in vivo*, may be responsible for some of the reports of virus particles containing non-viral nucleic acid (MICHEL et al., 1967; SIEGEL, 1971; WINOCOUR, 1968), and of phenotypically mixed particles. The latter are obtained from cells infected with two or more viruses, and are particles that contain the proteins of one virus and the genome of another. Phenotypic mixing was first found in experiments with T-even bacteriophages (DELBRÜCK and BAILEY, 1946), but seems most common among the mammal viruses that have enveloped particles (see, for example, BROWN et al., 1967; BURGE and PFEFFERKORN, 1966; CHOPPIN and COMPANS, 1970; VOGT, 1967) suggesting that mixing may occur most commonly during the envelopment of the nucleocapsid at the cell surface. However, phenotypic mixing is also found in non-enveloped viruses such as the picornaviruses (ITOH and MELNICK, 1959; LEDINKO and HIRST, 1961), and among plant viruses (KASSANIS and CONTI, 1971) where it may be important in the transmission of certain viruses by vectors (ROCHOW, 1972).

4. Viral RNAs Not Found in Virus Particles

Virus infected cells usually contain various viral RNAs which are not found in the particles of the virus.

There are, for example, usually at least two types of double stranded RNAs in cells infected with those viruses which have single stranded RNA genomes (WEISSMANN et al., 1968). These are concerned with replication of the virus and have been discussed earlier in the chapter.

Other viral RNAs not found in virus particles include the many viral mRNAs produced by those viruses whose genome is not both genome and messenger. Little is known of the structure of these mRNAs, but there is no evidence that they differ much from host cell mRNA.

Virus specified tRNAs are found in cells infected with some of the large bacteriophages. T4 bacteriophage, for example, produces 6–10 tRNA species that are specific for at least six of the amino acids (WEISS et al., 1968; DANIEL et al., 1970; TILLACK and SMITH, 1968). The role of these viral tRNAs during infection is not known (WILSON and KELLS, 1972; WILSON and ABELSON, 1972).

C. Concluding Remarks

We hope that this brief review has shown not only the various ways that RNAs are used by viral genomes, but also the great fascinating diversity of those types of genome we call viruses.

The study of viruses is clearly not only of importance as a way of gaining information that will help us control viruses or ameliorate their effects on their hosts, but has also contributed greatly to many branches of science and, in particular, to the biologists' syndrome called molecular biology.

Viruses have been used as tools for studying the cellular processes of, in particular, prokaryotes, as they can be used as purifiable genes which the cells translate into recognizable products. Now that attention is focusing more on studying the same but more complex processes in eukaryotes (Laskey et al., 1972), viruses will be increasingly used in the study of their development and control.

Glossary

This glossary lists alphabetically the viruses and virus groups named in the chapter and gives some information on them to help non-virologists. More information about individual viruses can be obtained from various sources, but two useful compilations are "Viruses of Vertebrates" by Andrewes and Pereira (3rd Edition, 1973; Baillière, Tindell and Cassell), and the C.M.I./A.A.B. "Descriptions of Plant Viruses" published by the Commonwealth Agricultural Bureaux, Slough, England since 1970.

Note that most of the names of animal viruses are either pseudo-Classical and derived from some characteristic feature of the virus (e. g. myxoviruses which have slimy particles, from *myxa* Gr. = slimy), or are acronyms or sigla (e. g. picornaviruses which have small [pico] RNA-containing particles), though togaviruses are named from the locality they were found (e. g. Semliki Forest virus). By contrast, most plant and insect viruses are named by the disease they cause in their commonest host or that from which they were first isolated (e. g. tobacco mosaic virus), and most group names are derived by contracting the name of the type member (e. g. *tobacco mosaic*-like *viruses* to tobamoviruses). The bacteriophages are given the code lettering and numbering used when they were isolated (e. g. T$_2$ or ϕX174), though many of the male-specific 'phages have codes containing f (e. g. f2, fr, fd).

In these notes, some of the information is given in cryptograms, the meanings of the symbols used in the cryptograms are as in Wildy (1971) and are as follows.

Each cryptogram consists of four pairs of symbols (e. g. tobacco mosaic virus: R/1: 2/5: E/E: S/O) with the following meanings:

1st pair   Type of nucleic acid/strandedness of nucleic acid.
Symbols for type of nucleic acid: R = RNA; D = DNA
Symbols for strandedness: 1 = single-stranded; 2 = double-stranded.

2nd pair   Molecular weight of nucleic acid (in millions)/percentage of nucleic acid in infective particles.
This term gives the composition of infective particles. The genome of some viruses is divided, when different pieces of the genome occur together in one type of particle, the symbol Σ indicates the total molecular weight of the pieces in the particles (e. g. clover wound tumour virus; R/2: Σ15/20: S/S: S,I/Au), but when the pieces occur in different particles the composition of each particle type is listed separately (e. g. tobacco rattle virus: R/1: 2.3/5 + (0.6–1.3)/5: E/E: S/Ne).

3rd pair   Outline of particle/outline of 'nucleocapsid' (the nucleic acid plus the protein most closely in contact with it).
Symbols for both properties:
S = essentially spherical
E = elongated with parallel sides, ends not rounded
U = elongated with parallel sides, end(s) rounded
X = complex or none of above.

4th pair   Kinds of host infected/kinds of vector.

  Symbols for kinds of host:
  A = actinomycete   P = pteridophyte
  B = bacterium   S = seed plant
  F = fungus   V = vertebrate
  I = invertebrate

  Symbols for kinds of vector:
  Ac = mite and tick (Acarina, Arachnida)
  Al = white-fly (Aleyrodidae, Hemiptera, Insecta)
  Ap = aphid (Aphididae, Hemiptera, Insecta)
  Au = leaf-, plant-, or tree-hopper (Auchenorrhyncha, Hemiptera)
  Cc = mealy-bug (Coccoidae, Hemiptera)
  Cl = beetle (Coleoptera, Insecta)
  Di = fly and mosquito (Diptera, Insecta)
  Fu = fungus (Chytridiales and Plasmodiophorales, Fungi)
  Gy = mirid, piesmid, or tingid bug (Gymnocerata, Hemiptera)
  Ne = nematode (Nematoda)
  Ps = psylla (Psyllidae, Hemiptera)
  Si = flea (Siphonaptera, Insecta)
  Th = thrips (Thysanoptera, Insecta)
  Ve = vectors known but none of above
  0 = spreads without a vector via a contaminated environment.

In all instances
  *   = property of the virus is not known
  ( ) = enclosed information is doubtful or unconfirmed
  [ ] = enclosed cryptogram gives information about a virus group.

Adenoviruses   [D/2: 20–25/12–14: S/S: V/O]

A large group of viruses that cause respiratory or more general diseases in many wild and domesticated mammals and birds throughout the world. They cause colds, sore throats (hence their name, from *adena* Gr. a gland), pneumonia, "swimming-pool" conjunctivitis or more serious diseases in man. When experimentally transmitted to different hosts, some cause tumours. They multiply in cell nuclei. Their particles are elegant complex icosahedra around 80 nm in diameter, they contain the genome and are made of five different proteins.

Adeno-Associated Virus or *Adeno-Satellite Virus*   D/1: 1.5/25: S/S: V/*

A virus found in cultured tissues, and in children, infected with adenoviruses, in whose absence it is unable to replicate. Particles of the virus are isometric and about 20 nm in diameter. They contain single-stranded DNA, but when this is liberated from the particles it becomes double-stranded as the DNA in some particles is complementary to, and denser than, that in the others; the dense strand is that from which viral mRNA is transcribed.

Alfalfa Mosaic Virus   R/1: 1.3/18 + 1.1/18 + 0.9/18 + (0.3/1): U/U: S/Ap

A virus found worldwide, often in legumes. It is readily transmitted by sap inoculation and by aphids (in the non-persistent manner) to a very wide range of dicotyledonous plants. Causes mild or severe mosaic symptoms. Its particles are bacilliform and of different lengths. Probably related to cucumber mosaic virus.

Arenaviruses   [R/1: Σ3.2/*: S/*: V, I/O]

This group includes many that cause severe, often fatal, haemorrhagic fevers in man and other mammals. Most have been isolated in the Americas, but lymphocytic choriomeningitis virus, which is the best known, is common, latent and congenitally-transmitted in rodents throughout the world. They are transmitted in infected dust, and perhaps also by various arthropods and other parasites. Their particles are irregularly isometric about 120 nm in diameter with a lipid-containing envelope and an amorphous core, which contains host-ribosomes, which appear as granules in the core of the particles (hence the group name; from *arenosus* L. sandy). The RNA genome is perhaps in two segments.

Baculoviruses   [D/2: 100/10: U/U: I/O]

This large cosmopolitan group includes the nuclear polyhedrosis and granulosis viruses of insects. They cause diseases of the larvae, principally of Lepidoptera, and there is increasing interest in using them as biological control agents in place of chemical insecticides. The baculoviruses infect via the gut, and multiply in nuclei so that the body contents of diseased larvae become a suspension of crystalline proteinaceous inclusions, which, when the body ruptures, spread over food plants. The inclusions contain one (granuloses) or more (nuclear polyhedroses) complex large rod-shaped particles (hence the group name, from *baculum* L. a stick). Their genome is double-stranded DNA, perhaps circular. Inclusions retain infectivity for very long periods, but when dissolved and the particles released, either chemically, or in the alimentary canal of an insect, the particles become much less stable and soon lose infectivity.

Broad Bean Mottle Virus   R/1: (1.1/22 + 1.0/20): S/S: S/*

A bromovirus of legumes. It has a narrow host range and was isolated once in England.

Brome Mosaic Virus   R/1: 1.1/22 + 1.0/20 + 0.75/(15): S/S: S/*

Type member of the bromoviruses. Found rarely in North America and Europe, narrow host range.

Bromoviruses   [R/1: 1.1/23 + 1.0/21 + 0.75/(16): S/S: S/*]

A group of three seemingly rare plant viruses with restricted host ranges. Sap transmitted but no vector known. Have given much interesting information on the reassembly of virus particles *in vitro*.

Cauliflower Mosaic Virus   D/2: 5/15: S/S: S/Ap

Type member of the caulimoviruses. A common virus of crucifers throughout the world.

Caulimoviruses   [D/2: 5/15: S/S: S/Ap]

A widespread group of about six plant viruses, which cause mosaic symptoms in a narrow-range of hosts. Transmitted by sap and by aphids in the non-persistent manner. Its type member, cauliflower mosaic virus, was the first plant virus shown to have a DNA genome, which is circular and double-stranded. Their isometric particles are 50 nm in diameter, and resemble those of the papovaviruses of mammals in composition and appearance.

Comoviruses   [R/1: 2.6/35 + 1.7/28: S/S: S/Cl]

A cosmopolitan group of about eight common plant viruses of restricted host ranges, many found in legumes. Transmitted by beetles, sap and, sometimes, seed. Cause mosaic and necrotic symptoms. Their angular isometric particles are of three types; one, just the protein shell, the others containing the two segments of the divided genome.

Cowpea Chlorotic Mottle Virus   R/1: 1.1/24 + 1.0/22 + 0.8/(18): S/S: S/*

An uncommon legume virus of North America. A bromovirus.

Cowpea Mosaic Virus   R/1: 2.5/33 + 1.5/24: S/S: S/Cl

Type member of the comoviruses. A common and economically important virus of cowpeas throughout the tropics.

Clover Wound Tumour Virus   R/2: Σ16/22: S/S: S, I/Au

This virus has been isolated only once and from a leafhopper. It infects certain agallian leafhoppers and a wide range of plants, causing no symptoms in the former and in the latter, tumours of pseudo-phloem cells in the vascular bundles. This virus is one of a group of perhaps six plant viruses, which will perhaps be named the clowtuviruses. All are transmitted by leafhoppers and characteristically cause microtumours of vascular tissue to give "rough leaf" symptoms, though rice dwarf virus, one of the best known of the group, causes no tumours. The clowtuviruses resemble closely the reoviruses of mammals and the cytoplasmic polyhedrosis viruses of insects, they all have isometric particles about 70 nm in diameter and their genome is segmented double-stranded RNA.

Cucumber Mosaic Virus   R/1: (1/18): S/S: S/Ap

A cosmopolitan virus with a particularly wide range of plant hosts in which it causes mosaics and mottles. Transmitted readily by sap, by aphids in the non-persistent manner, and, occasionally, by seed. Its isometric particles (diameter 30 nm) are like those of the bromoviruses in that they sediment as one component and contain a genome divided between different particles, however in composition, but not appearance, they resemble the particles of alfalfa mosaic virus, to which cucumber mosaic virus is probably related.

Cytoplasmic Polyhedrosis Viruses   [R/2: Σ13–18/16–30: S/S: I/O]

Viruses found in the larvae of Lepidoptera and Neuroptera throughout the world. Their ecology seems similar to that of the baculoviruses, but they replicate in the cytoplasm, and their polyhedral inclusions form there (hence their name). Scattered throughout their inclusions are many isometric particles, which resemble closely in appearance and composition, the particles of clover wound tumour virus and the reoviruses.

Encephalomyocarditis Virus   R/1: 2.6/31: S/S: V/O

A typical enterovirus, found throughout the world in various mammals, especially rodents, and once isolated from mosquitoes. Usually causes an inapparent gut infection, but may cause fevers often with myocarditis and encephalitis (i.e. inflammation of heart muscle or brain respectively); hence the name.

Enteroviruses   [R/1: 2.5/30: S/S: V, I/O]

A subgroup of the picornaviruses. Their particles are isometric about 30 nm in diameter, are acid-stable and, in CsCl solutions, have a density around 1.34. Usually infect cells lining the gut (hence their name, from enteron Gr. the intestines), but may spread to other tissues, especially nerves, causing inflammation and necrosis and, often, paralysis. So far found to be common throughout the world in mammals, insects and birds.

Entomopoxviruses   [D/*: */*: X/X: I/O]

Another polyhedra-producing group of insect viruses found in Coleoptera and Lepidoptera. The polyhedra contain particles which very closely resemble those of the poxviruses in structure and composition.

fd Bacteriophage   D/1: 1.7/12: E/E: B/O

A bacteriophage with filamentous particles 870 nm long and 5.5 nm in diameter. The genome is circular single-stranded DNA. Infects by attaching to the end of the f sex pili of male E. coli, thus like the ribophages it is a male-specific venereal disease. Does not lyse the host like most phages; infected bacteria contrive to grow and divide while releasing virus. Many similar phages have been found throughout the world; M13 is closely similar to fd, but others have longer particles, or other hosts, or infect via the I sex pili.

Foot and Mouth Disease Virus   R/1: 2.8/30: S/S: V/O

A virus of ungulates found in most parts of the world except North America and Australasia. Causes severe fever and vesicular lesions on various parts of the body. It is a picornavirus and has isometric particles about 30 nm in diameter, but, unlike those of the enteroviruses, they are unstable at pH 3 and, in CsCl solutions, have a density of 1.38–1.43.

φX174 Bacteriophage   D/1: 1.7/25: S/S: B/O

Another phage that infects E. coli. It has icosahedral particles about 30 nm in diameter that have prominent vertices. Much used in studies on nucleic acid biochemistry.

Galleria Densonucleosis Virus   D/1: 2.2/38: S/S: I/*

A virus that causes a lethal disease of Galleria (waxmoth) larvae, grows in cell nuclei. It has isometric particles about 20 nm in diameter. The single-stranded DNA in its particles behaves, when deproteinized, in the same way as that of adeno-associated virus.

Iridoviruses   [D/2: 130/5: S/S: I/*

A group of viruses found throughout the world in an increasing number of diverse hosts. They have characteristic complex isometric particles, which are around 130 nm in diameter and which were the first virus particles to be shown to be icosahedra. When centrifuged, the particles form crystalline pellets which give Bragg reflection of light and so appear a beautiful iridescent blue, hence the group name. Some of the iridovirus-like viruses from mosquitoes do not give iridescent pellets and have particles about 200 nm in diameter; they are perhaps a distinct group. The original iridoviruses were found in insect larvae, but similar particles have also been found in reptiles, amphibia, fungi and even green algae, and the particles of African swine fever virus, which is tick-borne, are similar.

Kilham's Rat Virus   D/1: 1.5/34: S/S: V/*

A latent virus of rats with isometric particles about 20 nm in diameter. It is the type member of the parvoviruses or picornaviruses, so named because of their smallness. This group includes the minute mouse virus and the adeno-associated viruses, but both the mouse and rat viruses have particles which contain a single type of DNA molecule, and thus their DNA does not become double-stranded when deproteinized.

Leukoviruses   [R/1: Σ10–13/1–2: S/*: V/O]

A large worldwide group of viruses with pleomorphic enveloped particles about 100 nm in diameter. Common in various warm blooded vertebrates and includes many viruses famous in cancer research, hence the group name which is derived from leukaemia, a cancer of the white blood cells (leukos Gr. white).

M13 Bacteriophage   D/1: 1.7/12: E/E: B/O

A filamentous phage perhaps closely related to fd.

Minute Mouse Virus   D/1: 1.7/*: S/S: V/O

A parvovirus found as a contaminant of a mouse adenovirus isolate. Its particles are similar to those of the adeno-associated viruses, both in size and appearance, but unlike them it is able to replicate without the help of adenovirus. Closely resembles Kilham's rat virus.

MS2 Bacteriophage   R/1: 1.1/32: S/S: B/O

One of the best known ribophages.

Orthomyxoviruses   [R/1: Σ5/1: S/E: V/O]

These are the influenza viruses and are common throughout the world in mammals and birds; they frequently cause major worldwide epidemics in man, perhaps as the result of genetic recombination between human strains and strains from other animals. These viruses infect the respiratory tract, and very rarely spread to other organs. Particles are pleomorphic about 100 nm in diameter, and have a lipid-containing envelope that contains virus-specified neuraminidase and haemagglutinin subunits, and that surrounds the RNA-containing nucleocapsid, which consists of filaments about 9 nm in diameter.

Papovaviruses   [D/2: 3–5/7–15: S/S: V/O]

These are viruses of mammals that cause warts, and occasionally other diseases. They are usually divided into two sub-groups the papillomaviruses and the polyomaviruses, the latter including the vacuolating viruses; it is a contraction of these three names that gives the group name, they are not found in Pope's eggs! Their double-stranded genome is circular and replicates in nuclei. Their particles are 'skew' icosahedra with 72 subunits, and are about 50 nm in diameter.

Paramyxoviruses   [R/1: 7/1: S/E: V/O]

This is a large group of economically important viruses of mammals, the best known being mumps, measles, dog distemper, rinderpest and Newcastle disease virus. Their particles are similar to those of orthomyxoviruses, and for a long time they were all grouped together as myxoviruses. However, the paramyxovirus nucleocapsid is 18 nm in diameter, twice that of the orthomyxoviruses, and its genome is in one piece. They cause various diseases, some serious, and are not confined to the respiratory tract.

Picornaviruses   [R/1: 2.5/30: S/S: V/O]

A very large group of animal viruses that have small ether-resistant isometric particles. The group contains the enteroviruses, which are viruses of the gut and have acid-resistant particles, also those with acid-labile particles such as the rhinoviruses (common cold viruses) and foot and mouth disease viruses, and the caliciviruses, which have intermediate properties, and include exanthema virus of pigs.

Poliovirus   R/1: 2.5/30: S/S: V/O

The best known enterovirus of man. Once caused widespread epidemics of infantile paralysis of people in developed countries of the world, but has been controlled by vaccines.

Potato Virus X   R/1: 2.1/6: E/E: S/O, (Fu)

The best known of the potex viruses. Once so common in cultivated potatoes that it was known as the healthy potato virus, as plants with only this virus were considered healthy.

Potexviruses   [R/1: 2.2–2.4/5.6: E/E: S/*, O, (Fu)]

A common and widespread group of plant viruses with restricted host ranges. The group includes potato virus X and white clover mosaic virus. Most are very infectious, and, unlike most plant viruses, they spread when infected and healthy plants rub together.

Poxviruses   [D/2: 160/5–7.5: X/*: V/O]

A large group of viruses of mammals and birds, usually divided into six subgroups. The group includes such famous viruses as vaccinia and smallpox, which was so named in the Middle Ages to distinguish it from the great, or French, pox which was syphilis. Most are transmitted by contact or through the alimentary canal, but some are transmitted by arthropods, for example rabbit myxomatosis virus which is transmitted by fleas and mosquitoes. Poxviruses cause skin lesions or more general diseases. They multiply in the cytoplasm and many form cytoplasmic inclusions, like the related entomopox-viruses. They have complex brick-shaped particles about 250 nm in diameter and 300 nm long, and which contain a complex convoluted nucleoid. The large linear DNA genome is not infectious when isolated.

Qβ Bacteriophage   R/1: 1.0/25: S/S: B/O

One of the best known male-specific ribophages.

R17 Bacteriophage   R/1: 1.1/32: S/S: B/O

Another well-known ribophage. Closely related to MS2 not Qβ.

Reoviruses   [R/2: Σ15/15: S/S: V/O, Di]

A group of widespread and common viruses of birds and mammals, even bats. They have a segmented double-stranded RNA genome which is contained in a double-shelled isometric particle about 80 nm in diameter. Some are common in man, but are not specifically correlated with any disease, hence their name which is an acronym from *respiratory enteric orphan virus*. Some of the reoviruses with more fragile particles are placed in a separate group, the orbiviruses, and include sheep blue tongue virus. The reoviruses are clearly similar to plant viruses like clover wound tumour and also the cyto-plasmic polyhedroses of insects.

Rhabdoviruses   [R/1: 3.5/2: U/E: V, I, S/O, Ac, Ap, Au, Di]

A worldwide group of viruses of plants, insects and animals, even fish. The group includes rabies, vesicular stomatitis, *Drosophila* σ, lettuce necrotic yellows, potato yellow dwarf and trout Egtved viruses. Many of the group that are found in plants are transmitted by, and infect, leafhopper or aphid vectors. All have similar complex bullet-shaped or bacilliform particles, hence the group name which is derived from *rhabdos* Gr., a rod. The particles are 130–220 nm long and 70 nm wide and consist of a lipid-containing envelope around the helically constructed nucleocapsid.

Rice Dwarf Virus   R/2: Σ10/11: S/S: S, I/Av

A rice virus confined to Japan and Korea. It is clearly related to clover wound tumour and the reo-viruses. Takata reported in 1895 that rice dwarf disease was transmitted in crops by leafhoppers; the first report of a vector-borne plant virus, and this virus was also the first to be shown to be transmitted by a vector to its progeny.

Satellite Virus   R/1: 0.4/20: S/S: S/Fu

The smallest RNA virus known. It multiplies only in plants also infected with tobacco necrosis virus (i.e. a virus of a virus), and both are transmitted by zoospores of the fungus *Olpidium*. This virus has been found so far only in Europe and North America, though tobacco necrosis virus is much more widespread.

Sendai Virus   R/1: 4–8/1: S/E: V/O

A paramyxovirus isolated from mice in Sendai, Japan, otherwise known as the haemagglutinating virus of Japan, but also found elsewhere in the world.

Sindbis Virus   R/1: Σ3/6: S/S: V, I/Di

Type member of the alphaviruses (Group A arboviruses), which are a sub-group of the togaviruses. Transmitted by, and replicates in, mosquitoes. Found in birds, man and other mammals in most tropics except the American tropics.

T-even Bacteriophages   [D/2: 137/40: X/X: B/O]

A group of closely related phages of *E. coli*, T is for type, not tailed nor Twort, the discoverer of phages. They have extremely complex particles which consist of an angular rounded head of diameter 100 nm, and a complex contractile tail of the same length. Fibres on the end of the tail stick to the bacterium, the tail sheath contracts, an inner tube of the tail penetrates the host and the genome is injected into the bacterium. The linear DNA genome is large and the particles alone contain at least twenty different proteins. Some of the cytosine of the DNA is hydroxymethylated presumably to thwart unfriendly host enzymes.

T-odd Bacteriophages

These are not all closely related to one another as are the T-even phages, but they all infect *E. coli*. The serologically related T3 and T7 [D/2: 19/50; S/S: B/O] have particles which have a head of about 60 nm diameter and a short non-contractile tail only about 10 nm long. By contrast T5; D/2: 80/39: X/X: B/O has particles with a head of diameter 90 nm and a non-contractile tail 200 nm long.

Tobacco Mosaic Virus   R/1: 2/5: E/E: S/O

One of the best known viruses. In 1898 the pathogens causing tobacco mosaic and foot and mouth disease were both shown to pass through bacteria-proof filters, thus establishing the new group of pathogens called the filterable viruses. Tobacco mosaic virus particles attain a large concentration in infected plants and are stable and readily purified, thus they have been used in much of the pioneering work on the structure and composition of virus particles. The virus has a wide host range, and is found throughout the world, its many strains fall into well defined groups some of which have distinct host ranges. Tobacco mosaic virus is one of the few plant viruses transmitted by contact, not only when plants rub together, but also on feeding animals and on tools, and the hands of cigarette smokers. Much is known of the structure and assembly of its helically constructed tubular particles which are 300 nm long and 18 nm in diameter.

Tobacco Necrosis Virus   R/1: 1.5/19: S/S: S/Fu

A worldwide virus transmitted in nature by the zoospores of the chytrid fungus *Olpidium* to roots of a wide range of plants; rarely infects plants systemically. Is sometimes accompanied by satellite virus. Its isometric particles are 26 nm in diameter.

Tobacco Rattle Virus   R/1: 2.3/5 + (0.6–1.3)/5: E/E: S/Ne

One of the tobraviruses. Has a particularly wide range of hosts. Found in many parts of the world.

Tobraviruses   [R/1: 2.3/5 + (0.6–1.3)/5: E/E: S/Ne]

These two viruses, tobacco rattle and pea early browning, are transmitted by soil-inhabiting nematodes (*Trichodorus* spp) to a wide range of hosts. Their tubular particles are about 20 nm wide, some are 50–100 nm long and contain one segment of the viral genome, the others are 180–200 nm long and contain the other part.

Togaviruses   [R/1: Σ3–5/4–8: S/S: V, I/Di, Ac]

This very large group of viruses is the largest part of a group once called the arboviruses (short for *a*rthropod-*bo*rne viruses). Togaviruses occur in all parts of the world and cause many severe and lethal diseases of man and other animals. The togaviruses are now divided into the alphaviruses (group A), which include Sindbis, Semliki Forest and Venezuelan equine encephalitis viruses, the flaviviruses (group B), which include dengue and yellow fever viruses (the group is named after the latter), and also various smaller groups of which the Bunyamwera supergroup is the largest. All replicate in their arthropod vectors. Most are transmitted by mosquitoes and these are found mostly in the tropics and sub-tropics, whereas some of the flavoviruses from areas with cold winters are transmitted by ticks. All have isometric particles 50–100 nm in diameter with a ribonucleoprotein core and a lipid-containing envelope, hence the name togavirus from *toga* L. a cloak.

Turnip Crinkle Virus   R/1: 2/25: S/S: S/Cl

A beetle-borne virus of crucifers found in Scotland. Its isometric particles are about 30 nm in diameter and are all of one type.

Turnip Yellow Mosaic Virus   R/1: 1.9/34: S/S: S/Cl

A beetle-borne virus of brassicas found in western Europe. Type member of the tymoviruses.

Tymoviruses   [R/1: 2.1/35: S/S: S/Cl]

A worldwide group of about ten beetle-borne viruses of plants. Tymovirus particles are isometric, about 28 nm in diameter with obviously grouped subunits, and are of two types; some are nucleoprotein and are infectious, the others are empty protein shells which are not infectious, and the demonstration by MARKHAM and SMITH in 1949 of these differences was the first clear evidence that the nucleic acid was the viral genome.

Vaccinia Virus   D/2: 160/5: X/*: V/O

Type member of the poxviruses. This domesticated virus was derived either from smallpox (variola) or cowpox, probably the latter (hence its name), and has been used to vaccinate (immunize) against smallpox since the experiments reported by JENNER, a Gloucestershire village doctor, in 1798. It was also the first virus to be grown in cultured cells.

Vesicular Stomatitis Virus   R/1: 3.5/2: U/E: V/O

Type member of the rhabdoviruses. Causes disease in cattle and horses in the New World, occasionally found in Europe and Africa.

References

ADAMS, J.M., JEPPESEN, P.G.N., SANGER, F., BARRELL, B.G.: Nature **223**, 1009 (1969).
AMMAN, J., DELIUS, H., HOFSCHNEIDER, P.H.: J. Mol. Biol. **10**, 557 (1964).
ANONYMOUS: Nature New Biol. **231**, 33 (1971).
ANONYMOUS: Nature New Biol. **239**, 223 (1972).
ARNOTT, S., HUTCHINSON, F., SPENCER, M., WILKINS, M.H.F., FULLER, W., LANGRIDGE, R.: Nature **211**, 227 (1966).
BACHRACH, H.L.: J. Mol. Biol. **8**, 348 (1964).
BALL, L.A.: J. Theoret. Biol. **36**, 313 (1972).
BALTIMORE, D.: Nature **226**, 1209 (1970).
BANCROFT, J.B.: Adv. Virus Res. **16**, 99 (1970).
BANCROFT, J.B., KAESBERG, P.: Biochim. Biophys. Acta **39**, 519 (1960).
BARWISE, A.H., WALKER, I.O.: FEBS Lett. **6**, 13 (1970).
BAWDEN, F.C., PIRIE, N.W.: Biochem. J. **34**, 1278 (1940).
BELLETT, A.J.D.: J. Virol. **1**, 245 (1967).
BERNARDI, A., SPAHR, P.-F.: Proc. Nat. Acad. Sci. U.S. **69**, 3033 (1972).
BILLETER, M.A., DAHLBERG, J.E., GOODMAN, H.M., HINDLEY, J., WEISSMANN, C.: Nature **224**, 1083 (1969).
BISHOP, D.H.L., ROY, P., BEAN, W.J., SIMPSON, R.W.: J. Virol. **10**, 689 (1972).
BLUMENTHAL, T., LANDERS, T.A., WEBER, K.: Proc. Nat. Acad. Sci. U.S. **69**, 1313 (1972).
BOL, J.F., VLOTEN-DOTING, L. VAN: Virology **51**, 102 (1973).
BOL, J.F., VLOTEN-DOTING, L. VAN, JASPARS, E.M.J.: Virology **46**, 73 (1971).
BOS, L., JASPARS, E.M.J.: C.M.I./A.A.B. Descrip. Plant Viruses **46** (1971).
BRATT, M.A., ROBINSON, W.S.: J. Mol. Biol. **23**, 1 (1967).
BRECK, L.O., GORDON, M.P.: Virology **40**, 397 (1970).
BROWN, F., MARTIN, S.J., CARTWRIGHT, B., CRICK, J.: J. Gen. Virol. **1**, 479 (1967).
BURGE, B.W., PFEFFERKORN, E.R.: Nature **210**, 1397 (1966).
BURGESS, R.R.: Ann. Rev. Biochem. **40**, 711 (1971).

BURGESS, R. R., TRAVERS, A. A., DUNN, J. J., BAUTZ, E. K. F.: Nature 221, 43 (1969).
BUTLER, P. J. G., KLUG, A.: Nature New Biol. 229, 47 (1971).
BUTLER, P. J. G., KLUG, A.: Proc. Nat. Acad. Sci. U.S. 69, 2950 (1972).
CANAANI, E., DUESBERG, P.: J. Virol. 10, 23 (1972).
CARPENTER, J. M., KLECZKOWSKI, A.: Virology 39, 542 (1969).
CASPAR, D. L. D., KLUG, A.: Cold Spring Harbor Symp. Quant. Biol. 27, 1 (1962).
CHAMBERLAIN, M., MCGRATH, J., WASKELL, L.: Nature 228, 227 (1970).
CHOPPIN, P. W., COMPANS, R. W.: J. Virol. 5, 609 (1970).
CONTRERAS, R., YSEBAERT, M., MIN JOU, W., FIERS, W.: Nature New Biol. 241, 99 (1973).
COOPER, P. D., GEISSLER, E., SCOTTI, P. D., TANNOCK, G. A.: In: Strategy of the viral genome, Ciba
    Found. Symp. (WOLSTENHOLME, G. E. W., O'CONNOR, M., eds.), p. 75. London: Churchill 1971.
CRICK, F. H. C., WATSON, J. D.: In: The nature of viruses, Ciba Found. Symp. (WOLSTENHOLME, G. E. W.,
    ed.), p. 5. Boston: Little and Brown 1957.
DANIEL, V., SARID, S., LITTAUER, U. V.: Science 167, 1682 (1968).
DARBY, G., MINSON, A. C.: J. Gen. Virol. 14, 199 (1972).
DELBRÜCK, M., BAILEY, W. T.: Cold Spring Harbor Symp. Quant. Biol. 11, 33 (1946).
DIENER, T. O.: Virology 50, 606 (1972).
DOBROV, E. N., KUST, W. V., TIKCHONENKO, T. I.: J. Gen. Virol. 16, 161 (1972).
EIKHOM, T. S., SPIEGELMAN, S.: Proc. Nat. Acad. Sci. U.S. 57, 1833 (1967).
EISERLING, F. A., DICKSON, R. C.: Ann. Rev. Biochem. 41, 467 (1972).
FINCH, J. T., GIBBS, A. J.: J. Gen. Virol. 6, 141 (1970).
FRAENKEL-CONRAT, H.: The chemistry and biology of viruses, p. 76. London and New York: Academic
    Press 1969.
FRAENKEL-CONRAT, H., WILLIAMS, R. C.: Proc. Nat. Acad. Sci. U.S. 41, 690 (1955).
FRANKLIN, R. E., KLUG, A., HOLMES, K. C.: In: The nature of viruses, Ciba Found. Symp. (WOLSTEN-
    HOLME, G. E. W., ed.), p. 39. Boston: Little and Brown 1957.
FRANKLIN, R. M.: Proc. Nat. Acad. Sci. U.S. 55, 1504 (1966).
GIBBS, A.: Advan. Virus Res. 14, 263 (1969).
GILLESPIE, D., MARSHALL, S., GALLO, R. C.: Nature New Biol. 236, 227 (1972).
GILLIES, S., BULLIVANT, S., BELLAMY, A. R.: Science 174, 694 (1972).
GLITZ, D. G., EICHLER, D.: Biochim. Biophys. Acta 238, 224 (1971).
GOLDBERG, M.: Tohoku Math. J. 43, 104 (1937).
GOMATOS, P. J., TAMM, I., DALES, S., FRANKLIN, R. M.: Virology 17, 441 (1962).
GREEN, M.: Ann. Rev. Biochem. 39, 701 (1970).
GREEN, M., CARTAS, M.: Proc. Nat. Acad. Sci. U.S. 69, 791 (1972).
GRONER, Y., POLLACK, Y., BERISSI, H., REVEL, M.: Nature New Biol. 239, 16 (1972a).
GRONER, Y., SCHEPS, R., KAMEN, R., KOLAKOFSKY, D., REVEL, M.: Nature New Biol. 239, 19 (1972b).
GUILLEY, M. H., STUSSI, C., HIRTH, L.: Compt. Rend. 272, 1181 (1971).
HARRIS, W. F.: Nature 240, 294 (1972).
HARRISON, B. D.: C.M.I./A.A.B. Descrip. Plant Viruses 12 (1970).
HAYASHI, M., HAYASHI, M. N., SPIEGELMAN, S.: Proc. Nat. Acad. Sci. U.S. 50, 664 (1963).
HORST, J., CONTENT, J., MANDELES, S., FRAENKEL-CONRAT, H., DUESBERG, P.: J. Mol. Biol. 69, 209
    (1972).
HUANG, A. S., BALTIMORE, D., BRATT, M.: J. Virol. 7, 389 (1971).
HUANG, A. S., BALTIMORE, D., STAMPFER, M.: Virology 42, 946 (1970).
ITOH, H., MELNICK, J. L.: J. Exp. Med. 109, 393 (1959).
JACOBSON, M. F., ASSO, J., BALTIMORE, D.: J. Mol. Biol. 49, 656 (1970).
JOHNSTON, M. C., MARTIN, S. J.: J. Gen. Virol. 11, 71 (1971).
JOHNSTON, R. E., BOSE, H. R.: Proc. Nat. Acad. Sci. U.S. 69, 1514 (1972).
JOKLIK, W. K., SKEHEL, J. J., ZWEERINK, H. J.: Cold Spring Harbor Symp. Quant. Biol. 35, 791 (1970).
KACIAN, D. L., WATSON, K. F., BURNY, A., SPIEGELMAN, S.: Biochim. Biophys. Acta 246, 365 (1971).
KADO, C. I., KNIGHT, C. A.: Proc. Nat. Acad. Sci. U.S. 55, 1276 (1966).
KADO, C. I., KNIGHT, C. A.: J. Mol. Biol. 36, 15 (1968).
KALMAKOFF, J., LEWANDOWSKI, L. J., BLACK, D. R.: J. Virol. 4, 851 (1969).
KAMEI, T., RUBIO-HUERTOS, M., MATSUI, C.: Virology 37, 507 (1969).
KAMEN, R.: Nature 228, 527 (1970).
KAPER, J. M.: In: Molecular basis of virology (FRAENKEL-CONRAT, H., ed.), A.C.S. Monograph 164,
    p. 1. New York: Reinhold Book Corporation 1968.

KASSANIS, B.: C.M.I./A.A.B. Descrip. Plant Viruses 15 (1971).
KASSANIS, B., CONTI, M.: J. Gen. Virol. 13, 361 (1971).
KATES, J.: Cold Spring Harbor Symp. Quant. Biol. 35, 743 (1970).
KENNEDY, S. I. T.: Biochem. Biophys. Res. Commun. 48, 1254 (1972).
KINGSBURY, D. W.: J. Mol. Biol. 18, 195 (1966).
KONDO, M., GALLERANI, R., WEISSMANN, C.: Nature 228, 525 (1970).
KUSHNER, D. J.: Bacteriol. Rev. 33, 302 (1969).
LANE, L. C., KAESBERG, P.: Nature New Biol. 232, 40 (1971).
LANGRIDGE, R., GOMATOS, P. J.: Science 141, 694 (1963).
LASKEY, R. A., GURDON, J. B., CRAWFORD, L. V.: Proc. Nat. Acad. Sci. U.S. 69, 3665 (1972).
LEBERMAN, R.: In: The molecular biology of viruses, Symp. Soc. Gen. Microbiol. (CRAWFORD, L. V.,
    STOKER, M. G. P., eds.), 18, p. 183. London: Cambridge University Press 1968.
LEDINKO, N., HIRST, G. K.: Virology 14, 207 (1961).
LISTER, R. M., BRACKER, C. E.: Virology 37, 262 (1969).
LITTAUER, U. Z., DANIEL, V., SARID, S.: In: Strategy of the viral genome (WOLSTENHOLME, G. E. W.,
    O'CONNOR, M., eds.), p. 169. London: Churchill 1971.
LITVAK, S., TARRAGO, A., TARRAGO-LITVAK, L., ALLENDE, J. E.: Nature New Biol. 241, 88 (1973).
MARTELLI, G. P., QUACQUARELLI, A., RUSSO, M.: C.M.I./A.A.B. Descrip. Plant Viruses 69 (1971).
MICHEL, M. R., HIRST, B., WEIL, R.: Proc. Nat. Acad. Sci. U.S. 58, 1381 (1967).
MIN JOU, W., HAEGEMAN, G., YSEBAERT, M., FIERS, W.: Nature 237, 82 (1972).
MIZUTANI, S., BOETTIGER, D., TEMIN, H. M.: Nature 228, 424 (1970).
MUNDRY, K. W.: Molec. Gen. Genet. 105, 361 (1969).
NATHANS, D., OESCHGER, M. P., POLMAR, S. K., EGGEN, K.: J. Mol. Biol. 39, 279 (1969).
NICHOLS, J. L., BELLAMY, A. R., JOKLIK, W. K.: Virology 49, 562 (1972).
ODA, K., JOKLIK, W. K.: J. Mol. Biol. 27, 395 (1967).
ÖBERG, B., PHILIPSON, L.: J. Mol. Biol. 58, 725 (1971).
ÖBERG, B. F., SHATKIN, A. J.: Proc. Nat. Acad. Sci. U.S. 69, 3589 (1972).
OHNO, T., INOUE, II., OKADA, Y.: Proc. Nat. Acad. Sci. U.S. 69, 3680 (1972).
OKADA, Y., OHASHI, Y., OHNO, T., NOZU, Y.: Virology 42, 243 (1970).
PEDERSEN, I. R.: Nature New Biol. 234, 112 (1971).
PETTIJOHN, D. E.: Nature New Biol. 235, 204 (1972).
PHILIPSON, L., WALL, R., GLICKMAN, G., DARNELL, J. E.: Proc. Nat. Acad. Sci. U.S. 68, 2806 (1971).
PINCK, M., YOT, P., CHAPEVILLE, F., DURANTON, H. M.: Nature 226, 954 (1970).
POGO, B. G. T., DALES, S., BERGOIN, M., ROBERTS, D. W.: Virology 43, 306 (1971).
PONS, M. W.: Virology 47, 823 (1972).
PRICE, R., PENMAN, S.: J. Mol. Biol. 70, 435 (1972a).
PRICE, R., PENMAN, S.: J. Virol. 9, 621 (1972b).
PRIDGEN, C., KINGSBURY, D. W.: J. Virol. 10, 314 (1972).
PROCHIANTZ, A., HAENNI, A. L.: Nature New Biol. 241, 168 (1973).
RALPH, R. K., BERGQUIST, P. L.: In: Methods in virology (MARAMOROSCH, K., KOPROWSKI, H., eds.),
    vol. 2, p. 463. London-New York: Academic Press 1967.
ROBERTS, J. W.: Nature 224, 1168 (1969).
ROBERTSON, H. D., JEPPESEN, P. G. W.: J. Mol. Biol. 68, 417 (1972).
ROCHOW, W. F.: Ann. Rev. Phytopath. 10, 101 (1972).
ROEDER, R. G., REEDER, R. H., BROWN, D. D.: Cold Spring Harbor Symp. Quant. Biol. 35, 727 (1970).
ROEDER, R. G., RUTTER, W. J.: Nature 224, 234 (1969).
ROEDER, R. G., RUTTER, W. J.: Proc. Nat. Acad. Sci. U.S. 65, 675 (1970).
ROSENBERG, H., DISKIN, B., ORON, L., TRAUB, A.: Proc. Nat. Acad. Sci. U.S. 69, 3815 (1972).
RUSHIZKY, G. W., KNIGHT, C. A.: Virology 11, 236 (1960).
SÄNGER, H. L.: J. Virol. 3, 304 (1969).
SAKUMA, S., WATANABE, Y.: J. Virol. 10, 628 (1972).
SALSER, W., RICARD, B.: Cold Spring Harbor Symp. Quant. Biol. 35, 19 (1970).
SCHACHTER, E. M., BENDET, I. J., LAUFFER, M. A.: J. Mol. Biol. 22, 165 (1966).
SCHEKMAN, R., WICKNER, W., WESTERGAARD, O., BRUTLAG, D., GEIDER, K., BERTSCH, L. L., KORN-
    BERG, A.: Proc. Nat. Acad. Sci. U.S. 69, 2691 (1972).
SHATKIN, A. J.: In: Molecular basis of virology (FRAENKEL-CONRAT, H., ed.), A.C.S. Monograph 164,
    p. 351. New York: Reinhold Book Corporation 1968.
SHEPHERD, R. J., WAKEMAN, R. J., ROMANKO, J. J.: Virology 36, 150 (1968).

Siegel, A.: Virology **46**, 50 (1971).
Simmons, D.T., Strauss, J.H.: J. Mol. Biol. **71**, 599 (1972).
Singer, B., Fraenkel-Conrat, H.: Virology **39**, 395 (1969).
Sinsheimer, R.L.: J. Mol. Biol. **1**, 37 (1959).
Sinsheimer, R.L., Starmann, B., Nagler, C., Guthrie, S.: J. Mol. Biol. **4**, 142 (1962).
Skehel, J.J.: Virology **45**, 793 (1971).
Skehel, J.J.: Virology **49**, 23 (1972).
Smith, R.E., Zweerink, H.J., Joklik, W.K.: Virology **39**, 791 (1969).
Staples, D.H., Hindley, J.: Nature New Biol. **234**, 211 (1971).
Staples, D.H., Hindley, J., Billeter, M.A., Weissmann, C.: Nature New Biol. **234**, 202 (1971).
Stavis, R.L., August, J.T.: Ann. Rev. Biochem. **39**, 527 (1970).
Steitz, J.A.: Nature New Biol. **236**, 71 (1972).
Sugiyama, T., Korant, B.D., Lonberg-Holm, K.K.: Ann. Rev. Microbiol. **26**, 467 (1972).
Summers, D.F., Maizel, J.V.: Proc. Nat. Acad. Sci. U.S. **59**, 966 (1968).
Summers, D.F., Maizel, J.V.: Proc. Nat. Acad. Sci. U.S. **68**, 2852 (1971).
Summers, D.F., Maizel, J.V., Darnell, J.E.: Proc. Nat. Acad. Sci. U.S. **54**, 505 (1965).
Summers, M.D., Anderson, D.L.: J. Virol. **9**, 710 (1972).
Suzuki, J., Haselkorn, R.: J. Mol. Biol. **36**, 47 (1968).
Symons, R.H., Rees, M.W., Short, M.N., Markham, R.: J. Mol. Biol. **6**, 1 (1963).
Taber, R., Rekosh, C., Baltimore, D.: J. Virol. **8**, 395 (1971).
Temin, H.M.: Virology **20**, 577 (1963).
Temin, H., Mizutani, S.: Nature **226**, 1211 (1970).
Thouvenel, J.-C., Guilley, H., Stussi, C., Hirth, L.: FEBS Lett. **16**, 204 (1971).
Tillack, T.W., Smith, D.E.: Virology **36**, 212 (1968).
Tinoco, I., Uhlenbeck, O.C., Levine, M.D.: Nature **230**, 362 (1971).
Travers, A.A.: Nature **225**, 1009 (1970).
Travers, A.A.: In: Strategy of the viral genome (Wolstenholme, G.E.W., O'Connor, M., eds.),
    p. 155. London: Churchill 1971.
Travers, A.A., Kamen, R.I., Schleif, R.F.: Nature **228**, 748 (1970).
Van Kammen, A.: Virology **34**, 312 (1968).
Van Ravenswaay Claasen, J.C., van Leeuwen, A.B.J., Duijts, L., Bosch, L.: J. Mol. Biol. **23**, 535
    (1967).
Van Vloten-Doting, L., Bol, J.F., Dingjan-Versteegh, A., Jaspars, E.M.J.: Int. Virol. 2, 2nd Int.
    Congr. Virol. Budapest, p. 232. Basel: Karger 1972.
Van Vloten-Doting, L., Jaspars, E.M.J.: Virology **33**, 684 (1967).
Van Vloten-Doting, L., Jaspars, E.M.J.: Virology **48**, 699 (1972).
Varma, A., Gibbs, A.J., Woods, R.D., Finch, J.T.: J. Gen. Virol. **2**, 107 (1968).
Verhoef, N.J., Bosch, L.: Virology **45**, 75 (1971).
Verhoef, N.J., Lupker, J.H., Cornelissen, M.C.E., Bosch, L.: Virology **45**, 85 (1971).
Verma, I.M., Menth, N.L., Baltimore, D.: J. Virol. **10**, 622 (1972).
Vigier, P.: Progr. Med. Virol. **12**, 240 (1970).
Vogt, P.K.: Virology **32**, 708 (1967).
Wagner, G.W., Bancroft, J.B.: Virology **45**, 321 (1971).
Walter, G., Seifert, W., Zillig, W.: Biochem. Biophys. Res. Commun. **30**, 240 (1968).
Watanabe, Y., Millward, S., Graham, A.F.: J. Mol. Biol. **36**, 107 (1968).
Weber, K., Rosenbusch, J., Harrison, S.C.: Virology **41**, 763 (1970).
Weiss, S.B., Hsu, W.-T., Foft, J.W., Scherberg, N.H.: Proc. Nat. Acad. Sci. U.S. **61**, 114 (1968).
Weissmann, C., Feix, G., Slor, H.: Cold Spring Harbor Symp. Quant. Biol. **33**, 83 (1968).
Wildy, P.: Classification and nomenclature of viruses, Monographs in Virology (Melnick, J.L., ed.),
    vol. 5. Basel: Karger 1971.
Wilson, J.H., Abelson, J.N.: J. Mol. Biol. **69**, 57 (1972).
Wilson, J.H., Kells, S.: J. Mol. Biol. **69**, 39 (1972).
Wimmer, E.: J. Mol. Biol. **68**, 537 (1972).
Wimmer, E., Chang, A.Y., Clark, J.M., Reichmann, M.E.: J. Mol. Biol. **38**, 59 (1968).
Winocour, E.: Virology **34**, 571 (1968).
Wrigley, N.G.: J. Gen. Virol. **5**, 123 (1969).
Zillig, W., Zechel, L., Rabussay, D., Schachner, M., Sethi, V.S., Palm, P., Heil, A., Seifert, W.:
    Cold Spring Harbor Symp. Quant. Biol. **35**, 47 (1970).

CHAPTER 11

Isolation, Purification and Fractionation of RNA

ROZANNE POULSON

Introduction

The isolation of RNA requires an effective means of cell disruption, a procedure for separating the nucleic acid from the protein or lipoprotein with which it is intimately associated, and a method for purification. To avoid degradation or denaturation, the RNA must be protected throughout from liberated nucleases, strong mechanical forces, high temperatures and extremes of pH and ionic strength. A number of methods which have been reported appear to meet these requirements when applied to one or a number of plants, animals or micro-organisms. However, none of them are universally applicable, due to a number of tissue and species specific differences. For example, variations in cell or organelle structure dictate different methods of disruption. Differences between tissues in the pH optima and inhibitor sensitivity of their nucleases necessitate procedures adapted to each material which minimize enzymic breakdown. Other variables of which a satisfactory method must take account include the resistance to denaturation of the RNA to be recovered and the nature of the principal contaminants. As a guide to the isolation of RNA from tissues for which no satisfactory procedure has been reported, a number of methods of cell disruption, nucleoprotein dissociation and RNA purification are described and evaluated.

Unlike methods for isolation, techniques for the separation of RNA into its component species are generally applicable irrespective of the tissue from which it is derived. As most of these fractionations have been described in detail elsewhere, only an outline of those most widely used is provided here.

A. Tissue Disruption

1. Extraction Medium

Denaturation and breakdown of RNA is most likely to occur during the initial stages of extraction when precise control of the ionic composition and strength of the extract is impossible, and before contaminating nucleases have been removed. Enzymic degradation can be minimized by performing all operations between 0° and 4° if possible, and by the use of RNAase inhibitors. These and other constituents of the extraction medium which affect the integrity of the RNA are considered below.

a) Total Ionic Strength. The secondary structure of RNA is dependent upon ionic strength (*I*). When *I* is low, RNA undergoes some degree of denaturation and this greatly increases its susceptibility to enzymic hydrolysis. NaCl is generally used to maintain the ionic strength of the extraction medium. The presence of NaCl is also an advantage during a phenol extraction as it depresses the solubility of the phenol in the aqueous phase.

b) Divalent Cations. In certain instances, a divalent cation, usually Mg^{2+}, is added to the extraction medium to preserve the secondary structure of RNA. More frequently, however, the presence of Mg^{2+} results in low yields of RNA due to aggregation and insolubilization of the RNA (SPORN and DINGMAN, 1963; STANLEY and BLOCK, 1965). In such cases the presence of a suitable chelating agent (e.g. EDTA) is essential.

c) Hydrogen Ion Concentration. To preserve the secondary structure of RNA, the pH of the extraction medium must be held between 5.0 and 9.0 by the use of a buffer (COX et al., 1956; LITTAUER and EISENBERG, 1959). A pH above 5.0 is also essential when phenol is included in the extraction medium to prevent solubilization of RNA in the phenol. Whenever possible a pH which provides maximum inhibition of RNAase action is chosen.

d) RNAase Inhibitors. RNA must be protected from nucleases from both the tissue and extraneous sources. The latter is achieved by wearing sterile gloves and using autoclaved glassware and solutions. The activity of nucleases within the tissue which are liberated during extraction can be minimized by including a suitable inhibitor in the extraction medium.

1) Bentonite is a montmorillonite clay, $Al_2O_3 \cdot 4SiO_2 \cdot H_2O$, which possesses many charged sites capable of binding metal ions and basic proteins such as RNAase (FRAENKEL-CONRAT et al., 1961; SINGER and FRAENKEL-CONRAT, 1961). It may be prepared for use by suspending the crude powder in distilled water (50 mg/ml). The fraction sedimenting between 6,000 g for 10 minutes and 10,000 g for 15 minutes is resuspended in 0.1 M Na_2-EDTA pH 7.5 (10 mg/ml) and stirred for 48 hours at 25° to remove metal ions. The suspension is centrifuged differentially and the 10,000 g pellet is washed repeatedly with water to remove the EDTA. It is then made up as a 5% suspension in 0.01 M sodium acetate pH 6.0.

The suspension of uniformly sized particles is added to the extraction medium and subsequently is removed, with bound RNAase, by centrifugation. At pH 6.0 bentonite (35 µg/ml) is an effective inhibitor of RNAases from yeast, vascular plants and pancreas; but at pH 7.4 a concentration of 3.5 mg/ml is required for complete inhibition (LITTAUER and SELA, 1962). It should be noted that bentonite adsorbs RNA as well as protein (BLANTON and BARNETT, 1969; FRAENKEL-CONRAT et al., 1961; WALTERS et al., 1970), although to a lesser extent (probably no more than 50 µg RNA/mg bentonite). However, the amount adsorbed varies with the source of RNA and it is advisable to determine the minimum amount of bentonite required to inhibit RNAase action.

2) A copolymer of L-tyrosine and L-glutamic acid (TG) is an effective inhibitor of RNAases at pH 5.0 but not at pH 7.4 (LITTAUER and SELA, 1962).

3) Diethyl pyrocarbonate (DEP), $C_2H_5-O-CO-O-CO-O-C_2H_5$, is a potent inhibitor of RNAase activity (FEDORCSÁK and EHRENBERG, 1966; SOLYMOSY et al., 1968). It causes enzyme inactivation by denaturing proteins (ROSÉN and

FEDORCSÁK, 1966; WOLF et al., 1970). DEP is highly unstable in the presence of water and on contact with polystyrene (ÖBERG, 1970). At 20° it has a half-life of 75 minutes, decomposing into ethanol and carbon dioxide. DEP should not, therefore, be added to extraction media prior to use. To prevent breakdown during storage it should be kept at 0° under nitrogen.

4) Macaloid is a purified hectorite (sodium magnesium lithofluorosilicate). It is used as a suspension (1 mg/ml) of fine particles which, being negatively charged, adsorb RNAase. The macaloid and bound RNAase are then removed from the extract by centrifugation. It may remove RNAase more effectively than bentonite since it is reputed to permit isolation of the rRNAs from purified yeast mitochondrial ribosomes in equimolar ratio (GRIVELL et al., 1971), while bentonite does not (YU et al., 1972).

The macaloid is prepared for use by dispersing the powder in hot 0.05 M Tris-HCl pH 7.6 containing 0.01 M magnesium acetate and homogenizing in a high shear mixer for 5 minutes. The suspension is cooled and centrifuged, the supernatant is discarded and the gelatinous layer of the pellet separated from the coarser particles. The gel is suspended in 0.05 M Tris-HCl pH 7.5 to give a final concentration of 1 % (MARCUS and HALVORSON, 1968).

5) Polyvinylsulfate (PVS) has been used as an RNAase inhibitor (BERNFIELD et al., 1960; HOWELLS and WYATT, 1969). However, it is less effective than bentonite even at high concentrations (LITTAUER and SELA, 1962). It is also very difficult to separate from RNA preparations except by column chromatography. This substance is not, therefore, recommended for general use.

6) Anionic detergents inhibit RNAase activity. They are also used to dissociate nucleoprotein complexes as described below.

2. Disruption of Cells and Organelles

a) Viruses. Viruses are readily separated from the host tissue by one of the standard techniques (GOODING and HERBERT, 1967; GREEN and PIÑA, 1963; LEVINTON and DARNELL, 1960; ROBINSON et al., 1965; STACE-SMITH, 1966; STEERE, 1956). The main problem is the dissociation of the protein coat of the virion without degrading the RNA. The most commonly used methods are treatment with sodium dodecyl sulphate (SDS) or phenol or a combination of both. The lipid-rich coats of both the influenza virus and the Rous sarcoma virus are rapidly dissolved when incubated at 37° for a few seconds in the presence of 1 % SDS (NOLL and STUTZ, 1968; ROBINSON et al., 1965). Tobacco mosaic virus may be disrupted with a 1 % solution of SDS if the mixture is heated at 50° for 5 minutes (FRAENKEL-CONRAT et al., 1957). Alternatively, it can be ruptured by vigorously shaking the viral suspension with an equal volume of 88 % phenol at 0–20° for 5 minutes (GIERER and SCHRAMM, 1956; SCHUSTER et al., 1956). The poliovirus is disrupted by vigorously shaking with phenol at 60° for 5 minutes (WARNER et al., 1963). Hot (60°) phenol is also used to extract infectious RNA from purified equine encephalomyelitis virus (WECKER, 1959).

b) Bacteria. Various mechanical devices have been used to disrupt bacterial cells e. g. Servall omni-mixer, French pressure cell, glass beads and Hughes press.

However, if non-mechanical methods are effective, they are preferable as shearing of the RNA is avoided.

Some bacterial cell walls can be ruptured by adding 1% SDS to the cell suspension, the suspension becoming clear when cell lysis is complete. In other cases the bacterial cells must be treated with lysozyme before they can be lysed with SDS (BUTLER and GODSON, 1963; MARMUR, 1961; REPASKE, 1958). RNA may also be extracted directly from bacterial cells with a phenol-SDS mixture (SCHERRER and DARNELL, 1962), but in most cases rRNA is not extracted unless the cell wall is removed before the phenol is added. However, the restricted pores of bacterial and yeast cell walls are useful in retaining large (ribosomal) RNAs and allowing the efflux of small (4–6S) RNAs when treated with phenol or other membrane disrupting solutes. Use of this method reduces the difficulty of later separation of RNAs of various sizes [cf. C and D below].

Bacterial spores are especially difficult to disrupt and a method has been described (TAKAHASHI, 1968) which includes treatment with a sulfhydryl compound to sensitize the spores to the action of lysozyme. The sensitized spores are then incubated with 0.5 mg/ml lysozyme at 37° for 1 hour and the suspension is centrifuged at 12,000 g for 20 minutes and the supernatant used as the source of RNA.

c) Plant Tissues. Plant tissues are usually disrupted mechanically. This inevitably causes some RNA degradation. Homogenization should not, therefore, be continued until 100% cell breakage is achieved unless quantitative recovery of RNA is essential. Tissues, which may be frozen in liquid nitrogen, can be ground with a pestle in a mortar, or, more conveniently, with a VirTis '45' homogenizer, Waring blendor, Lourdes or Servall omni-mixer. The volume of extraction medium should be 3–10 times that of the tissue. If the tissue is difficult to disrupt rapidly it should first be dehydrated. This may be done by extracting the tissue in a Waring blendor in either ethanol or ethylene glycol and washing the extract with ethanol, ethanol:ether (1:3), ethanol:ether (1:1) and anhydrous ether and drying in a stream of nitrogen.

d) Animal Tissues. Animal tissues can be mechanically disrupted more easily than most plant materials. In many cases they are readily homogenized with a motor-driven teflon pestle in a glass homogenizer of the POTTER-ELVEHJEM type, using the minimum number of vertical strokes to achieve adequate cell breakage. The efficiency of cell breakage may be improved by using an all glass homogenizer.

e) Chloroplasts. Chloroplasts are suspended in 0.01 M Tris-HCl pH 7.5 containing 0.1 M NaCl and 0.5% SDS (EISENSTADT and BRAWERMAN, 1964). Lysis occurs as the chloroplast suspension is shaken with liquefied phenol at 0°. Alternatively, chloroplasts may be lysed by a 10 minute incubation at 0° with 1% Triton X-100 (SPENCER and WHITFELD, 1966).

f) Nuclei. Nuclei are suspended in 0.001 M potassium phosphate pH 6.8 containing 0.002 M $MgCl_2$ and 0.32 M sucrose. SDS or Triton X-100 are added to a final concentration of 0.1% or 0.3% respectively, and the mixture stirred at 20° until the solution is clear. Another method is to shake the nuclei for a few seconds in a mixture of 10% sodium deoxycholate and 10% Tween 40 (1:2) (PENMAN, 1966). Unless treated with 50 µg/ml DNAase, the lysed nuclei cannot be used directly for phenol extraction because the DNA forms a nucleohistone gel which traps some of the nuclear RNA. Further details of RNA extraction from nuclei are given in Chapter 3.

g) Mitochondria. Mitochondria are suspended in 0.01 M Tris-HCl pH 7.5 containing 0.1 M NaCl, 0.1 % macaloid and 2 % SDS or 1 % sodium deoxycholate. Lysis occurs as the suspension is shaken with phenol at 0° (GRIVELL et al., 1971; RIFKIN et al., 1967).

h) Ribosomes. Ribosomes are suspended in 1 volume of 0.01 M Tris-HCl pH 7.5 containing 0.1 M NaCl and 0.001 M $MgCl_2$; SDS is added to a final concentration of 0.5 % for 80S ribosomes or 2 % for 70S ribosomes. The mixture is incubated at 37° for 10 seconds and the lysate immediately chilled in ice (NOLL and STUTZ, 1968). Alternatively, ribosomes can be suspended in 0.1 M Tris-HCl pH 7.5 containing 0.001 M $MgCl_2$ and mixed with an equal volume of cold buffer-saturated phenol. The mixture is stirred at 20° for 1 hour to ensure the complete release of RNA.

B. Nucleoprotein Dissociation and Deproteinization of Released RNA

1. Materials

a) Dissociating Agents. A dissociating agent, usually an anionic detergent, is included in the extraction medium to lyse viruses, bacteria and organelles, to free RNA from protein and lipoprotein structures and to inhibit RNAase activity. The dissociation of nucleoproteins is dependent upon the protein:detergent ratio and the duration of exposure (NOLL and STUTZ, 1968). Complete release of RNA with a minimum of degradation is achieved by using a high concentration of detergent and a brief exposure period.

1) Sodium 4-aminosalicylate is a strong detergent and at a concentration of 6 % disintegrates the chromosomal structure and releases DNA. If DNA is not required 0.5 % sodium naphthalene-1,5-disulfonate should be substituted (KIRBY, 1965; 1968). Sodium 4-aminosalicylate may be maintained in solution at 0–2° by the addition of 6 % 2-butanol (PARISH and KIRBY, 1966).

2) Sodium deoxycholate is comparable with SDS [cf. 4) below] in its capacity to disrupt nucleoprotein complexes, but it is more easily separated from the RNA; this is done by precipitating the RNA at pH 5.0 with 2 volumes of 95 % ethanol containing 0.2 M sodium acetate (BLOW and RICH, 1960).

3) Sodium dodecyl sarcosinate (Sarkosyl NL-97) is substituted for SDS [cf. 4) below] when CsCl gradients are used because, unlike SDS, it is soluble in concentrated CsCl.

4) Sodium or lithium dodecyl sulfate (SDS and LDS) at a concentration of 1 % extract DNA together with RNA into the aqueous phase of the phenol extraction system (see below); however, at lower concentrations (0.1 %) only RNA is extracted (DINGMAN and SPORN, 1962). The potassium salt of dodecyl sulfate has a very low solubility and K^+ should, therefore, be excluded from the extraction medium.

5) Sodium naphthalene-1,5-disulfonate is a hydrophilic salt which is able to release tRNA and rRNA into the aqueous phase of the phenol extraction system but not DNA or other RNA species (Kirby, 1956, 1968).

6) Sodium tri-isopropylnaphthalene sulfonate is a powerful detergent which can be used in greater concentrations than SDS without forming a single phase system in the phenol extraction procedure.

b) Deproteinizing Agents. One of a number of substances, generally referred to as deproteinizing agents, may be added to the extraction medium to precipitate proteins which have been released from the nucleoprotein complexes. Certain of these additives also inhibit RNAase activity and disrupt viruses and bacteria.

1) Chloroform containing 1–4% 1-pentanol has proved a more effective deproteinizing agent than phenol [cf. 4) below] in some tissues (Ingle and Burns, 1968).

2) Diethyl pyrocarbonate (DEP) is a very effective deproteinizing agent and RNAase inhibitor (Fedorcsák and Ehrenberg, 1966; Rosén and Fedorcsák, 1966).

3) Guanidium and lithium chloride denature the proteins of ribosomes and release free RNA which contains less than 1% protein (Barlow et al., 1963; Cox, 1968; Cox and Arnstein, 1963; Sela et al., 1957).

4) Phenol is a very efficient protein denaturant (Kirby, 1956, 1957; Pusztai, 1966). However, it does not completely inhibit RNAase activity even at high (60°) temperatures (Huppert and Pelmont, 1962; Kickhoffen and Burger, 1962; Rushizky et al., 1963). It also has the disadvantage that its high UV absorption interferes with spectrophotometric assays of RNA, although phenol can be removed from aqueous RNA preparations by extracting them several times with 2 volumes of diethyl ether. The most serious disadvantage of isolation procedures using phenol is that some RNA is invariably lost in precipitates of denatured proteins.

Phenol must be distilled (preferably under vacuum over zinc dust) to remove preservatives and oxidative products which cause the selective breakdown of 25S and 23S RNAs (Poulson, unpublished observations). The freshly distilled phenol is saturated with either water or buffer. (If the phenol is not equilibrated it will take up some of the aqueous phase.)

The addition of 0.1% 8-hydroxyquinoline to the phenol provides several advantages (Kirby, 1962). It prevents oxidation of the phenol, partially inhibits RNAase activity and chelates metal ions involved in binding RNA to proteins.

The freezing point of phenol—8-hydroxyquinoline mixtures can be lowered by the addition of 10% *m*-cresol (Kirby, 1965). *m*-Cresol also improves the ability of the mixture to deproteinize extracts.

2. Methods

a) Sodium Dodecyl Sulfate (SDS). SDS releases both RNA and DNA from protein complexes and partially inhibits RNAase action (Kay et al., 1952; Noll and Stutz, 1968). Its principal advantage is that it achieves quantitative recovery of RNA. However, except for certain cases, e.g. cytoplasm, ribosomes and some viruses, the RNA preparations are heavily contaminated with proteins. It is also unsuitable for the isolation of RNA from materials rich in DNA, such as nuclei, unless subsequent measures are taken to remove the DNA (DI GIROLAMO et al., 1964).

Extraction medium:
A. 0.005 M Tris-HCl pH 7.5   B. 10% SDS
   0.010 M NaCl

1) Ribosomes or viral particles are suspended in 1 volume of cold solution A.

2) To the suspension (or directly to the cytoplasmic extract) add solution B to a final concentration of 0.5% SDS for cytoplasmic extracts and 80S ribosomes, 2% for 70S ribosomes and 0.5-1% for viral particles.

3) The mixture is incubated at 37° for 1 minute, and then chilled in ice.

4) The lysate may be used immediately for characterization studies, or, if it is heavily contaminated with protein, it may be deproteinized with phenol (cf. below).

b) Phenol. Unlike SDS which only dissociates nucleoprotein complexes, phenol also precipitates the dissociated proteins from the aqueous phase (KIRBY, 1956). However, some RNA complexed with denatured protein is precipitated together with all of the DNA. For this reason the method is rarely quantitative but it has been used for qualitative studies on RNA from yeasts, reticulocytes and some micro-organisms and also for the extraction of tRNA and DNA-free RNA from nuclei. Below 20°, tRNA and rRNA are preferentially extracted. At 50° and 60° respectively, ribosomal precursor RNA and rapidly labelled polydisperse RNA are also released (GEORGIEV and MONTIEVA, 1962; GEORGIEV et al., 1963; LERMAN et al., 1965). Thus, by means of sequential extraction at progressively higher temperatures, a preliminary fractionation of RNA may be achieved. A single extraction at 60° to obtain whole cell RNA is not recommended as there is evidence of rRNA degradation at high temperatures (PENE et al., 1968; PLAGE-MANN, 1970).

Extraction medium:
A. 0.05 M Sodium acetate pH 6.0   B. Phenol:acetate buffer:8-hydroxyquinoline
   0.14 M NaCl   (50:10:0.05 v/v/w)
   0.5% Bentonite

1) Tissue is suspended in 1 volume of solution A.

2) An equal volume of solution B is added.

3) The mixture is shaken vigorously at 0-5° for 30 minutes.

4) The mixture is centrifuged at 8,000 g for 10 minutes.

5) The interfacial precipitate may be re-extracted at 50° and 60° by adding equal volumes of solutions A and B to release ribosomal precursor RNA and rapidly labelled polydisperse RNA respectively. (As an alternative to extraction at high temperatures, the ribosomal precursor RNA and rapidly labelled RNA may be released by re-extracting the interfacial precipitate and phenol layer with a solution containing 0.5% SDS. The detergent-released RNA will be contaminated with DNA which can be removed by treatment with RNAase-free DNAase.)

6) Each aqueous phase is washed twice with 3 volumes of diethyl ether to remove the phenol.

7) RNA is precipitated from each of the aqueous phases with 2 volumes of 95% ethanol containing 0.2 M sodium acetate.

Precipitation of RNA from the aqueous phase with ethanol may take from 1 to 16 hours at −20°. If no acetate is added, complete RNA precipitation may take longer. The RNA is collected by centrifugation at 12,000 g for 20 minutes. Traces of ethanol are removed by recentrifuging the packed pellet. The RNA preparation may be freed from contaminating phenol and detergents other than SDS by dissolving it in 0.15 M sodium acetate pH 6.0 containing 0.5% SDS at 20° and reprecipitating the RNA with 2 volumes of 95% ethanol at −20°.

The crude RNA preparation may be stored, prior to purification, without incurring any damage. The RNA pellet is washed with absolute ethanol and can either be dried under vacuum over $CaCl_2$ or P_2O_5 and stored at $-20°$, or dissolved in a small volume of 0.15 M sodium acetate pH 6.0 containing 0.5% SDS as an RNAase inhibitor and stored in the presence of 2 volumes of ethanol at $-20°$.

c) Phenol-SDS. The SDS method yields RNA preparations heavily contaminated with protein while the phenol method does not dissociate all nucleoprotein complexes. Extraction with a combination of phenol and SDS gives a preparation containing all the RNA species with little protein contamination (DINGMAN and SPORN, 1962; ITOH and HIRAI, 1966). However, as with SDS alone, DNA is extracted, and as with phenol alone, the method is not entirely quantitative. Contamination with DNA can be avoided by extraction at 60° (SCHERRER and DARNELL, 1962), though as mentioned above, degradation of rRNA may occur. Despite these shortcomings the method has been widely used for comparative studies on RNA from plants, animals and micro-organisms.

Extraction medium:
A. 0.01 M Tris-HCl pH 7.5   B. Phenol: water: 8-hydroxyquinoline: *m*-cresol
   1% Bentonite   (50:10:0.05:7 v/v/w/v)
   1% SDS
1) Tissue is extracted in 3–10 volumes of solutions A and B in a ratio of 1:1.
2) The homogenate is shaken or stirred vigorously at 0–4° for 10 minutes.
3) The mixture is centrifuged at 10,000 g for 5 minutes.
4) NaCl is added to the aqueous phase to give a final concentration of 0.5 M.
5) An equal volume of solution is added and the mixture stirred and centrifuged.
6) Step 5 is repeated until no additional protein is precipitated.
7) The aqueous phase is washed twice with 3 volumes of peroxide-free diethyl ether to remove phenol. The phases can be separated rapidly by centrifugation at 5,000 g for 2 minutes.
8) RNA is precipitated from the aqueous phase with 2 volumes of 95% ethanol containing 0.2 M sodium acetate.
Sodium deoxycholate may be substituted for SDS (CLICK and HACKETT, 1966). If the tissue is extracted with 1 part of sodium glycinate buffer (0.1 M pH 8.0 containing 0.1 M NaCl, 0.01 M Na_2-EDTA, 1% bentonite and 1% sodium deoxycholate) plus 2 parts of phenol reagent (phenol saturated with Na_2-EDTA) only a single deproteinization is necessary (i.e. omit steps 4–6) (POULSON and BEEVERS, 1970). Three or 10 volumes of extraction medium should be used for plant and animal, and bacterial cells respectively.

d) Phenol—Naphthalene-1,5-disulfonate—Tri-isopropylnaphthalene Sulfonate. As with cold phenol, tRNA and rRNA are extracted by this procedure while the rapidly labelled RNA and DNA remain in association with precipitated proteins (KIRBY, 1965; 1968). However, it is more quantitative than the phenol method for the following reasons. First, t- and rRNA are released before treatment with phenol so that protein bound rRNA is not lost; second, the detergents are effective RNAase inhibitors which eliminates the need to use bentonite which may adsorb RNA.

Extraction medium:
A. 0.01 M Tris-HCl pH 7.5   B. Phenol: water: 8-hydroxyquinoline: *m*-cresol
   0.05 M NaCl   (50:10:0.05:7 v/v/w/v)
   0.5% Sodium naphthalene-1,5-disulfonate
1) Tissue is extracted in 10 volumes of solution A.
2) An equal volume of solution B is added to the homogenate.
3) The mixture is stirred at 20° for 20 minutes.
4) The mixture is centrifuged at 6,000 g for 20 minutes.

5) The aqueous phase is made 5% with respect to sodium tri-isopropylnaphthalene sulfonate (a more thorough deproteinization is afforded by making the aqueous phase 1% with respect to tri-isopropylnaphthalene sulfonate and 6% with respect to 4-aminosalicylate) and 0.5 volumes of solution B are added.

6) The mixture is stirred at 20° for 15 minutes.

7) The mixture is centrifuged at 8,000 g for 10 minutes.

8) NaCl is added to the aqueous phase to give a final concentration of 0.5 M.

9) An equal volume of solution B is added and the mixture stirred and centrifuged.

10) RNA is precipitated from the aqueous phase with 2 volumes of 95% ethanol containing 0.2 M sodium acetate.

e) Phenol—Tri-isopropylnaphthalene Sulfonate—4-Aminosalicylate. This procedure isolates all RNA species as does hot phenol or cold phenol-SDS (KIRBY, 1965; 1968). As in the latter method, DNA is also extracted. It is more quantitative than either of these methods, however, because the nucleoproteins are dissociated before phenol is added and the detergents eliminate the need for bentonite as an RNAase inhibitor [cf. discussion of method d)].

Extraction medium:

A. 0.01 M Tris-HCl pH 7.5
   0.05 M NaCl
   1% Sodium tri-isopropylnaphthalene
     sulfonate
   6% Sodium 4-aminosalicylate
   6% 2-Butanol

B. Phenol: water: 8-hydroxyquinoline: *m*-cresol
   (50:10:0.05:7 v/v/w/v)

1) Tissue is extracted in 10 volumes of solution A.

2) An equal volume of solution B is added to the homogenate.

3) The mixture is stirred at 20° for 20 minutes.

4) The mixture is centrifuged at 6,000 g for 20 minutes.

5) NaCl is added to the aqueous phase to give a final concentration of 0.5 M and an equal volume of solution B is added.

6) The mixture is stirred at 20° for 15 minutes.

7) The mixture is centrifuged at 8,000 g for 10 minutes.

8) The aqueous phase is re-extracted with an equal volume of solution B.

9) RNA is precipitated from the aqueous phase with 2 volumes of 95% ethanol containing 0.2 M sodium acetate.

f) Phenol-Chloroform. Chloroform is more effective than phenol for deproteinizing extracts from some tissues, e.g. *Pinus* species and some fern gametophytes (INGLE and BURNS, 1968). Where this is the case, the preceding method e), modified after step 4 as described below, has been used successfully (INGLE and BURNS, 1968).

1)–4) [cf. preceding method e)].

5) The aqueous phase is incubated at 20° for 10 minutes.

6) NaCl is then added to the aqueous layer to give a final concentration of 0.5 M and an equal volume of chloroform: 1-pentanol (25:1) is added.

7) The mixture is shaken at 20° for 15 minutes.

8) The mixture is centrifuged at 8,000 g for 10 minutes.

9) The aqueous phase is re-extracted twice with the chloroform mixture.

10) RNA is precipitated from the aqueous phase with 2 volumes of 95% ethanol containing 0.2 M sodium acetate.

A phenol-chloroform mixture has also been used for the extraction of RNA from isolated nuclei (PENMAN, 1966).

1) Lysed nuclei are shaken with 1 volume of water-saturated phenol.

2) The mixture is heated to 60° and shaken again.

3) An equal volume of chloroform containing 1% 1-pentanol is added.

4) The mixture is heated at 60°, shaken and centrifuged. A large flocculent precipitate containing nuclear RNA accumulates at the interface and sinks into the phenol phase unless chloroform is present.

5) The aqueous phase and interfacial precipitate are re-extracted 3 times with the chloroform—1-pentanol mixture.

6) RNA is precipitated from the aqueous phase with 2 volumes of 95% ethanol containing 0.2 M sodium acetate.

g) Diethyl Pyrocarbonate (DEP)-SDS. DEP is a potent RNAase inhibitor and has been used with SDS instead of phenol (SOLYMOSY et al., 1968, 1970). Evidence that a less degraded product is thereby obtained is indicated by reports that chloroplast rRNAs are recovered in an equimolar ratio with DEP-SDS while most 23S rRNA is lost when phenol-SDS is used. It has also been found that, compared with the phenol-SDS method, higher yields of RNA are obtained by this procedure. DEP-SDS has been used successfully for the isolation of RNA from a number of sources including whole plant and animal tissues, yeast mitochondria and bacteria (ABADON and ELSON, 1970; GRIVELL et al., 1971; FEDORCSÁK et al., 1969; FORRESTER et al., 1970; SOLYMOSY et al., 1968, 1970).

Extraction medium: 0.05 M Tris-HCl pH 7.5
                              0.005 M $MgCl_2$
                              1% SDS

1) Tissue is extracted with 10 volumes of extraction medium.

2) DEP (0.02 ml/ml) may be added at the beginning of the extraction, however, improved preparations have been obtained if it is added immediately after the cells have been disrupted.

3) The homogenate is incubated at 37° for 5 minutes with occasional stirring.

4) The mixture is centrifuged at 8,000 g for 5 minutes at 4°.

5) The supernatant is removed and NaCl is added (0.1 g/ml supernatant) and dissolved at 20° with stirring.

6) The mixture is incubated at 37° for 5 minutes.

7) The mixture is centrifuged at 10,000 g for 20 minutes.

8) RNA is precipitated from the supernatant with 2 volumes of 95% ethanol containing 0.2 M sodium acetate.

This method may be simplified for the isolation of bacterial RNA by using a single deproteinization step (SUMMERS, 1970).

1) Bacterial cells are extracted with 10 volumes of extraction medium containing DEP (0.03 ml/ml).

2) The mixture is incubated at 37° for 5 minutes.

3) The mixture is chilled in ice and 0.05 volumes of saturated NaCl are then added to precipitate protein and SDS, which are removed by centrifugation at 10,000 g for 10 minutes.

4) RNA is precipitated from the supernatant with 2 volumes of 95% ethanol containing 0.2 M sodium acetate.

h) Guanidium Chloride. This method, which has not been used extensively, is reported to be suitable for the isolation of RNA from ribosomes and some viruses (COX, 1966, 1968; COX and ARNSTEIN, 1963; GRINNAN and MOSHER, 1951; REICHMANN and STACE-SMITH, 1959; SELA et al., 1957; VOLKIN and CARTER, 1951). Precipitation of ^{32}P-labelled RNA from guanidium chloride (steps 5–9) can also be used to remove contaminating ^{32}P-labelled nucleoside 5′-phosphates.

Extraction medium:
A. 0.05 M Tris-HCl pH 7.5   B. 6.0 M Guanidium chloride
      0.025 M KCl   (reagent grade guanidium chloride
      0.005 M $MgCl_2$ neutralized with concentrated HCl)
      0.25 M Sucrose

1) Ribosomes are suspended (10 mg/ml) in solution A.

2) Five volumes of solution B are added (the final concentration of guanidium chloride must be at least 4.0 M).

3) After chilling the mixture in ice for 15 minutes, 0.5 volumes of ethanol are added.

4) The mixture is cooled to $-20°$ for 30 minutes.

5) The mixture is centrifuged at 3,000 g for 5 minutes.

6) The precipitate containing RNA is quickly suspended in cold 4.0 M guanidium chloride buffered with 0.001 M EDTA pH 7.0.

7) The RNA is then precipitated with 0.5 volumes of ethanol and collected by centrifugation.

8) The precipitate is redissolved in guanidium chloride and re-precipitated with ethanol and collected by centrifugation.

9) The RNA pellet is extracted with dilute neutral salt solution (e.g. 0.01 M Tris-HCl or 0.002 M EDTA pH 7.0) for 30 minutes. The insoluble residue is separated by centrifugation and re-extracted; the process being repeated until the precipitate is completely dissolved.

10) RNA in the combined supernatants is converted to the K^+ salt by the addition of 2 volumes of ethanol containing 0.2 M potassium acetate. Alternatively, it may be chromatographed on Amberlite IR-120 in the K^+ form at pH 7.0.

i) Lithium Chloride (LiCl). Previous applications of this method have been confined to the extraction of RNA from ribosomes and reticulocytes (BARLOW et al., 1963; SPITNIK-ELSON, 1965). The procedure is probably as quantitative as the SDS method and yields RNA less heavily contaminated with protein.

Extraction medium:

A. 0.05 M Tris-HCl pH 7.5   B. 4.0 M Lithium chloride
   0.025 M KCl
   0.005 M $MgCl_2$
   0.25 M Sucrose

1) Ribosomes are suspended (10 mg/ml) in solution A.

2) An equal volume of solution B is added and the mixture shaken.

3) The mixture is stored at 2° for 12–16 hours to allow complete precipitation of the RNA.

4) The mixture is centrifuged at 3,000 g for 5 minutes.

5) The RNA pellet is washed twice with 2.0 M LiCl to remove tRNA and other impurities soluble in 2.0 M LiCl.

C. Purification

RNA preparations obtained by any of the methods described above may be contaminated by water soluble materials such as DNA, polysaccharides, sugar phosphates and nucleoside 5'-mono-, di- and triphosphates. Purification may be achieved using one, or a combination of the procedures described below.

1. Separation of RNA from Polysaccharides

a) Selective Precipitation of RNA with Cetyltrimethylammonium Bromide (CTA). RNA (and DNA if present) is precipitated as the insoluble cetyltrimethyl-ammonium salt and recovered by centrifugation while most contaminating polysaccharides, sugar phosphates and nucleoside 5'-phosphates remain in solution (BELLAMY and RALPH, 1968; RALPH and BELLAMY, 1964). The CTA procedure removes almost all the contaminating ^{32}P-labelled phosphate esters from ^{32}P-labelled RNA preparations.

Crude RNA is dissolved in 1 ml of 0.05 M Tris-HCl pH 7.5 and 0.1 ml of 5% CTA is added slowly with stirring. The mixture is chilled in ice for 5 minutes and then centrifuged for 10 minutes at 12,000 g. To transform the RNA to the sodium salt, the precipitate is washed 3 times with 70% ethanol containing 0.2 M sodium acetate and recovered each time by centrifugation. The precipitate is dissolved in Tris buffer and the process repeated if necessary.

b) Partitioning of Aqueous RNA Solutions with 2-Methoxyethanol. Purification of RNA by precipitation with CTA does not completely remove the polysaccharides and sugar phosphates. However, this can be accomplished by extracting RNA solutions in phosphate buffer with 2-methoxyethanol (BELLAMY and RALPH, 1968; RALPH and BELLAMY, 1964). The purified RNA can be recovered from the methoxyethanol layer by precipitation with CTA.

Plant RNA is dissolved in 0.025 M Tris-HCl pH 8.0 containing 0.025 M NaCl and mammalian RNA is dissolved in 0.1 M sodium acetate pH 5.1. Equal volumes of 2.5 M potassium phosphate pH 8.0 and 2-methoxyethanol are added and the mixture shaken vigorously for 2 minutes. The upper organic phase, containing the RNA, is removed carefully to avoid contamination with the interphase material and mixed with an equal volume of 1% CTA. The suspension is chilled in ice for 5 minutes and then centrifuged at 5,000 g for 5 minutes. The purified CTA-RNA is washed 3 times with 70% ethanol containing 0.2 M sodium acetate to convert the insoluble CTA-RNA to the water soluble sodium salt.

c) Partitioning of 2,2'-Diethyldihexylammonium-RNA Salts Between an Aqueous and Organic Phase. This method exploits the solubility of alkylammonium salts of RNA in organic solvents. RNA but not polysaccharides is extracted into the organic phase from an aqueous solution (KIDSON et al., 1963). The extraction is quantitative providing the Na^+ concentration is less than 0.05 M.

An aliquot of RNA solution is shaken with 0.55 volumes of 2,2'-diethyldihexylammonium acetate:glacial acetic acid:1-butanol mixture (11.7:2:100 v/v/v). RNA is recovered from the organic phase by precipitation with 2 volumes of ethanol containing 0.2 M sodium acetate. This method is convenient for recovering radioactive RNA from sucrose gradients and column fractions. The organic phase containing the RNA can be added directly to a toluene-based scintillation liquid and radio-activity determined.

d) Selective Precipitation of Polysaccharides from Ethanolic Solutions with Tetraethylammonium Bromide (or Triethylammonium Acetate). This purification depends on the differential solubility of polysaccharides and the tri- and tetra-ethylammonium salts of RNA in 50% ethanol (KIRBY, 1964).

Crude RNA (50—100 mg) is dissolved in 5 ml of 0.01 M tetramethylammonium bromide and 20 ml of 0.2 M tetraethylammonium bromide in 50% ethanol are added. The addition of 5 ml of ethanol to this mixture precipitates the polysaccharides which are easily removed by centrifugation at 10,000 g for 5 minutes. The supernatant is collected and 25 ml of ethanol and 0.5 g of sodium benzoate in 6 ml of 70% ethanol are added to convert the RNA back to the sodium salt which is precipitated and collected by centrifugation at 12,000 g for 20 minutes.

2. Separation of rRNA from tRNA and DNA

a) Selective Precipitation of rRNA with 10% m-Cresol. Precipitation of rRNA with 10% m-cresol removes contaminating polysaccharides and tRNA which remain in solution (KIRBY, 1965).

The aqueous extract containing crude RNA is made 20% with respect to sodium benzoate (necessary because m-cresol forms a 2-phase system with water but only a single phase if sodium benzoate is present) 3% with respect to NaCl (necessary to maintain ionic strength and preserve the secondary structure of RNA in the presence of *m*-cresol) and 10% with respect to *m*-cresol. The mixture is chilled in ice for 1 hour and the precipitated rRNA (and DNA if present) is recovered by centrifugation at 1,000 g for 10 minutes. The precipitate is washed successively with the sodium benzoate:NaCl: *m*-cresol mixture, 4% NaCl in 75% ethanol, 75% ethanol and finally with absolute ethanol.

b) Purification of rRNA by Extraction with 3.0 M Sodium Acetate. Since rRNA is insoluble in strong salt solutions while low molecular weight DNA, tRNA and polysaccharides are soluble, a simple purification can be achieved (HASTINGS and KIRBY, 1966).

Crude RNA is extracted 3 times with 25 ml of cold 3.0 M sodium acetate pH 6.0. The residual rRNA is recovered each time by centrifugation at 8,000 g for 10 minutes. The rRNA pellet is finally washed with 75% ethanol containing 0.2 M sodium acetate.

c) Separation of rRNA and Highly Polymerized DNA by Selective Precipitation with NaCl. Separation of rRNA and high molecular weight DNA may be effected in the following way (PARISH and KIRBY, 1966).

Two volumes of ethanol containing 10% *m*-cresol are added to the crude aqueous extract to precipitate the rRNA and DNA. After chilling the mixture in ice for 1 hour the precipitated rRNA and DNA are collected by a 10 minute centrifugation at 8,000 g. The pellet is washed twice with 75% ethanol containing 0.2 M sodium acetate and then dissolved in 0.1 M sodium acetate pH 6.0. The solution is made 3.0 M with respect to NaCl and stored at −5° for 12 hours. The precipitated rRNA, free of DNA, is recovered by centrifugation at 12,000 g for 10 minutes and washed with 3.0 M sodium acetate pH 6.0 to remove NaCl. The precipitate is finally washed with 75% ethanol containing 0.2 M sodium acetate and then with absolute ethanol.

3. Separation of tRNA from rRNA and DNA

a) Selective Precipitation of High Molecular Weight Nucleic Acid Contaminants from tRNA Preparations with m-Cresol.
Concentrated solutions of *m*-cresol precipitate rRNA, DNA and polysaccharides while leaving tRNA in solution (KIRBY, 1968).

The aqueous extract containing crude RNA is made 20% with respect to sodium benzoate and 20% with respect to *m*-cresol. The mixture is chilled in ice and the precipitate removed by centrifugation at 12,000 g for 10 minutes. The tRNA is then precipitated from the supernatant by the addition of 2 volumes of 95% ethanol.

b) DEAE-Cellulose Chromatography. DEAE-cellulose chromatography is the most widely used method for separating tRNA from contaminating rRNA, DNA and polysaccharides (HOLLEY et al., 1961).

The DEAE-cellulose column (Whatman DE-32 or DE-52) is equilibrated with 0.02 M Tris-HCl pH 7.5 containing 0.25 M NaCl, 0.01 M MgCl$_2$, 0.001 M EDTA, 0.002 M Na$_2$S$_2$O$_3$ and 0.01% isoamyl acetate. The same buffer is used for applying the sample, washing the column and eluting with a linear gradient of NaCl from 0.65 M to 1.5 M. Fractions containing the desired tRNA or tRNAs are combined and precipitated with 2 volumes of 95% ethanol, washed with ethanol, dissolved in water and freeze dried.

c) Gel Filtration on Sephadex G-100 (or G-200). High molecular weight rRNA can be separated from low molecular weight tRNA by molecular sieve chromatography (WETTSTEIN and NOLL, 1965).

The Sephadex G-100, or G-200 column (2.5 × 40 cm), is equilibrated with 0.01 M Tris-HCl pH 7.5 containing 1.0 M NaCl and 0.001 M EDTA. The same buffer is used for applying the crude RNA sample (100 mg) and eluting the column. The rRNA elutes with the void volume of the column while the tRNA is retarded. The tRNA is recovered from the column fractions as described for DEAE-cellulose [cf. b) above].

d) Selective Precipitation of High Molecular Weight Nucleic Acid Contaminants from tRNA Preparations with 2-Propanol. DNA and rRNA are precipitated out of 0.3 M sodium acetate by the addition of 2-propanol leaving the tRNA in solution (Zubay, 1962).

Crude RNA is dissolved in 0.3 M sodium acetate pH 7.0 and 0.54 volumes of 2-propanol are added slowly with stirring. The temperature is adjusted to 20° and the mixture centrifuged at 8,000 g for 5 minutes. The supernatant containing the tRNA is saved and the precipitate is redissolved in sodium acetate and reprecipitated by the addition of 0.54 volumes of 2-propanol. The tRNA is precipitated from the combined supernatants by the addition of a further 0.5 volumes of 2-propanol.

e) Purification of tRNA by Extraction with NaCl. Contaminating rRNA and DNA may be removed from aqueous tRNA preparations by precipitation with NaCl (Zubay, 1962).

The aqueous extract containing crude RNA is made 1.0 M with respect to NaCl. The mixture is chilled in ice for 1 hour and then centrifuged at 15,000 g for 30 minutes and the sediment re-extracted with 1.0 M NaCl. The tRNA is precipitated from the combined saline supernatants with 2 volumes of ethanol. Further purification of the tRNA is generally required and this may be achieved by 2-propanol fractionation or chromatography on DEAE-cellulose [cf. b) and d) above].

D. Fractionation

1. Zone Velocity Centrifugation in Sucrose Gradients

Ribonucleic acids can be separated according to their sedimentation velocity by centrifugation through sucrose gradients. This method may be used to separate rRNA from tRNA and to characterize rRNA species. However, resolution is restricted to rRNA species which differ by more than 2S.

Preparation of materials—The sucrose gradients are prepared in buffer to maintain the secondary structure of the RNA. In general, the conditions optimal for the isolation of a particular RNA will also be optimal during its fractionation on sucrose gradients. Magnesium ions should be omitted to prevent intermolecular aggregation and it is advisable to include EDTA to ensure complete removal of Mg^{2+} and other divalent cations which may accompany the RNA. Sucrose solutions can be pretreated with bentonite or DEP to remove traces of RNAase.

Preparation of sucrose gradients—Devices for the production of linear, concave and convex exponential gradients have been described (Britten and Roberts, 1960; McConkey, 1968). Linear gradients, usually 5–20% or 15–30%, are the most frequently used although isokinetic gradients (15–30%) have a higher resolving power (Noll, 1967). To minimize RNAase activity, the gradients are prechilled and run at 1–4°. If they are run at higher temperatures it is advisable to add 0.1% SDS as a nuclease inhibitor.

Operating procedure—The RNA sample is dissolved in the same buffer as the sucrose and carefully layered on the top of the gradient. At least 1 mg of RNA in 0.5–1 ml of buffer can be fractionated on a 30 ml gradient. Centrifugation at 63,850 g (average) for 16 hours achieves maximum resolution of the rRNA species on a 5–20% linear sucrose gradient in a 1 × 3 in. tube.

2. Methylated Albumin Kieselguhr (MAK) Chromatography

MAK chromatography may be used to separate RNA from DNA and to segregate various classes of RNA (MANDELL and HERSHEY, 1960; MURAKAMI, 1968; OSAWA and SIBATANI, 1968). Transfer RNAs can be fractionated by MAK chromatography but there is considerable overlapping of some species (YAMANE and SUEOKA, 1963). Both plant and bacterial rRNA species are resolved on MAK columns but mammalian rRNAs are not. The recovery of some classes of RNA is low if elution is confined to a NaCl gradient. For example, if rapidly labelled RNA is eluted from a MAK column with a NaCl gradient (0.5–2.0 M), less than half is eluted. The remainder of the rapidly labelled RNA (mainly DNA-like RNA) can be recovered by elution with 1.0 M ammonia (BILLING et al., 1969; ELLEM and SHERIDAN, 1964; KUNZ et al., 1970; LINGREL, 1967).

Preparation of materials—Methylated serum albumin is prepared by suspending 5 g of bovine serum albumin in 500 ml of absolute methanol and adding 4.2 ml of 12.0 M HCl. The mixture is stored in the dark at 20° for 3 days with occasional shaking; during this time the protein dissolves and after methylation is re-precipitated. The precipitate is collected by centrifugation and washed with methanol until the washings are colorless. It is then washed twice with anhydrous ether and dried *in vacuo* over KOH. The dried esterified protein is powdered and prepared as a 1% solution in water.

Hyflo Supercel (kieselguhr) is prepared for use by suspending 100 g in 200 ml of 1.0 M NaOH and filtering the mixture. The kieselguhr is washed successively with 300 ml of 1.0 M NaOH, 500 ml of distilled water, 500 ml of 1.0 M HCl and then washed with water until the washings are near neutral. The washings should be decanted as this removes fines from the kieselguhr. To ensure complete removal of fines, the kieselguhr is suspended in 500 ml of distilled water in a cylinder and after allowing to settle for several minutes (longer times give slower flow rates) the fines are decanted. The kieselguhr is dried at 40°.

Methylated albumin coated kieselguhr (MAK) is prepared by boiling (to expel air) and cooling a suspension of 20 g of washed kieselguhr in 100 ml of 0.1 M NaCl in 0.05 M potassium phosphate pH 6.7. Five ml of 1% methylated albumin solution are added while stirring, plus an additional 20 ml of 0.1 M buffered saline.

Preparation of column—The methylated albumin kieselguhr slurry is poured into a 2 cm diameter column. A more elaborate method designed to prevent channelling has been described (MANDELL and HERSHEY, 1960), but this is unnecessary provided that the kieselguhr is sufficiently coarse to permit a satisfactory flow rate under gravity.

Operating procedure—The RNA sample is dissolved in 0.1 M buffered saline and applied to the column. At least 5 mg of RNA applied in 20 ml of buffer can be fractionated on a 50 ml column. After sample application is complete, the column is washed with 1 bed volume of 0.1 M buffered saline. The RNA species are then eluted from the column with a linear gradient of NaCl in 0.05 M potassium phosphate buffer pH 6.7 and finally with 1.0 M ammonia. The quantity and composition of the salt gradient vary with the source and type of RNA under investigation. Ribosomal RNA and tRNA species are eluted using NaCl gradients of approximately 0.1–1.2 M and 0.3–0.9 M respectively.

3. Polyacrylamide Gel Electrophoresis

Fractionation of RNA by gel electrophoresis is more rapid and surpasses in resolution sucrose gradient centrifugation and MAK chromatography (LOENING, 1967, 1968, 1969, 1970; PEACOCK and DINGMAN, 1967). The method is particularly suitable for the fractionation of ribosomal and messenger RNA species. RNA species differing in molecular weight by only 2–3 nucleotides in a molecular weight range of 10,000–50,000 can be clearly separated using 7.5% acrylamide gels. In the high molecular weight range, excellent separation of rRNA species can be obtained using 2.4% acrylamide gels (Fig. 1).

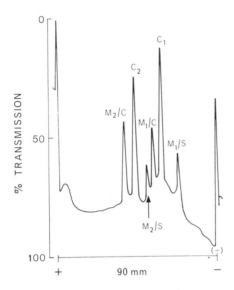

Fig. 1. Gel electrophoresis of a mixture of rRNA species. Nucleic acids were extracted from the cytoplasm and mitochondria of *Candida parapsilosis* and *Saccharomyces cerevisiae* (Yu et al., 1972). Mixtures of the nucleic acids were applied to 2.4% acrylamide gels and subjected to electrophoresis for 4 hours at 22°. Further details are given under 'D3; Polyacrylamide Gel Electrophoresis'. After washing, the gels were scanned at 265 nm. Symbols are as follows: M_1/S and M_2/S = mitochondrial rRNA species of *Saccharomyces cerevisiae*; M_1/C and M_2/C = mitochondrial rRNA species of *Candida parapsilosis*; C_1 and C_2 = cytoplasmic rRNAs from both yeasts (not resolved)

Preparation of materials—Acrylamide is recrystallized from chloroform (50 g/l) and methylene-bis-acrylamide is recrystallized from acetone (10 g/l).

Stock acrylamide solutions:

15 g recrystallized acrylamide,

0.75 g recrystallized bis-acrylamide for 2.4–5% gels,

or 0.375 g recrystallized bis-acrylamide for 5–7.5% gels.

Water to 100 ml.

Stock buffer solutions:

A. 0.12 M Tris-acetate pH 7.8   or B. 0.108 M Tris-phosphate pH 7.6

   0.06 M Sodium acetate   0.09 M NaH_2PO_4

   0.003 M Na_2-EDTA   0.003 M Na_2-EDTA

The Tris buffers are made up with acetate or phosphate rather than with HCl to avoid the formation of hypochlorite; Tris-phosphate is used in buffer B because it has a better buffering capacity at the lower pH. Sodium ions are included to help preserve the secondary structure of RNA and a chelating agent has been found necessary to prevent the high molecular weight RNA sticking to the gel surface.

Preparation of polyacrylamide gels—Ribosomal RNA species are separated on 2.4% gels while tRNA and other low molecular weight species are separated on 5–7.5% gels. Gels with a concentration of less than 2.4% can be strengthened by combining 0.5% agarose with the acrylamide (Dingman and Peacock, 1968). The polyacrylamide gel solutions are prepared as outlined below:

2.4% gels: 5 ml acrylamide stock solution + 6.25 ml buffer + 19.7 ml water,

5.0% gels: 5 ml acrylamide stock solution + 3.00 ml buffer + 6.7 ml water,

7.5% gels: 5 ml acrylamide stock solution + 2.00 ml buffer + 2.7 ml water.

The solutions are mixed and de-aerated under vacuum for 30 seconds. Twenty-five µl of TEMED (*N,N,N',N'*-tetramethylethylenediamine) and 0.25 ml of freshly prepared aqueous 10% ammonium persulfate are added and the solution gently mixed. It is then immediately pipetted into the gel tubes to a height of 9 cm. Plexiglass is preferred to glass for the electrophoresis tubes because the acrylamide gels are more easily removed after electrophoresis. To prevent the gels sliding out of the Plexiglass tubes during electrophoresis, the lower end is covered with dialysis membrane. During polymerization of the acrylamide solution, a serum cap covers the dialysis membrane. After polymerization of the acrylamide, which occurs within 10–15 minutes, the serum caps are removed.

Operating procedure—The tubes are placed in the electrophoresis unit and diluted stock buffer (stock buffer solution diluted 3 times with water) is added to each electrolyte container. The volume of buffer should be at least 400 ml to minimize rise in temperature and change in pH. If electrophoresis is carried out above 4°, SDS (0.1%) is added to the buffers to inhibit RNAase activity. The gels are pre-run for 1 hour at a constant current setting of 5 mA/gel (7.5 v/cm length of gel) to remove free acrylamide, catalysts and other impurities.

The RNA is dissolved in diluted stock buffer containing 10% sucrose (2 mg/ml) and 25 µl samples layered onto the gels without removing the electrolyte buffer. Electrophoresis on 2.4% gels is continued at 5 mA/gel for $2\frac{1}{4}$ hours for separation of whole cell RNA, or 4 hours for maximum resolution of rRNA species.

Analysis of polyacrylamide gels—After electrophoresis, the dialysis membranes are removed and the gels allowed to slide out of the electrophoresis tubes into test tubes containing distilled water. The water is replaced several times over a period of 30 minutes to wash out the high UV absorbing material which accumulates at both ends of the gel during electrophoresis. The gels are placed in a 1 cm light-path cuvette and scanned at 260 nm. Alternatively, they may be stained by the following procedure (PEACOCK and DINGMAN, 1967). The gels are rinsed for 10 minutes in 1.0 M acetic acid to lower their pH and then stained for 1 hour in 0.2% toluidine blue or methylene blue in 0.4 M sodium acetate pH 4.7. Excess stain is removed by repeated washing with water. The position of the bands may then be measured. The RNA can be estimated quantitatively by scanning the stained gels at 660 nm with a densitometer.

If labelled RNA is used, the distribution of radioactivity in the gels can be determined in the following way. Gels are placed in aluminum boats and frozen in dry ice. The frozen gel is sectioned into 0.5–1 mm slices. The slices are placed in vials and 0.5 ml of 10% piperidine in 0.001 M EDTA is added and the gels incubated at 40° for 4 hours. The dried gel slices are allowed to swell in 0.5 ml of water for 2 hours before 10 ml of scintillation liquid is added and radioactivity determined.

References

ABADON, P. N., ELSON, D.: Biochim. Biophys. Acta 199, 528 (1970).

BARLOW, J.J., MATHIAS, A. P., WILLIAMSON, R.: Biochem. Biophys. Res. Commun. 13, 61 (1963).

BELLAMY, A.R., RALPH, R.K.: In Methods in enzymology (GROSSMAN, L., and MOLDAVE, K., eds.), vol. 12, part B, p. 156. New York and London: Academic Press 1968.

BERNFIELD, P., NISSELBAUM, J.S., BERKELEY, B.J., HANSON, J.: J. Biol. Chem. 235, 2852 (1960).

BILLING, R.J., INGLIS, A. M., SMELLIE, R. M.S.: Biochem. J. 113, 571 (1969).

BLANTON, M. V., BARNETT, B.: Anal. Biochem. 32, 150 (1969).

BLOW, D.M., RICH, A.: J. Amer. Chem. Soc. 82, 3566 (1960).

BRITTEN, R.J., ROBERTS, R.B.: Science 131, 32 (1960).

BUTLER, J.A.V., GODSON, G.N.: Biochem. J. 88, 176 (1963).

CLICK, R.E., HACKETT, D.P.: Biochim. Biophys. Acta 129, 74 (1966).

COX, R.A.: Biochem. Prep. 11, 104 (1966).

COX, R.A.: In Methods in enzymology (GROSSMAN, L., and MOLDAVE, K., eds.), vol. 12, part B, p. 120. New York and London: Academic Press 1968.

COX, R.A., ARNSTEIN, A.R.V.: Biochem. J. 89, 574 (1963).

COX, R.A., JONES, A.S., MARSH, G.E., PEACOCK, A.R.: Biochim. Biophys. Acta 21, 576 (1956).

DI GIROLAMO, A., HENSHAW, E.C., HIATT, A. H.: J. Biol. Chem. 8, 749 (1964).

DINGMAN, C.W., PEACOCK, A.R.: Biochemistry 7, 659 (1968).

DINGMAN, C.W., SPORN, M.B.: Biochim. Biophys. Acta 61, 164 (1962).

EISENSTADT, J., BRAWERMAN, G.: J. Mol. Biol. 10, 392 (1964).

ELLEM, K.A.O., SHERIDAN, J.W.: Biochem. Biophys. Res. Commun. 16, 505 (1964).

FEDORCSÁK, I., EHRENBERG, L.: Acta Chem. Scand. 20, 107 (1966).

FEDORCSÁK, I., NATARAJAN, A.T., EHRENBERG, L.: Eur. J. Biochem. 10, 450 (1969).

Forrester, I.T., Nagley, P., Linnane, A.W.: FEBS Lett. **11**, 59 (1970).
Fraenkel-Conrat, H., Singer, B., Tsugita, A.: Virology **4**, 54 (1961).
Fraenkel-Conrat, H., Singer, B., Williams, R.C.: Biochim. Biophys. Acta **25**, 87 (1957).
Georgiev, G.P., Montieva, V.L.: Biochim. Biophys. Acta **61**, 153 (1962).
Georgiev, G.P., Samarina, O.P., Lerman, M.T., Smirnov, M.N., Severtzov, A.N.: Nature **28**, 1291 (1963).
Gierer, A., Schramm, G.: Nature **177**, 702 (1956).
Gooding, G.V., Herbert, T.T.: Phytopathology **57**, 1285 (1967).
Green, M., Piña, M.: Virology **20**, 199 (1963).
Grinnan, E.L., Mosher, W.A.: J. Biol. Chem. **191**, 719 (1951).
Grivell, L.A., Reijnders, L., Borst, P.: Biochim. Biophys. Acta **247**, 91 (1971).
Hastings, J.R.B., Kirby, K.S.: Biochem. J. **100**, 532 (1966).
Holley, R.W., Apgar, J., Doctor, B.P., Farrow, J., Marini, M.A., Merrill, S.H.: J. Biol. Chem. **236**, 200 (1961).
Howells, A.J., Wyatt, G.R.: Biochim. Biophys. Acta **174**, 86 (1969).
Huppert, J., Pelmont, J.: Arch. Biochem. Biophys. **98**, 214 (1962).
Ingle, J., Burns, R.G.: Biochem. J. **110**, 605 (1968).
Itoh, T., Hirai, T.: Ann. Phytopathol. Soc. Japan **32**, 227 (1966).
Kay, E.R.M., Simmons, N.S., Dounce, A.L.: J. Amer. Chem. Soc. **74**, 1724 (1952).
Kickhoffen, B., Burger, M.: Biochim. Biophys. Acta **65**, 190 (1962).
Kidson, C.S., Kirby, K.S., Ralph, R.K.: J. Mol. Biol. **7**, 312 (1963).
Kirby, K.S.: Biochem. J. **64**, 405 (1956).
Kirby, K.S.: Biochem. J. **66**, 495 (1957).
Kirby, K.S.: Biochim. Biophys. Acta **55**, 545 (1962).
Kirby, K.S.: In Progress in nucleic acid res. and mol. biol. (Davidson, J.N., and Cohn, W.E., eds.), vol. 3, p. 1. New York and London: Academic Press 1964.
Kirby, K.S.: Biochem. J. **96**, 266 (1965).
Kirby, K.S.: In Methods in enzymology (Grossman, L., and Moldave, K., eds.), vol. 12, part B, p. 87. New York and London: Academic Press 1968.
Kunz, W., Niessing, J., Schneiders, B., Sekeris, C.E.: Biochem. J. **116**, 563 (1970).
Lerman, M.I., Viladimirtseva, E.A., Tersikh, V.V., Georgiev, G.P.: Biokhimiya **30**, 322 (1965).
Levinton, L., Darnell, J.E.: J. Biol. Chem. **235**, 70 (1960).
Lingrel, J.B.: Biochim. Biophys. Acta **142**, 75 (1967).
Littauer, U.Z., Eisenberg, H.: Biochim. Biophys. Acta **32**, 320 (1959).
Littauer, U.Z., Sela, M.: Biochim. Biophys. Acta **61**, 609 (1962).
Loening, U.E.: Biochem. J. **102**, 251 (1967).
Loening, U.E.: In Chromatographic and electrophoretic techniques (Smith, I., ed.), vol. 2, p. 437. London: W. Heinemann Ltd. 1968.
Loening, U.E.: Biochem. J. **113**, 131 (1969).
Loening, U.E.: In Methods in medical res. (Olson, R.E., ed.), vol. 12, p. 359. Chicago: Year Book Medical Publ. Inc. 1970.
Mandell, J.D., Hershey, A.D.: Anal. Biochem. **1**, 66 (1960).
Marcus, L., Halvorson, H.O.: In Methods in enzymology (Grossman, L., and Moldave, K., eds.), vol. 12, part A, p. 498. New York and London: Academic Press 1968.
Marmur, J.: J. Mol. Biol. **3**, 208 (1961).
McConkey, E.H.: In Methods in enzymology (Grossman, L., and Moldave, K., eds.), vol. 12, part A, p. 620. New York and London: Academic Press 1968.
Murakami, W.T.: In Methods in enzymology (Grossman, L., and Moldave, K., eds.), vol. 12, part A, p. 634. New York and London: Academic Press 1968.
Noll, H.: Nature **215**, 360 (1967).
Noll, H., Stutz, E.: In Methods in enzymology (Grossman, L., and Moldave, K., eds.), vol. 12, part B, p. 129. New York and London: Academic Press 1968.
Öberg, B.: Biochim. Biophys. Acta **204**, 430 (1970).
Osawa, S., Sibatani, A.: In Methods in enzymology (Grossman, L., and Moldave, K., eds.), vol. 12, part A, p. 678. New York and London: Academic Press 1968.
Parish, J.H., Kirby, K.S.: Biochim. Biophys. Acta **129**, 554 (1966).
Peacock, A.R., Dingman, C.W.: Biochemistry **6**, 1818 (1967).

PENE, J.J., KNIGHT, E., DARNELL, J.E.: J. Mol. Biol. **33**, 609 (1968).
PENMAN, S.: J. Mol. Biol. **17**, 117 (1966).
PLAGEMANN, P.G.W.: Biochim. Biophys. Acta **224**, 451 (1970).
POULSON, R., BEEVERS, L.: Plant Physiol. **46**, 509 (1970).
PUSZTAI, A.: Biochem. J. **99**, 93 (1966).
RALPH, R.K., BELLAMY, A.R.: Biochim. Biophys. Acta **87**, 9 (1964).
REICHMANN, M.E., STACE-SMITH, R.: Virology **9**, 710 (1959).
REPASKE, R.: Biochim. Biophys. Acta **30**, 225 (1958).
RIFKIN, M.R., WOOD, D.D., LUCK, D.J.L.: Proc. Nat. Acad. Sci. U.S. **58**, 1025 (1967).
ROBINSON, W.S., PITKANEN, A., RUBIN, H.: Proc. Nat. Acad. Sci. U.S. **54**, 857 (1965).
ROSÉN, G.C., FEDORCSÁK, I.: Biochim. Biophys. Acta **130**, 401 (1966).
RUSHIZKY, G.W., GRECO, A.E., HARTLEY, R.W., SOBER, H.A.: Biochem. Biophys. Res. Commun. **10**, 311 (1963).
SCHERRER, K., DARNELL, J.E.: Biochem. Biophys. Res. Commun. **7**, 486 (1962).
SCHUSTER, H., SCHRAMM, G., ZILLIG, W.: Z. Naturforsch. **11**b, 339 (1956).
SELA, M., ANFINSEN, C.B., HARRINGTON, W.F.: Biochim. Biophys. Acta **26**, 506 (1957).
SINGER, B., FRAENKEL-CONRAT, H.: Virology **4**, 59 (1961).
SOLYMOSY, F., FEDORCSÁK, I., GULYÁS, A., FARKAS, G.L., EHRENBERG, L.: Eur. J. Biochem. **5**, 520 (1968).
SOLYMOSY, F., LAZAR, G., BAGI, G.: Anal. Biochem. **38**, 40 (1970).
SPENCER, D., WHITFELD, P.R.: Arch. Biochem. Biophys. **117**, 337 (1966).
SPITNIK-ELSON, P.: Biochem. Biophys. Res. Commun. **18**, 557 (1965).
SPORN, M.B., DINGMAN, C.W.: Biochim. Biophys. Acta **68**, 389 (1963).
STACE-SMITH, R.: Virology **29**, 240 (1966).
STANLEY, W.M., BOCK, R.M.: Biochemistry **4**, 1302 (1965).
STEERE, R.L.: Phytopathology **46**, 60 (1956).
SUMMERS, W.C.: Anal. Biochem. **33**, 459 (1970).
TAKAHASHI, T.: In Methods in enzymology (GROSSMAN, L., and MOLDAVE, K., eds.), vol. 12, part B, p.99. New York and London: Academic Press 1968.
VOLKIN, E., CARTER, C.F.: J. Amer. Chem. Soc. **73**, 1516 (1951).
WALTERS, R.A., GURLEY, L.B., SAPONARA, A.G., ENGER, M.D.: Biochim. Biophys. Acta **199**, 525 (1970).
WARNER, J., MADDEN, M.J., DARNELL, J.E.: Virology **19**, 393 (1963).
WECKER, E.: Virology **7**, 241 (1959).
WETTSTEIN, F.O., NOLL, H.: J. Mol. Biol. **11**, 35 (1965).
WOLF, B., LESNAW, J.A., REICHMANN, M.E.: Eur. J. Biochem. **13**, 519 (1970).
YAMANE, T., SUEOKA, N.: Proc. Nat. Acad. Sci. U.S. **50**, 1093 (1963).
YU, R., POULSON, R., STEWART, P.: Mol. Gen. Genetics **114**, 313 (1972).
ZUBAY, G.: J. Mol. Biol. **4**, 347 (1962).

Subject Index

Molecular Biology, Biochemistry and Biophysics
Molekularbiologie, Biochemie und Biophysik

Editors: A. Kleinzeller, G. F. Springer, H. G. Wittmann

Volume 1
J. H. van't Hoff: Imagination in Science

Translated into English with notes and a General Introduction by G. F. Springer. With 1 portrait
VI, 18 pages. 1967
DM 6,60; US $2.40

Volume 2
K. Freudenberg and A. C. Neish: Constitution and Biosynthesis of Lignin

With 10 figures
IX, 129 pages. 1968
Cloth DM 28,− ;US $9.90

Volume 3
T. Robinson: The Biochemistry of Alkaloids

With 37 figures
X, 149 pages. 1968
Cloth DM 39,−
US $13.80

Volume 4
A. S. Spirin and L. P. Gavrilova: The Ribosome

With 26 figures
X, 161 pages. 1969
out of print

Volume 5
B. Jirgensons: Optical Rotatory Dispersion of Proteins and Other Macromolecules

With 65 figures
XI, 166 pages. 1969
Cloth DM 46,−
US $16.30

Volume 6
F. Egami and K. Nakamura: Microbial Ribonucleases

With 5 figures
IX, 90 pages. 1969
Cloth DM 28,− ;US $9.90

Volume 7
F. Hawkes: Nucleic Acids and Cytology

With 34 figures
Approx. 288 pages
in preparation

Volume 8
Protein Sequence Determination

A Sourcebook of Methods and Techniques
Edited by
S. B. Needleman
With 77 figures
XXI, 345 pages. 1970
Cloth DM 84,−
US $29.70

Volume 9
R. Grubb: The Genetic Markers of Human Immunoglobulins

With 8 figures
XII, 152 pages. 1970
Cloth DM 42,−
US $14.90

Volume 10
R. J. Lukens: Chemistry of Fungicidal Action

With 8 figures
XIII, 136 pages. 1971
Cloth DM 42,−
US $14.90

Volume 11
P. Reeves: The Bacteriocins

With 9 figures
XI, 142 pages. 1972
Cloth DM 48,−
US $17.00

Distribution rights for U. K., Commonwealth, and the Traditional British Market (excluding Canada):
Chapman & Hall Ltd., London

Prices are subject to change without notice

■ Prospectuses on request

Springer-Verlag Berlin · Heidelberg · New York
München London Paris Sydney Tokyo Wien

A Glossary of Genetics and Cytogenetics

Rieger
Michaelis
Green

Classical and Molecular

With 90 figures
507 pages. 1968
Cloth DM 66, —

Third completely revised edition of "Genetisches und Cyto-genetisches Wörterbuch" by R. Rieger and A. Michaelis, both Genetische Abteilung, Institut für Kulturpflanzenforschung der Deutschen Akademie der Wissenschaften in Gatersleben, and M. M. Green, Department of Genetics, University of California, Davis, Calif.

U. K. and
Commonwealth
Publishers:
Allen & Unwin Ltd.,
London

The aim of the book is to list, explain or define, and collate the special terminology which has grown up over the past 100 years in the fields of genetics, cytology and cytogenetics. Some 2500 terms have been catalogued in alphabetical order, their originator named, their synonymity with related terms given, and their meaning explained in detail. Thus the text is not simply a dictionary of terms but rather a short encyclopedia designed to give the reader a thorough understanding of each term's significance and relevance. This glossary will be most useful to both specialists and non-specialists. The terminology ranges from expressions of merely historical interest to the latest developments in molecular genetics and biology.

■ Prospectus
on request!

Springer-Verlag
Berlin · Heidelberg · New York